Air Quality Assessment Standards and Sustainable Development in Developing Countries

Air Quality Assessment Standards and Sustainable Development in Developing Countries

Special Issue Editors

Weixin Yang
Guanghui Yuan

MDPI • Basel • Beijing • Wuhan • Barcelona • Belgrade

MDPI

Special Issue Editors
Weixin Yang
University of Shanghai for Science and
Technology
China

Guanghui Yuan
Shanghai University of Finance and
Economics
China

Editorial Office
MDPI
St. Alban-Anlage 66
4052 Basel, Switzerland

This is a reprint of articles from the Special Issue published online in the open access journal *Sustainability* (ISSN 2071-1050) from 2018 to 2019 (available at: https://www.mdpi.com/journal/sustainability/special_issues/Air_Quality_Assessment_Sustainable_Development)

For citation purposes, cite each article independently as indicated on the article page online and as indicated below:

LastName, A.A.; LastName, B.B.; LastName, C.C. Article Title. *Journal Name* **Year**, *Article Number, Page Range.*

ISBN 978-3-03928-014-8 (Pbk)
ISBN 978-3-03928-015-5 (PDF)

Contents

About the Special Issue Editors

Weixin Yang received his Ph.D. in Economics in 2006. He is currently teaching at University of Shanghai for Science and Technology, and the Development Institute of Fudan University. He has published over 100 papers, book chapters, and conference proceedings. He is also invited as a reviewer and academic editor of well-known journals. His main research interests include development economics, game theory, resource and environmental management and policy analysis.

Guanghui Yuan received his Ph.D. degree from Shanghai University of Finance and Economics. He is currently teaching at Shanghai University of Finance and Economics. He has published more than 50 research papers. His main research interests include system modeling, information economy, data mining, and algorithm design.

Preface to "Air Quality Assessment Standards and Sustainable Development in Developing Countries"

Air pollution is a critical challenge faced by the world, especially in developing countries. During the rapid economic growth and remarkable pace of industrialization of typical developing countries, such as China and India, massive amounts of fossil fuels, including coal, petroleum, and natural gas, have been consumed. In addition to the usual pollution from fossil fuels and sandstorm, fine atmospheric pollutants (such as $PM_{2.5}$, PM_{10}, and O_3) continue to emerge and are compounded to create new hazards. Air pollution not only affects the public health and quality of life in developing countries, but also poses a threat to the sustainable development of their economies and societies. Quick and efficient control of air pollution in developing countries has become a common concern in academia.

In order to scientifically measure and improve air quality, the U.S. established the Environmental Protection Agency (EPA) in the 1970s and published the first Pollution Standards Index. In 1999, the EPA added the measurement of daily average PM2.5 concentration into their PSI and thereby formed an independent Air Quality Index (AQI). After that, major developed countries constructed their own air quality assessment systems. The World Health Organization (WHO) and various scientific research institutions have also established their own air quality measurement standards, which are constantly adjusted. Therefore, researchers from various universities have conducted in-depth studies on air quality measurement and have offered amendment suggestions to the current measurement indicator system.

Therefore, we have edited and published this book to discuss the effectiveness of current air quality assessment standards in developing countries, the impact of air quality governance policies based on those standards, and the improvement of those standards in order to effectively achieve sustainable development. Scholars have delved into the issues of air quality assessment standards and sustainable development in developing countries in 13 excellent academic papers which provide excellent materials for students studying at the BSc, MSc, and Ph.D. levels as well as researchers to understand air quality assessment standards and sustainable development in developing countries.

Weixin Yang, Guanghui Yuan
Special Issue Editors

sustainability

MDPI

Article

Evaluating China's Air Pollution Control Policy with Extended AQI Indicator System: Example of the Beijing-Tianjin-Hebei Region

Guanghui Yuan [1],[†] and Weixin Yang [2],[*],[†]

[1] School of Information Management and Engineering, Fintech Research Institute, Shanghai University of Finance and Economics, Shanghai 200433, China; guanghuiyuan@outlook.com or flame_yuan@163.sufe.edu.cn

[2] Business School, University of Shanghai for Science and Technology, Shanghai 200093, China

[*] Correspondence: iamywx@outlook.com; Tel.: +86-21-5596-0082

[†] Guanghui Yuan and Weixin Yang are joint first authors. They contributed equally to this paper.

Received: 14 January 2019; Accepted: 7 February 2019; Published: 12 February 2019

Abstract: This paper calculated and evaluated the air quality of 13 cities in China's Beijing-Tianjin-Hebei (BTH) region from February 2015 to January 2018 based on the extended AQI (Air Quality Index) Indicator System. By capturing the heterogeneous information in major pollutant indicators and the standardization process, we depicted the important effect of other relevant features of pollutant indicators beyond single-point data. Based on that, we further calculated the assessment value of the air quality of different cities in the BTH region by using the Collaborative Filtering Backward Cloud Model to construct differentiated weights of different indicators. With help of the Back Propagation (BP) Neutral Network, we simulated the effect of the pollution control policies of the Chinese government targeting air pollution since March 2016. Our conclusion is: the pollution control policies have improved the air quality of Beijing by 55.74%, and improved the air quality of Tianjin by 34.38%; while the migration of polluting enterprises from Beijing and Tianjin has caused different changes in air quality in different cities of Hebei province—we saw air quality deterioration by 58.60% and 38.68% in Shijiazhuang and Handan city respectively.

Keywords: AQI indicators; air pollution; collaborative filtering; Beijing-Tianjin-Hebei region

1. Introduction

Among the environmental challenges China is facing now, air pollution is one of the key issues that draw the attention of academic circles [1–4]. In order to scientifically measure air quality and better prevent and control air pollution, China has officially launched the Technical Regulation on Ambient Air Quality Index (on trial) (HJ 633-2012) in 2016 [5].

Air pollution refers to the circumstances where the concentration of certain substances in the atmosphere reaches a certain level that it can harm the ecosystem as well as humans and other species living in it, and threaten the survival of human beings [6]. Currently, the pollutants that China has covered in regular monitor and air quality evaluation include sulfur dioxide (SO_2), nitrogen dioxide (NO_2), carbon monoxide (CO), inhalable particles (PM_{10} and $PM_{2.5}$), and ozone (O_3) [5,7]. Above pollutants all cause serious threats to the sustainable development and health of human beings.

According to the two National Standards on Air Quality Measurement published by the Chinese Ministry of Environmental Protection on 29 February 2012—Ambient air quality standards (GB 3095-2012) and Technical Regulation on Ambient Air Quality Index (on trial) (HJ 633-2012)—that became effective on January 1st, 2016, the air quality measurement of China mainly relies on the calculation of AQI (Air Quality Index), with the method of [5]:

- First, calculate the Individual Air Quality Index of certain pollutant ($IAQI_P$):

$$IAQI_P = \frac{IAQI_{Hi} - IAQI_{Lo}}{BP_{Hi} - BP_{Lo}}(C_P - BP_{Lo}) + IAQI_{Lo} \tag{1}$$

In equation above, C_P represents the mass concentration of pollutant P; BP_{Hi} is the higher threshold of pollutant concentration near C_P corresponding to specified IAQI (Individual Air Quality Index) regulated by government policy; BP_{Lo} is the lower threshold of pollutant concentration near C_P regulated by government; $IAQI_{Hi}$ is the corresponding IAQI to BP_{Hi}; while $IAQI_{Lo}$ is the corresponding IAQI to BP_{Lo}.

- Then, take the largest number from all $IAQI_P$ to calculate the AQI:

$$AQI = max\{IAQI_1, IAQI_2, IAQI_3, \cdots, IAQI_n\} \tag{2}$$

Nevertheless, there are some issues to be further discussed in above calculation method:

(1) The final AQI only reflects one pollutant—only the pollutant with the highest $IAQI$. Although it is further defined in "AQI Technical Specifications (Trial Use)" that "when AQI is above 50, the pollutant with the highest $IAQI$ is the "primary pollutant"; if there are more than one pollutant with the same highest $IAQI$, then all of such pollutants are classified as "primary pollutants"; all pollutants whose $IAQI$ is above 100 should be classified as "pollutants exceeding limits" [5]. However, even based on such definitions, we are unable to capture the impact of pollutants other than the one with the highest $IAQI_P$ on air quality.

(2) As regulated by government, the threshold of pollutant concentration corresponding to specific $IAQI$ is 500 for average PM$_{2.5}$ within 24 hours, and 600 for average PM$_{10}$ in 24 hours [5]. However, recently in our actual air quality monitoring practice, sometimes the concentration of certain pollutants (such as PM$_{2.5}$) in certain regions reached far higher than the threshold that it went "off the charts" [8,9]. Because BP_{Hi} and BP_{Lo} in Equation (1) above is subject to the range of (0, 500) and (0, 600), this calculation method cannot reflect the exact AQI.

(3) Given above issues, it is difficult for us to accurately measure and assess the air quality of different cities, not to mention comparing the effect of air pollution control policies across the cities. In current research practice, the assessment and comparison of air quality across provinces and cities is usually simplified to be based on PM$_{2.5}$ data, which is not helpful in identifying the whole picture of pollutant sources and creates more challenges for the design and evaluation of air pollution control policies.

Hence, this paper has selected 13 cities across the Beijing-Tianjin-Hebei (BTH) region—the region with the heaviest air pollution in China and ranking top among the 13 target regions assigned by the government for air pollution control [10]. Our study has covered the two municipalities directly under the Central Government, Beijing and Tianjin. Meanwhile, since the launch of "Beijing-Tianjin-Hebei Integration Policy" in 2014, the 13 cities in this region have shown stronger synergy in terms of policy design and execution. Therefore, the air pollution conditions as well as the effectiveness of government policy in the BTH region have valuable implications for wider areas of China.

A number of academic studies have also been conducted on the air quality problem in the BTH region. Lang et al. studied the vehicular emissions trends in the BTH region from 1999 to 2010 by the COPERT IV model. They showed that vehicular emissions of CO and VOC (Volatile Organic Compounds) have decreased while emissions of NO$_X$ and PM$_{10}$ have kept increasing in Tianjin and Hebei [11]. Xu et al. studied the health risks caused by SO$_2$ emissions in the different cities in the BTH region. Using the Community Multi-scale Air Quality (CMAQ) modeling system, they simulated the fate and transport of SO$_2$ in the BTH region. They discovered that a risk-based approach should be preferred because it will help improve the efficiency in resource utilization [12]. Zhao et al. collected more than 400 PM$_{2.5}$ samples in Beijing, Tianjin, Shijiazhuang, and Chengde over four seasons from 2009 to 2010. They indicated that the characteristics of carbonaceous aerosol pollution were spatially similar and season-dependent in the plain area of the BTH region [13]. Sheng et al. compared the air quality of the BTH region just before and after Asia-Pacific Economic Cooperation (APEC) meetings of 2014. They showed that the APEC emission reduction measures have effectively improved the air quality of the BTH, especially in Beijing [14]. Miao et al. used the Weather Research and Forecasting Model and the Flexible-particle Dispersion Model to investigate the pollutant transport mechanisms of a haze event in 2011 over the BTH region. They suggested that the penetration by sea-breeze could strengthen the vertical dispersion in BTH and carry the local pollutants to the downstream areas [15]. Zhou et al. investigated the ammonia emission inventory for the BTH region with the updated source-specific emission factors and the county-level activity data. They found that higher ammonia emission was concentrated in the areas with more rural and agricultural activity of Shijiazhang, Handan, Xingtai, Tangshan and Cangzhou than other cities in BTH [16]. Han et al. studied the intense air pollution occurred in the BTH region in January 2013. By multisatellite datasets, air sounding and surface meteorological observations, they showed that there was a vertical overlap of fog and aerosol layers during the foggy haze episodes, which would worsen the regional air quality and have notable effects on the radiation balance [17]. Guo et al. investigated the reduction potentials of PM_{10}, NO_x, CO and HC under different control policies in the BTH region during 2011–2020. They showed that the emission standards updating policy would achieve a substantial reduction of all the pollutants, while the eliminating high-emission vehicles policy can reduce emissions more effectively in short-term than in long-term, especially in Beijing [18]. Chen et al. used Voronoi spatial interpolation method to estimate the $PM_{2.5}$ concentration in the BTH region. They showed that up to 14,051 deaths and 6574 million yuan loss would be avoided when the $PM_{2.5}$ concentration fell by 25% in BTH [19]. Zhu et al. studied the spatial impacts of foreign direct investment (FDI) on SO_2 emissions in the BTH region by spatial panel data from 2000 to 2013. They found that the increase in FDI inflows would also increase air pollution levels and influence the air quality of surrounding cities [20]. Wang et al. developed a modified inter-regional and sectoral model to study the embodied emission flows based on the input-output table of the BTH region. They showed that the transfer pattern of the most significant pollutant flow was the same for SO_2, Soot, Dust and NO_x, which accounted for 35.7% to 42.0% of the total embodied emissions of pollutants from Hebei province to Beijing [21]. Zhang et al. calculated the intended maximum emission levels in the BTH region by modelling the relationship between $PM_{2.5}$ concentration and other air pollutant emissions. They indicated that the $PM_{2.5}$ concentrations in BTH was influenced by local air pollutant emissions, wind speed, lagged $PM_{2.5}$ concentrations, and $PM_{2.5}$ concentrations in adjacent cities [22].

However, none of the above studies considered the combined effects of the six pollutants. In fact, the severity of each pollutant is not all the same in cities in the BTH region (see Figure 1) [23].

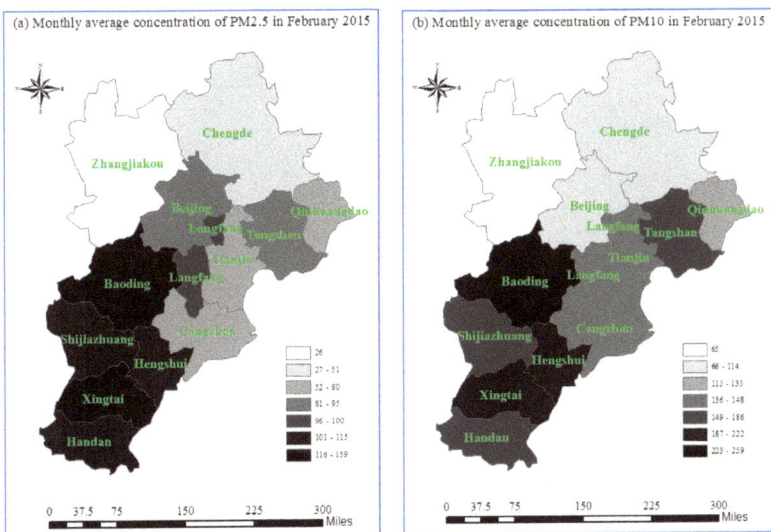

Figure 1. *Cont.*

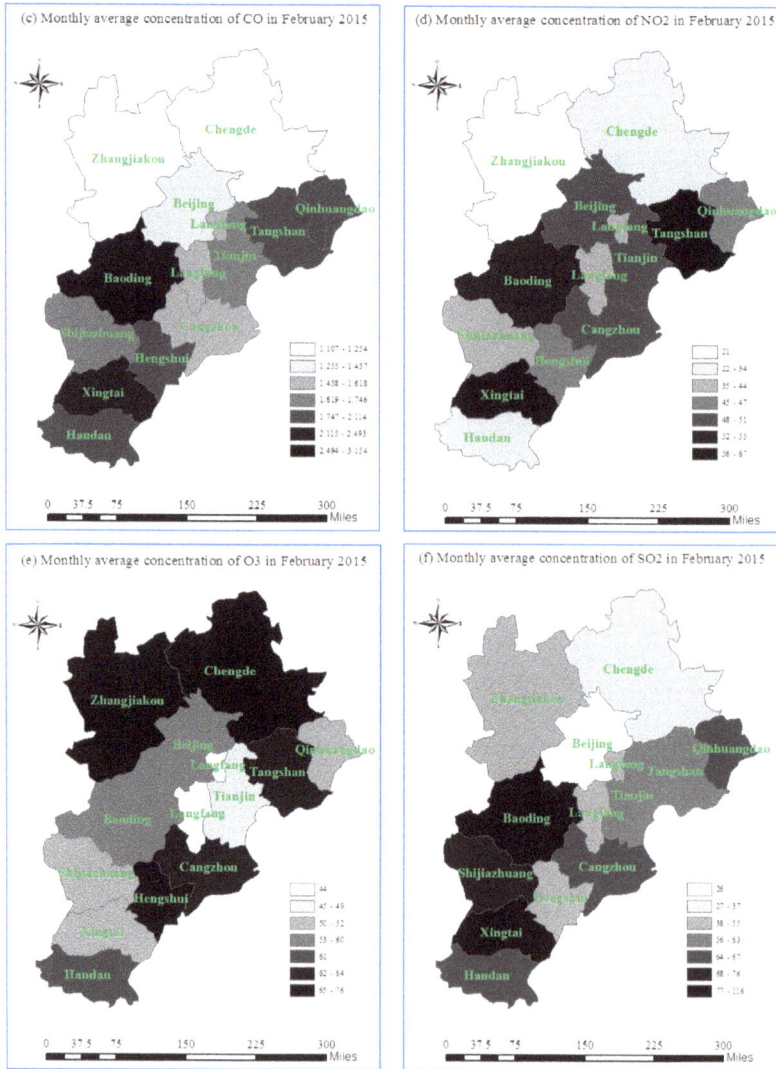

Figure 1. Monthly average concentration of six pollutants in the BTH Region (February 2015): (**a**) PM$_{2.5}$ (unit: $\mu g/m^3$); (**b**) PM$_{10}$ (unit: $\mu g/m^3$); (**c**) CO (unit: mg/m^3); (**d**) NO$_2$ (unit: $\mu g/m^3$); (**e**) O$_3$ (unit: $\mu g/m^3$); (**f**) SO$_2$ (unit: $\mu g/m^3$).

Therefore, this paper has extended the indicators defined in "AQI Technical Specifications (Trial Use)" (HJ 633-2012). First, we built an indicator system that covers all six main air pollutants and based on the interval analysis method [24,25], we further constructed three kinds of heterogeneous information (information with different dimensions, such as interval number, mean value, and variance) as well as calculated the standardization form of each pollutant indicator in order to capture the effect of each pollutant on air quality within our study period. Then we adopted the Collaborative Filtering Backward Cloud Model to obtain the different weights for calculation of air quality assessment score of each city based on air pollutant concentration data of the BTH region from February 2015 to January 2018. Furthermore, we simulated the Background Trend Line of the dynamic change in air quality in absence of air pollution control policies in cities of the BTH region with help of the Back Propagation (BP) Neural Network method in order to quantify the influence of government policy on pollution control of different cities. Last but not least, we proposed tailored policy recommendations for air pollution control.

The structure of this paper is as follows: Part 2 introduced the methodology and data used in this paper. Part 3 illustrated our calculation results and analysis of the effect of air pollution control policies on various cities in BTH region since March 2016. Part 4 provided conclusions and related policy recommendations.

2. Materials and Methods

2.1. Data

The data adopted by this paper came from the official daily air quality data and pollutant monitoring data published by the Data Center of China's Ministry of Environmental Protection [23], the City Air Quality Publishing Platform of China's National Environmental Monitoring Center [26], as well as local governments of Beijing, Tianjin and Hebei. The data range from February 2015 to January 2018, and included the daily average concentration numbers of 6 main air pollutants ($PM_{2.5}$, PM_{10}, CO, NO_2, O_3, and SO_2).

2.2. Methods

Because this paper has extended the official AQI indicator system of the Chinese government to six main pollutants and 18 indicators, we first selected 3 variables of heterogeneous information (interval number, mean value, and mean variance) for each indicator, and then obtained the standardization form of these variables by common practice and calculated the distance between heterogeneous information and its positive thresholds (the corresponding minimum value of each attribute indicator during the observation period) and negative thresholds (the corresponding maximum value of each attribute indicator during the observation period). Then we further adopted the Collaborative Filtering Model that helps to sort and select the optimal assessment method given multiple indicators in order to determine the differentiated weights of different indicators. Finally, we calculated the air quality assessment scores of 13 cities by the Backward Cloud Model [27].

2.2.1. Construction and Standardization of Heterogeneous Information

We selected 3 variables of heterogeneous information for each main pollutant indicator—interval number, mean value, and mean variance. Among these variables, the mean value and mean variance can be written as real number $e_{ij}(j \in C_1)$. We first obtained the standardization form of e_{ij} as x_{ij}:

$$x_{ij} = \begin{cases} \dfrac{e_{ij}}{e_{maxj}}, & j \in C_1^b \\ 1 - \dfrac{e_{ij}}{e_{maxj}}, & j \in C_1^c \end{cases} \tag{3}$$

in which $e_{maxj} = \max\{e_{ij} | i = 1, 2, \ldots, m\}$.

The standardization form of the interval number $e_{ij} = \left[e_{ij}^-, e_{ij}^+ \right], j \in C_2$ can be written as:

$$x_{ij} = \begin{cases} \left[\dfrac{e_{ij}^-}{e_{maxj}^+}, \dfrac{e_{ij}^+}{e_{maxj}^+} \right], & j \in C_2^b \\ \left[1 - \dfrac{e_{ij}^+}{e_{maxj}^+}, 1 - \dfrac{e_{ij}^-}{e_{maxj}^+} \right], & j \in C_2^c \end{cases} \tag{4}$$

in which $e_{maxj}^+ = max\left\{ e_{ij}^+ | i = 1, 2, \ldots, m \right\}$.

2.2.2. Calculate the Distance between the Heterogeneous Information and Its Positive and Negative Thresholds

In order to compare different assessment methods, let x^+ and x^- be the positive and negative thresholds of the heterogeneous information respectively, i.e., the extremal solutions of the best case and worst case scenario. Therefore, if x_{ij} represents the jth attribute value of the ith indicator, its distance from its positive threshold value, $d(x_{ij}, x^+)$ can be calculated by:

$$d(x_{ij}, x^+) = \begin{cases} d\left(d_{ij}, d_i^+\right)^2, & j \in C_1 \\ \frac{1}{2}\left[\left(d_{ij}, e_i^{-+}\right)^2 + \left(d_{ij}, e_i^{++}\right)^2\right], & j \in C_2 \\ \frac{1}{3}\left[\left(a_{ij}, a_i^+\right)^2 + \left(b_{ij}, b_i^+\right)^2 + \left(c_{ij}, c_i^+\right)^2\right], & j \in C_3 \end{cases} \tag{5}$$

While its distance from its negative threshold value, $d(x_{ij}, x^-)$ can be calculated by:

$$d(x_{ij}, x^-) = \begin{cases} d\left(d_{ij}, d_i^-\right)^2, & j \in C_1 \\ \frac{1}{2}\left[\left(d_{ij}, e_i^{--}\right)^2 + \left(d_{ij}, e_i^{+-}\right)^2\right], & j \in C_2 \\ \frac{1}{3}\left[\left(a_{ij}, a_i^+\right)^2 + \left(b_{ij}, b_i^+\right)^2 + \left(c_{ij}, c_i^+\right)^2\right], & j \in C_3 \end{cases} \tag{6}$$

2.2.3. Decide Indicator Weights by Using Collaborative Filtering Algorithm

After obtaining the distance between the heterogeneous information from its positive and negative threshold values, we measured the differentiation between various indicators by taking the opposite number of their similarity value calculated by the MSD Similarity Formula.

$$d(y_i, y_j) = \frac{\sum_{h=1}^{card(S_{ij})}\left(b_{hi} - b_{hj}\right)^2}{card(S_{ij})} \tag{7}$$

in which $d(y_i, y_j)$ is the differentiation between indicator y_i and y_j; S_{ij} is the set of all assessment models that cover both y_i and y_j; b_{hi} is the standardized assessment score of indicator y_i by assessment model S_h.

The mean differentiation of indicator y_i with all other indicators, $\overline{d_i}$ can be written as:

$$\overline{d_i} = \frac{\sum_{j=1}^{n} d_{ij}}{n} \tag{8}$$

In equation (8), d_{ij} is the differentiation between the ith indicator y_i and the jth indicator y_j. The differential weight of indicator y_j, ω_i can be expressed as:

$$\omega_i = \frac{\overline{d_i}}{\sum_{i=1}^{n} \overline{d_i}} \tag{9}$$

2.2.4. The Backward Cloud Model

In order to combine quantitative and qualitative assessment, we selected Backward Cloud Model with no specific degrees to calculate the air quality score of different cities. First, we obtained the mean value of the air quality scores $(\overline{X})$ based on the information of n indicators (x_i).

$$\overline{X} = \frac{1}{n}\sum_{i=1}^{n} x_i \tag{10}$$

This average value $\overline{X}$ is the expected value of air quality of this city. This is the best indicator for qualitative assessment of a city's air quality.

Given the expected value, we can further calculate the entropy of air quality of different cities, En:

$$En = \sqrt{\frac{\pi}{2}} \times \frac{1}{n} \sum_{i=1}^{n} |x_i - \overline{X}| \tag{11}$$

This entropy (En) means the width of the information, which represents the uncertainty and ambiguity in one city's air quality score. The bigger the entropy's value is, the higher the uncertainty becomes.

We then further obtained the hyper entropy (He) of each city's air quality score through:

$$He = \sqrt{|S^2 - En^2|} \tag{12}$$

in which S^2 is the variance of various assessment models against their respective expectation value Ex.

This hyper entropy (He) reflects the uncertainty of various entropy values by showing the dispersion degree of fuzzy information. The bigger the hyper entropy value is, the more disperse a city's air quality score is, and the more randomness there is. A smaller hyper entropy value means less uncertainty and randomness, and better air quality of a city. The larger the evaluation value obtained, the worse was the air quality of the city at that time. Therefore, with help of the Backward Cloud Model, we obtained the qualitative result expressed by a certain number and realized the integration of quantitative scores and qualitative expression, able to qualitatively describe a city's air quality based on a quantitative number.

3. Results

With help of the Collaborative Filtering Backward Cloud Model discussed in 3.1 and the MATLAB algorithm we developed (refer to Appendix A) and based on the pollutant data listed in 3.2, we calculated the Air Quality Assessment Score of 13 cities in BTH region from February 2015 to January 2018 (1095 days) as shown below through Tables 1–4.

Table 1. Air quality assessment score of cities in the BTH region (2015.02–2015.10).

	2015-02	2015-03	2015-04	2015-05	2015-06	2015-07	2015-08	2015-09	2015-10
Baoding	0.8440	0.8561	0.7136	0.6942	0.6789	0.6942	0.7094	0.6473	0.7373
Beijing	0.4373	0.4329	0.4446	0.4723	0.5354	0.4845	0.4516	0.4808	0.4200
Cangzhou	0.5017	0.4831	0.3742	0.6043	0.4379	0.0800	0.4219	0.4058	0.2640
Chengde	0.3135	0.2663	0.1712	0.3941	0.1988	0.3590	0.2240	0.2055	0.1351
Handan	0.5969	0.3893	0.5457	0.5285	0.6499	0.5829	0.8147	0.7053	0.7513
Hengshui	0.7697	0.6868	0.5638	0.6405	0.6189	0.6357	0.6046	0.8389	0.8801
Langfang	0.3117	0.3767	0.5031	0.4415	0.4170	0.4393	0.5946	0.6149	0.5629
Qinhuangdao	0.3005	0.3197	0.3800	0.3050	0.1375	0.1286	0.2647	0.1929	0.1937
Shijiazhuang	0.4626	0.6096	0.5823	0.3846	0.4678	0.2947	0.4143	0.4639	0.4341
Tangshan	0.6236	0.8502	0.9725	0.9995	0.7670	0.7661	0.6697	0.7291	0.7173
Tianjin	0.3130	0.3760	0.0886	0.2781	0.4627	0.0627	0.2310	0.2190	0.2366
Xingtai	0.7590	0.6527	0.5737	0.5618	0.7026	0.6843	0.8039	0.7065	0.7335
Zhangjiakou	0.2422	0.2290	0.2675	0.4207	0.2968	0.1705	0.2582	0.1371	0.0819

Within our study period, the most important air pollution control policy by the Chinese government is the one announced by Prime Minister Li Keqiang in the "Government Work Report" (March 2016) that "we must prioritize the control of air pollution and water pollution with the goal of reducing chemical oxygen demand (COD) and ammonia-nitrogen emissions by 2%, reducing the emissions of sulfur dioxide and oxynitride by 3% and controlling the concentration of PM$_{2.5}$ in key areas" [28]. As the key area listed in the "Government Work Report", the BTH region has made great effort on air pollution control under the policy guidance of the central government since March 2016. In order to depict the effect of such air pollution control policy, we adopted the Back Propagation (BP) Neural Network method to simulate the Background Trend Line of the dynamic change in air quality in absence of these pollution control policies in cities of the BTH region, and compared with the actual numbers (especially since the air pollution control campaign that started in March 2016) from below Tables 2–5, in order to quantify the influence of policy on pollution control of different cities.

Table 2. Air quality assessment score of cities in the BTH region (2015.11–2016.07).

	2015-11	2015-12	2016-01	2016-02	2016-03	2016-04	2016-05	2016-06	2016-07
Baoding	0.7371	0.6991	0.6298	0.8750	0.7225	0.6364	0.6870	0.6127	0.6329
Beijing	0.4446	0.3742	0.1996	0.2896	0.5057	0.3786	0.5397	0.5779	0.5328
Cangzhou	0.7212	0.4504	0.3726	0.5587	0.5211	0.4021	0.4265	0.4103	0.4098
Chengde	0.2902	0.1759	0.3461	0.4294	0.2476	0.3135	0.3312	0.3481	0.3446
Handan	0.6415	0.7078	0.6554	0.8445	0.7648	0.7955	0.2903	0.5024	0.3518
Hengshui	0.7449	0.7871	0.7306	0.9171	0.7632	0.6172	0.5894	0.6422	0.6467
Langfang	0.5093	0.4317	0.3473	0.3294	0.4516	0.2237	0.2164	0.4799	0.5971
Qinhuangdao	0.2070	0.1576	0.0904	0.1124	0.2889	0.3187	0.3366	0.2894	0.1180
Shijiazhuang	0.6835	0.5144	0.5899	0.6014	0.7660	0.3267	0.5278	0.4469	0.5410
Tangshan	0.6982	0.4357	0.4623	0.6462	0.8139	0.7107	0.8880	0.9609	0.5922
Tianjin	0.4943	0.2709	0.2772	0.1672	0.4672	0.4470	0.3450	0.4395	0.0983
Xingtai	0.6978	0.6196	0.6253	0.7029	0.7446	0.5027	0.5205	0.5598	0.7750
Zhangjiakou	0.4137	0.2942	0.3691	0.3947	0.2122	0.1761	0.1561	0.0899	0.2541

Table 3. Air quality assessment score of cities in the BTH region (2016.08–2017.04).

	2016-08	2016-09	2016-10	2016-11	2016-12	2017-01	2017-02	2017-03	2017-04
Baoding	0.3092	0.6628	0.5853	0.5509	0.5631	0.6854	0.6255	0.6643	0.5634
Beijing	0.4877	0.2403	0.3245	0.2401	0.3293	0.3087	0.1616	0.3003	0.2197
Cangzhou	0.2731	0.5797	0.5798	0.5681	0.6649	0.3238	0.5536	0.6097	0.6751
Chengde	0.1787	0.0767	0.1436	0.2105	0.1548	0.2277	0.2727	0.2135	0.2898
Handan	0.5420	0.7057	0.4738	0.6708	0.7452	0.6672	0.7120	0.6064	0.7989
Hengshui	0.6088	0.6651	0.5344	0.6288	0.4565	0.5000	0.5819	0.6314	0.6944
Langfang	0.7345	0.3213	0.1966	0.2460	0.4123	0.3987	0.4081	0.4372	0.4947
Qinhuangdao	0.2415	0.2363	0.2490	0.2354	0.2881	0.4494	0.3947	0.4700	0.4536
Shijiazhuang	0.4370	0.6672	0.7702	0.7248	0.6939	0.6808	0.7176	0.7575	0.5961
Tangshan	0.4957	0.7255	0.6287	0.5398	0.5315	0.4429	0.5363	0.6436	0.7999
Tianjin	0.2715	0.4562	0.3333	0.3651	0.3652	0.3113	0.3510	0.6127	0.7321
Xingtai	0.3989	0.7404	0.5957	0.5495	0.6229	0.6827	0.8329	0.7001	0.5804
Zhangjiakou	0.2147	0.0795	0.2242	0.2973	0.2653	0.3020	0.2287	0.1595	0.3066

Table 4. Air quality assessment score of cities in the BTH region (2017.05-2018.01).

	2017-05	2017-06	2017-07	2017-08	2017-09	2017-10	2017-11	2017-12	2018-01
Baoding	0.7170	0.6288	0.6332	0.5885	0.7320	0.4273	0.7028	0.7288	0.6464
Beijing	0.4714	0.3771	0.4886	0.2573	0.3874	0.2164	0.1696	0.1455	0.2238
Cangzhou	0.6445	0.5570	0.4912	0.4590	0.4080	0.6839	0.5789	0.6079	0.5319
Chengde	0.2721	0.0333	0.0113	0.0156	0.0321	0.0345	0.1520	0.1651	0.2904
Handan	0.8597	0.7773	0.4710	0.8398	0.6658	0.6679	0.7722	0.7462	0.8278
Hengshui	0.6058	0.4576	0.3713	0.4443	0.3034	0.5248	0.5525	0.5562	0.6148
Langfang	0.5550	0.4469	0.5713	0.3136	0.4420	0.2623	0.1900	0.2319	0.2227
Qinhuangdao	0.1415	0.3120	0.2403	0.2963	0.3901	0.4707	0.4547	0.3096	0.2895
Shijiazhuang	0.8026	0.7218	0.7478	0.6380	0.7405	0.5206	0.6798	0.6535	0.7337
Tangshan	0.8857	0.8552	0.7815	0.7767	0.7740	0.9348	0.6063	0.5592	0.4841
Tianjin	0.5620	0.2684	0.3644	0.3660	0.4094	0.4932	0.3299	0.3422	0.3066
Xingtai	0.9449	0.8580	0.8058	0.8240	0.7968	0.7023	0.8016	0.7332	0.8120
Zhangjiakou	0.2200	0.1316	0.3945	0.1164	0.0076	0.1465	0.2841	0.3168	0.2854

By calculations under the Back Propagation (BP) Neural Network (refer to Appendix B for calculation principles and MATLAB algorithm), we obtained the Output Layer result of the Background Trend Line of the dynamic change in air quality in absence of pollution control policies in cities of the BTH region (see Tables 5 and 6).

Table 5. Air quality assessment score of cities in the BTH region in absence of pollution control policies simulated by BP Neural Network (2016.03–2017.02).

	2016-03	2016-04	2016-05	2016-06	2016-07	2016-08	2016-09	2016-10	2016-11	2016-12	2017-01	2017-02
Baoding	0.7141	0.6426	0.6936	0.5723	0.7663	0.7724	0.6455	0.5991	0.6180	0.7840	0.8399	0.6400
Beijing	0.3049	0.2277	0.3364	0.3493	0.4288	0.4765	0.3892	0.5081	0.3669	0.5614	0.4194	0.2388
Cangzhou	0.6544	0.4714	0.4608	0.6713	0.6385	0.5143	0.5513	0.8875	0.7668	0.5740	0.5653	0.8375
Chengde	0.2007	0.1839	0.2299	0.3231	0.3848	0.3714	0.0989	0.1431	0.1857	0.3122	0.3320	0.2900
Handan	0.7429	0.8698	0.5428	0.7207	0.5328	0.7528	0.6692	0.7502	0.6657	0.8170	0.6676	0.8009
Hengshui	0.8102	0.6837	0.6847	0.6510	0.7660	0.6932	0.6373	0.6334	0.5426	0.5414	0.6892	0.4523
Langfang	0.3677	0.2627	0.2616	0.4169	0.4979	0.5571	0.4164	0.4421	0.4618	0.4716	0.4404	0.4145
Qinhuangdao	0.2321	0.2816	0.2950	0.2332	0.1304	0.1523	0.1672	0.2275	0.2729	0.2912	0.2811	0.2364
Shijiazhuang	0.5435	0.2637	0.4381	0.4880	0.5605	0.3365	0.5877	0.5408	0.4270	0.6493	0.6235	0.7687
Tangshan	0.8113	0.7203	0.9273	0.6284	0.6987	0.6137	0.8433	0.4622	0.6677	0.6327	0.6958	0.4147
Tianjin	0.2904	0.4180	0.4251	0.4001	0.1298	0.2197	0.4675	0.3188	0.3793	0.3588	0.1361	0.1890
Xingtai	0.7526	0.6157	0.6479	0.6187	0.6765	0.6574	0.7214	0.5547	0.7381	0.6489	0.5686	0.8927
Zhangjiakou	0.3478	0.2511	0.2808	0.2428	0.3394	0.1671	0.1872	0.1969	0.1812	0.2373	0.4282	0.3372

Table 6. Air quality assessment score of cities in the BTH region in absence of pollution control policies simulated by BP Neural Network (2017.03–2018.01).

	2017-03	2017-04	2017-05	2017-06	2017-07	2017-08	2017-09	2017-10	2017-11	2017-12	2018-01
Baoding	0.5835	0.7104	0.7298	0.7840	0.6943	0.6023	0.6213	0.7601	0.8161	0.6563	0.5962
Beijing	0.4722	0.2262	0.3870	0.2695	0.1791	0.4292	0.1752	0.3647	0.3668	0.4232	0.5592
Cangzhou	0.8008	0.6312	0.6974	0.8698	0.5925	0.5720	0.6442	0.8901	0.7110	0.6378	0.7566
Chengde	0.1187	0.2083	0.2715	0.2756	0.2806	0.0710	0.0947	0.1265	0.3133	0.3152	0.2944
Handan	0.6248	0.8848	0.6933	0.8227	0.5450	0.7733	0.6853	0.7818	0.7096	0.8973	0.6596
Hengshui	0.6033	0.5944	0.5077	0.6647	0.4597	0.4780	0.4740	0.5312	0.5630	0.5836	0.5839
Langfang	0.3855	0.4036	0.4090	0.3492	0.4113	0.3605	0.4343	0.3678	0.3757	0.4592	0.4507
Qinhuangdao	0.2361	0.2571	0.1853	0.3129	0.2706	0.2246	0.2333	0.2393	0.2493	0.2748	0.3849
Shijiazhuang	0.5336	0.4076	0.4625	0.6518	0.6150	0.5515	0.5547	0.5504	0.3813	0.5402	0.6140
Tangshan	0.7088	0.7663	0.9073	0.5687	0.7585	0.7460	0.8443	0.5906	0.6523	0.5995	0.7192
Tianjin	0.5981	0.4961	0.3449	0.2995	0.2353	0.1183	0.3231	0.4765	0.4491	0.4385	0.0569
Xingtai	0.7000	0.5182	0.9210	0.7241	0.6244	0.8950	0.7228	0.5857	0.8659	0.6564	0.6443
Zhangjiakou	0.2997	0.2921	0.2385	0.1762	0.2177	0.1173	0.1909	0.2273	0.3975	0.4304	0.3504

Finally, we can make the comparison between air quality assessment scores simulated by BP Neural Network and actual air quality scores of cities in the BTH region (see Figure 2).

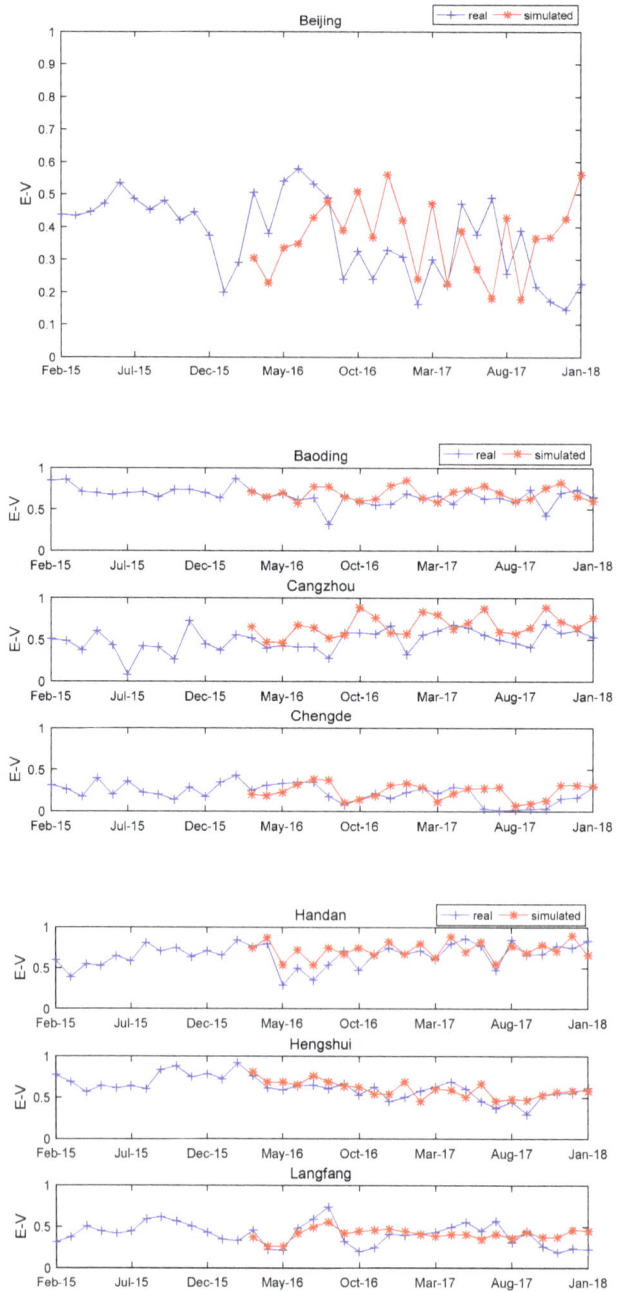

Figure 2. *Cont.*

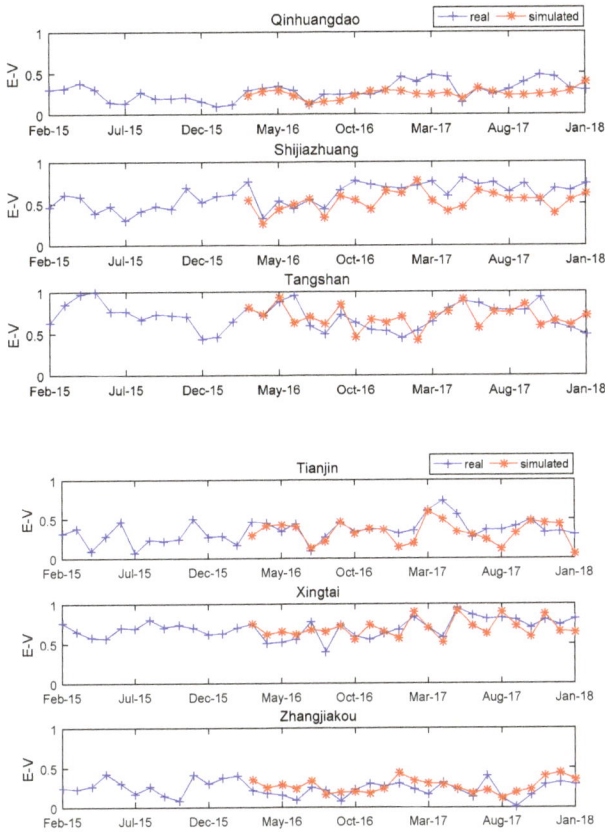

Figure 2. Comparison between air quality assessment scores simulated by BP Neural Network and actual air quality scores of cities in the BTH region.

4. Discussion

According to our research method, the larger the evaluation value obtained, the worse was the air quality of the city at that time. Through comparison between the air quality assessment scores simulated by BP Neural Network and the actual air quality scores, we found that the air pollution control policy since March 2016 has shown different effectiveness and impact on cities of the BTH region as below:

(1) Beijing's air quality scores have shown improvement since August 2016 after the pollution control policy was implemented, but have also experienced fluctuations before April 2017. Beijing has shown success in air pollution control since October 2017, and greatly improved its ranking in air quality among the 13 cities of BTH region from 2017 to 2018. It has even ranked top one for 3 months and ranked among the top three for 8 months in 2017. The air quality of Beijing has improved 55.74% from 0.5057 in March 2016 to 0.2238 in January 2018, and improved 48.82% since the beginning of our study period (February 2015). Behind this remarkable improvement, Beijing has made heavy investment and issued numerous administrative orders with Chinese characteristics.

- From 2016 to 2017, Beijing has invested as much as 34.78 billion RMB in air pollution control, which is almost 7 times of its investment in 2013 [29]. The biggest investment is on "Replacing Coal and Reducing Nitrogen Emission", i.e., facilitating the energy source change from coal to clean energy and reducing the annual coal consumption of Beijing from 22.7 million tons in 2013 to less than 6 million tons by end of 2017 through administrative orders and equipment upgrade [30]. One of the administrative orders with Chinese characteristics is from Beijing Construction Committee on September 15th 2017 that all road construction work (including any earthwork and house demolition) and hydraulic engineering projects must suspend from November 15th 2017 to March 15th 2018 (the heating season in Beijing) in order to completely eliminate construction dust [31]. Only one month after this administrative order (from October 2017 to January 2018), Beijing has achieved the best air quality score during the entire study period.

- Beijing has prioritized the policy control on high-emission vehicles by extending its forbidden area. In November 2016, Beijing government has issued its revised "Air Pollution Emergency Plan" which stipulates that on days of Air Pollution Orange Alert and Red Alert, all light-duty gasoline vehicles with National Level I and Level II Emission Standards are forbidden on the road in the whole city [32]. Since February 15th 2017, Beijing government has further forbidden cars with National Level I and Level II Emission Standards driving in areas within the 5th Ring Road during workdays (whole day) [33]. By end of 2017, Beijing government has forced the retirement of 2.17 million old motor vehicles, which accounted for 36.12% of its total motor vehicles [30].
- Beijing has implemented more strict elimination policies for polluting companies. The "Production Technology and Equipment Upgrade / Retirement List of Heavy Polluting Industries of Beijing" was officially effective in July 2017, which strictly requires to remove 115 production technologies and 57 production equipment of 11 industries including the steel industry, non-ferrous metals industry, building materials industry, chemical industry, textile printing and dyeing industry, papermaking industry, etc. within a specified time limit and forbids starting or extending any similar projects [34]. By the end of 2017, Beijing government has cleaned up around 11 thousand heavy pollution companies [30]. However, it's worth noticing that a large number of those companies (especially large industrial companies) simply moved from Beijing to a nearby city. The result is moving the pollution sources from Beijing to Hebei province.

(2) As the other municipality directly under the central government in this region, Tianjin has experienced large fluctuations during the study period. Although its air quality has improved by 34.38% after implementation of the pollution control policies, its air quality score has once dropped to the worst level of 0.7321 in April 2017 and gradually improved afterwards. Its air quality score in January 2018 only improved by 2.04% compared to its level in February 2015. According to the inspection result on Tianjin provided by the Environmental Protection Inspectorate sent by the central government in July 2017, the execution as well as effectiveness of Tianjin's air pollution control policy has large fluctuations with even worse air quality in several periods. With high concentration of heavy and chemical industries in the city and severe structural pollution, Tianjin still initialized or planned to initialize several thermal power projects without regard to the environment, which resulted in a large increase in the concentration of nitrogen dioxide in the atmosphere in 2016, and an increase of $PM_{2.5}$ by 27.5% in the first quarter of 2017 [35]. Although thee Tianjin government has taken a series of remedial measures including longer suspension period than Beijing—from October 2017 to March 2018, all road construction work, hydraulic engineering projects, earthwork, house demolition and cement mixing work are paused in Tianjin's urban area [36]. However, the policy has not shown much effectiveness so far.

(3) Baoding and Cangzhou have shown most overall policy effectiveness within the study period with their own characteristics. Although Baoding has achieved an improvement of 23.41% in air quality score in January 2018 compared with that of the beginning period, except for August 2016 and October 2017, its air quality score has been above 0.50 for most of the study period. Cangzhou has shown lower air quality score than other cities in the study period partly due to its geographical location close to the Bohai Sea. However, since October 2017, its air quality score has deteriorated to above 0.5, resulting in an air quality score in January 2018 that has declined by 6.02% compared with that of the beginning period. We have noticed that in June 2017, Cangzhou Bohai New Area planned a "Beijing Enterprise Zone" in order to receive the immigrating companies of non-capital functions

from Beijing. By the end of 2017, almost one thousand companies have settled down in this "Beijing Enterprise Zone", among which there are nearly 800 clothes manufacturing companies [37].

(4) The air pollution control policy of Handan, Hengshui, Xingtai and Zhangjiakou has shown low effectiveness within the study period with large differences in air quality scores. All these cities have experienced a decline of air quality scores when comparing the last period with the beginning period, except for Hengshui (the score of Handan has declined by 38.68%, that of Zhangjiakou declined by 17.84%, while that of Xingtai declined by 6.98%). We have noticed that these 4 cities have all received large numbers of polluting companies that migrated out of Beijing in the study period. In 2014, Beijing government decided to move its Lingyun Building Materials & Chemical Co.,Ltd. from Beijing to Handan, which was the first central-government-owned enterprise that was forced to migrate out of Beijing during our study period and received its production permit from Handan government in October 2015. Before that, this company emitted 400 thousand tons of carbon dioxide, 9 thousand tons of sulfur dioxide, and 10 thousand tons of dust and fume in Beijing every year [38]. In addition, as one of the leading textile printing and dyeing companies of Beijing, Victor's Clothing Company also migrated to Hengshui in 2015, only leaving its head office and design center in Beijing [39]. All these migrating companies plus the existing polluting companies in these cities such as Handan Iron and Steel Group Company, panel and plate processing companies in Xingtai, chemical plants in Hengshui, and emissions from the growing numbers of motor vehicles in Zhangjiakou in recent years—the various factors have offset the effects of the air pollution control policies.

(5) The air quality score of Langfang, which is located between the two municipalities directly under the central government—Beijing and Tianjin, dropped to the worst level of 0.7345 in August 2016, ranking bottom among the 13 cities in our study scope. However, after that, its air quality score has seen distinct improvement and reached its best level of 0.2227 in January 2018 (improved by 28.55% compared with its beginning level), ranking top among the 13 cities. Located in the ecological conservation area north-west of Beijing, Chengde has kept an outstanding air quality record of under 0.40. We noticed that the air quality score of Chengde first dropped to the worst level during June and October 2017 but then climbed up. Its air quality score has only improved by 7.37% when comparing that of the ending period with the beginning period. This result shows big fluctuations in air quality and policy effectiveness of these 2 cities in our study period and needs further enhancement in the future.

(6) Tangshan's case is a little special. Although its air quality in January 2018 has improved by 22.37% compared with the beginning of the period, its air quality score has ranked bottom in 9 months across the study period of 36 months, and has showed no sign of improvement until October 2017. In order to understand the reason behind, we must be aware that from 2014 to 2017, Tangshan received the most industries that migrated from Beijing and Tianjin among cities in Hebei province, with total investment of 575.1 billion RMB and 442 projects of investment over 100 million RMB, including large heavy-pollution industrial companies such as Capital Iron and Steel Company and Beijing Coking and Chemical Plant [40]. The Caofeidian District of Tangshan with large numbers of immigrating companies from Beijing and Tianjin is only one-hour drive from downtown Tangshan [41]. Therefore, the moving-in of industrial companies has caused huge impact on the air quality of Tangshan. That is why Tangshan government appropriated 66.70 million RMB from its fiscal income and constructed an air quality grid monitoring and decision-making support system with high accuracy for the purpose of air pollution monitoring and control which was officially launched in September 2017. This system has integrated resources from various government departments including the environmental protection department, public security department, housing development department and land department, and installed almost 600 miniaturized and integrated online monitoring devices with international standards in the urban area [42]. Moreover, Tangshan has put great emphasis on staggering peak production of iron and steel companies. Since November 15th 2017, Tangshan government has demanded that all of its 35 iron and steel companies adopt staggering-peak production [43]. For example, Tangshan Iron and Steel Co., Ltd. under Hebei Iron

and Steel Group has limited its steel production to 477.5 thousand tons of by suspending the operation of blast furnaces [44], which has greatly helped the improvement of air quality since October 2017.

(7) The air pollution control policy has achieved little effect in Shijiazhuang and Qinhuangdao. The air quality score of Shijiazhuang in January 2018 has deteriorated by 58.60% when compared with that of the beginning period. The possible reasons are: First, the geographic location of Shijiazhuang is very close to the Taihang Mountains, which blocks the wind or air circulation and causes air pollutants to linger above the city, creating difficulty for the clean-up of air pollution [45]. Moreover, apart from its own polluting industries including the iron and steel industry and cement industry, Shijiazhuang has also received large numbers of polluting industries from Beijing and Tianjin in recent years, including the building materials industry, leather manufacturing industry, pharmaceuticals industry, etc. [46]. Many of these polluting companies have set their new location to be between Tangshan and Qinhuangdao, which has impacted the air quality of Qinhuangdao and offset the effectiveness of air pollution control policies to some extent [47].

5. Conclusions

This paper calculated and assessed the air quality of 13 cities of the BTH region from February 2015 to January 2018 based on the extended AQI indicator system. By constructing and standardizing Heterogeneous Information of major pollutant indicators including interval number, mean value, and variance, we depicted the important effect of other relevant features of pollutant indicators beyond single-point data. Based on that, we further calculated the air quality scores of different cities in the BTH region by using the Collaborative Filtering Backward Cloud Model to construct differentiated weights of different indicators. With help of the Back Propagation (BP) Neutral Network, we simulated the effect of the pollution control policies of the Chinese government targeting air pollution since March 2016. Our conclusion is: the pollution control policies have improved the air quality of Beijing by 55.74%, and improved the air quality of Tianjin by 34.38%; while the migration of polluting enterprises from Beijing and Tianjin has caused different changes in air quality in different cities of Hebei province—we saw air quality deterioration by 58.60% and 38.68% in Shijiazhuang and Handan city respectively. Based on findings above, we provided below policy recommendations for air pollution control of the BTH region:

(1) Embrace more market measures than administrative orders in the battle against air pollution. Currently, most of the measures targeting the air pollution in BTH region are administrative orders and penalty. Although these administrative orders and penalty have achieved certain results, these tools are not efficient or sustainable enough and do not match with the requirement under market economy. Therefore, in the future battle against air pollution, apart from improving the accuracy of air quality measurement, we should also design more tax categories for specific pollutant emissions, such as carbon tax, sulfur dioxide tax, and $PM_{2.5}$ tax. At the same time, we should convert the current environmental protection fee to corresponding local tax; decrease the production of pollution products by income effect and substitution effect of tax; encourage companies to save energy [48,49] and cut emissions in order to solve the issue of pollution [50,51].

(2) Improve the compensation system for both economic and environmental loss during industry migration in the BTH region. During the air pollution control campaign of Beijing and Tianjin, large numbers of polluting companies moved to cities in Hebei province, including some heavy pollution companies such as the Capital Iron and Steel Company and Beijing Coking and Chemical Plant that moved to Tangshan, the Lingyun Building Materials & Chemical Co., Ltd. that moved to Handan, Beijing's No. 1 Machine Tool Plant that moved to Baoding, etc. This impacted the air pollution control work of cities in Hebei province to some extent. Therefore, the BTH region should establish and improve the compensation system for industry migration and industry upgrade in this region. Based on the overall industry plan of this region, the government should be fully aware of the economic development and environmental protection pressure on these destination cities of polluting industries, and offer sufficient compensation in terms of economic development and environmental protection

resources in order to realize a fair competition within this region and achieve synergy in regional economic development.

(3) Develop pollution control technologies and continuously improve air quality through technological advancement. On one hand, we should encourage colleges and scientific research institutions in this region to continue working on air pollution control technologies, and enhance the cleansing and control of industrial wastegas and motor vehicle exhaust. On the other hand, we should continuously develop and implement new energy technologies in this region; improve traffic management and green construction in the city; and further reduce pollution by encourage public transportation and other environmentally friendly methods such as walking and cycling.

Author Contributions: G.Y. and W.Y. are joint first authors. They contributed equally to this paper. Conceptualization, W.Y.; Methodology, G.Y. and W.Y.; Resources, G.Y. and W.Y.; Software, G.Y.; Validation, W.Y.; Formal Analysis, G.Y. and W.Y.; Data Curation, G.Y. and W.Y.; Writing—Original Draft Preparation, G.Y. and W.Y.; Writing—Review and Editing, G.Y. and W.Y.

Funding: Guanghui Yuan is financially supported by the National Natural Science Foundation of China (grant number 71271126) and the Graduate Innovation Fund of Shanghai University of Finance and Economics. Weixin Yang is financially supported by the Humanities and Social Sciences Research Fund of the University of Shanghai for Science and Technology, and the Decision-making Consultation Research Project of Shanghai Municipal Government. The authors gratefully acknowledge the above financial supports.

Conflicts of Interest: The authors declare no conflict of interest.

Appendix A. MATLAB algorithm for the Collaborative Filtering Backward Cloud Model

```
load('DATA.mat');
data=DATA;
L=5;
a1=max(data(:,1));
data(:,1)=1-data(:,1)/a1;
data(:,2)=data(:,2);
a2=max(max(data(:,4:5)));
data(:,4:5)=data(:,4:5)/a2;
a3=max(max(data(:,6:8)));
data(:,6:8)=data(:,6:8)/a3;
data1=xiangduizhengtiejindu(data,L);
MSD=chayi(data1); MSD_=mean(MSD,2);
MSDsum=sum(MSD_);
W=MSD_./MSDsum;
W=W';
 [Ex,En,He]=nixiangyun(data1,W);


function [MSD]=chayi(data)
[m,n]=size(data);
for i=1:n
for j=i:n
        AA=[data(:,i) data(:,j)];
        [z1,z2]=find(isnan(AA));
        AA(z1,:)=[[];
        [p,q]=size(AA);
        a=intersect(AA(1),AA(2));
        b=length(a);
        card(i,j)=1-b/(2*p-b);
        %card(i,j)=pdist(AA', 'jaccard');
```

```
            qiuhe(i,j)=mean((AA(:,1)-AA(:,2)).^2);
            %qiuhe(i,j)=sum((AA(:,1)-AA(:,2)).^2);
            msd(i,j)=qiuhe(i,j)./card(i,j);
        end
    end
    MSD=msd+msd';
    for i=1:n
        MSD(i,i)=0;
    end

    function dataZZ=xiangduizhengtiejindu(data,L)
    dataz=max(data);
    dataz(10)=max(data(:,10));
    dataf=min(data);
    dataf(10)=min(data(:,10));
    dataZ(:,1)=(data(:,1)-dataz(1)).^2;
    dataZ(:,2)=1/L.*((data(:,2)+data(:,3)-(dataz(2)+dataz(3))).^2);
    dataZ(:,3)=1/2.*(((data(:,4)-dataz(4)).^2+(data(:,5)-dataz(5)).^2));
    dataZ(:,4)=1/3.*((data(:,6)-dataz(6)).^2+(data(:,7)-dataz(7)).^2+(data(:,8)-dataz(8)).^2);
    dataZ(:,5)=1/3.*((data(:,9)-dataz(9)).^2+(data(:,10)-dataz(10)).^2+((data(:,9)+data(:,10))-
    (dataz(9)+dataz(10))).^2);
    dataF(:,1)=(data(:,1)-dataf(1)).^2;
    dataF(:,2)=1/L.*((data(:,2)+data(:,3)-(dataf(2)+dataf(3))).^2);
    dataF(:,3)=1/2.*(((data(:,4)-dataf(4)).^2+(data(:,5)-dataf(5)).^2));
    dataF(:,4)=1/3.*((data(:,6)-dataf(6)).^2+(data(:,7)-dataf(7)).^2+(data(:,8)-dataf(8)).^2);
    dataF(:,5)=1/3.*((data(:,9)-dataf(9)).^2+(data(:,10)-dataf(10)).^2+((data(:,9)+data(:,10))-
    (dataf(9)+dataf(10))).^2);
    dataZZ=dataZ./(dataZ+dataF);

    function [Ex,En,He]=nixiangyun(UU,W)
    UU=mapminmax(UU',0,1);
    UU=UU';
    [m,n]=size(UU);
    X_=W*UU';
    Ex=X_;
    sum1=zeros(1,m);
    En=zeros(1,m);
    for i=1:m
    for j=1:n
            BB=abs(UU(i,j)-X_(i));
            sum1(i)=sum1(i)+BB;
        end
        En(i)=(pi/2)^2*mean(sum1(i),2);
    end
    S2=zeros(1,m);
    for i=1:m
        S2(i)=var(UU(i,:));
    end
    He=zeros(1,m);
    for i=1:m
```

```
    He(i)=(abs(S2(i)-En(i)^2))^0.5;
end
```

Appendix B. Calculation Principles of BP Neural Network and MATLAB algorithm

Appendix B.1. Calculation Principles of BP Neural Network

In March 2016, Chinese Prime Minister Li Keqiang officially raised in the "Government Work Report" that we must prioritize the control of air pollution and water pollution with the goal of reducing chemical oxygen demand (COD) and ammonia-nitrogen emissions by 2%, reducing the emissions of sulfur dioxide and oxynitride by 3% and controlling the concentration of $PM_{2.5}$ in key areas including Beijing, Tianjin and Hebei [28]. Since March 2016, the Chinese government has made great effort in air pollution control under the aligned policy guidance of the central government. In order to depict the effect of such air pollution control policy, we adopted the Back Propagation (BP) Neural Network method to simulate the Background Trend Line of the dynamic change in air quality in absence of these pollution control policies in cities of the Beijing-Tianjin-Hebei (BTH) region in order to quantify the influence of policy on pollution control of different cities.

The BP Neural Network is a multilayered feedforward network, consisting of the Input Layer, Hidden Layer and Output Layer. In calculations and predictions related to policy analysis, the BP Neural Network with single hidden layer can approximate any continuous function in a bounded region with any specified precision [52]. In our model, the number of neurons (nodes) on the input and output layers of the BP Neural Network equals the number of dimensions of our input vector (pollutant data) and output vector (assessment score). Its topological structure is shown in Figure A1, in which $X = (x_1, x_2, \ldots, x_n)$ represents the Input Vector of pollutant data in the past n days while the expectation value of the $(n+1)^{th}$ day is y. Let the number of nodes on the hidden layer be m, the link weight between the input layer and hidden layer be $w_{ij}(i = 1, 2, \ldots, n; j = 1, 2, \ldots, m)$, the link weight between the hidden layer and output layer be $v_{1j}(j = 1, 2, \ldots, m)$, and the thresholds of nodes on the hidden layer and output layer be $\theta_j(j = 1, 2, \ldots, m)$ and γ respectively.

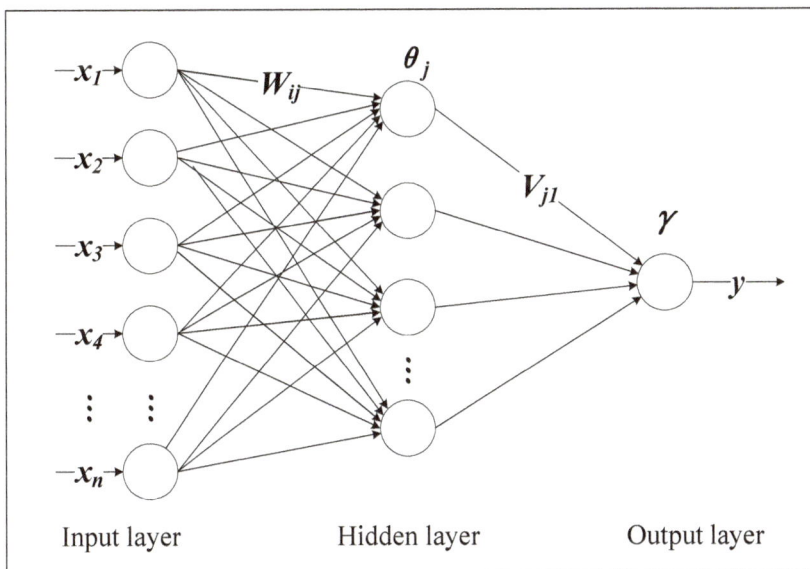

Figure A1. Structure of Single Hidden Layer, Single Output BP Neural Network.

Through forward propagating of input signals (pollutant data) and back propagation of error signals, we can complete the calculation process of such BP Neural Network: propagating the pollutant input vector x_i through the Input Layer, Hidden Layer and Output Layer, and obtaining the estimated output of $\hat{y}$ (assessment score) on the Output Layer by using the link weight of $\hat{y}$, w_{ij} and v_{1j} between different layers as well as randomly assigned threshold values of θ_j and γ and the activation function; propagating e—the error between the output value $\hat{y}$ and the expected value of y through the Input Layer, Hidden Layer and Output Layer, and modifying the link weights between different layers towards the direction of diminishing errors. Assume the number of learning samples is p, expressed by a vector of $(x^1, x^2, \ldots, x^p)$. After obtaining the output vector $y_j^p (j = 1, 2, \ldots, m)$ of the pth sample, we could calculate the error of the pth sample E_p by the square error function.

$$E_p = \frac{1}{2} \sum_{j=1}^{m} \left(t_j^p - y_j^p \right)^2 \tag{13}$$

In above equation, t_j^p is the expected output.
The global error with p samples is:

$$E = \frac{1}{2} \sum_{p=1}^{p} \sum_{j=1}^{m} \left(t_j^p - y_j^p \right)^2 \tag{14}$$

(1) Output Layer Weight Change
Using the BP algorithm of accumulative error to modify ω_{jk} in order to minimize the global error of E:

$$\Delta \omega_{ik} = \mu \frac{\partial E}{\partial \omega_{jk}} = \mu \frac{\partial}{\partial \omega_{jk}} \left(\sum_{p=1}^{p} E_p \right) = \sum_{p=1}^{p} \mu \frac{\partial E_p}{\partial \omega_{jk}} \tag{15}$$

The error signal in above equation is:

$$\Delta \omega_{jk} = \sum_{p=1}^{P} \sum_{j=1}^{m} \mu \left(t_j^p - y_j^p \right) f_2'(S_j) z_k \tag{16}$$

(2) Hidden Layer Weight Change

$$\Delta v_{jk} = \mu \frac{\partial E}{\partial v_{ki}} = \mu \frac{\partial}{\partial v_{ki}} \left(\sum_{p=1}^{p} E_p \right) = \sum_{p=1}^{p} \mu \frac{\partial E_p}{\partial v_{jk}} \tag{17}$$

The equation for weight adjustment of neural networks on the Hidden Layer is as follows:

$$\Delta v_{jk} = \sum_{p=1}^{P} \sum_{j=1}^{m} \mu \left(t_j^p - y_j^p \right) f_2'(S_j) \omega_{jk} f_1'(S_k) x_i \tag{18}$$

By repeating above processes until convergence of the algorithm, we achieved the output layer results of the Background Trend Line of the dynamic air quality change in absence of air pollution control policies in cities of the BTH region (as listed in the Tables 5 and 6).

Appendix B.2. MATLAB Algorithm

```
function [A]=BP(BB)
 [N,M]=size(BB);
for kk=1:N
    x=BB(kk,:);
 lag=6;
```

```
iinput=x;
n=length(iinput);
inputs=zeros(lag,n-lag);
for i=1:n-lag
    inputs(:,i)=iinput(i:i+lag-1)';
end
targets=x(lag+1:end);
hiddenLayerSize = 10;
net = fitnet(hiddenLayerSize);
net.divideParam.trainRatio = 70/100;
net.divideParam.valRatio = 15/100;
net.divideParam.testRatio = 15/100;
 [net,tr] = train(net,inputs,targets);
yn=net(inputs);
errors=targets-yn;
errorsA(kk,:)=errors;

fn=23;

f_in=iinput(n-lag+1:end)';
f_out=zeros(1,fn);
for i=1:fn
    f_out(i)=net(f_in);
    f_in=[f_in(2:end);f_out(i)];
end
A(kk,:)=f_out;
end
end
```

References

1. Australia State of the Environment National Air Quality Standards: Ambient Air Quality. 2016. Available online: https://soe.environment.gov.au/theme/ambient-air-quality/topic/2016/national-air-quality-standards (accessed on 18 December 2018).
2. Yang, W.; Li, L. Efficiency evaluation of industrial waste gas control in China: A study based on data envelopment analysis (DEA) model. *J. Clean. Prod.* **2018**, *179*, 1–11. [CrossRef]
3. Su, P.; Lin, D.; Qian, C. Study on Air Pollution and Control Investment from the Perspective of the Environmental Theory Model: A Case Study in China, 2005–2014. *Sustainability* **2018**, *10*, 2181. [CrossRef]
4. Yang, Y.; Yang, W. Does Whistleblowing Work for Air Pollution Control in China? A Study Based on Three-party Evolutionary Game Model under Incomplete Information. *Sustainability* **2019**, *11*, 324. [CrossRef]
5. Ministry of Environmental Protection of the People's Republic of China. *Technical Regulation on Ambient Air Quality Index (on Trial): HJ 633-2012*; China Environmental Science Press: Beijing, China, 2012.
6. ISO 4220. *Ambient Air—Determination of a Gaseous Acid Air Pollution Index—Titrimetric method with Indicator or Potentiometric End-Point Detection*; International Organization for Standardization: Geneva, Switzerland, 1983.
7. Ministry of Environmental Protection of the People's Republic of China. *Ambient Air Quality Standards: GB3095-2012*; China Environmental Science Press: Beijing, China, 2012.
8. Xinhuanet. Hebei's air quality "burst" and PM2.5 Index in Shijiazhuang is over 1000. Available online: http://www.xinhuanet.com/local/2016-12/19/c_129411383.htm (accessed on 21 April 2018).
9. People's Central Broadcasting Station What is the Feeling When the PM2.5 Index has Exceeded 1400? Look at Shenyang. Available online: http://news.cnr.cn/native/gd/20151109/t20151109_520447867.shtml (accessed on 21 April 2018).

10. Ministry of Environmental Protection of the People's Republic of China. *The "Twelfth Five-Year Plan" for Prevention and Control of Atmospheric Pollution in Key Regions*; Clean Air Alliance of China: Beijing, China, 2012.
11. Lang, J.; Cheng, S.; Wei, W.; Zhou, Y.; Wei, X.; Chen, D. A study on the trends of vehicular emissions in the Beijing-Tianjin-Hebei (BTH) region, China. *Atmos. Environ.* **2012**, *62*, 605–614. [CrossRef]
12. Xu, J.; Wang, X.; Zhang, S. Risk-based air pollutants management at regional levels. *Environ. Sci. Policy* **2013**, *25*, 167–175. [CrossRef]
13. Zhao, P.; Dong, F.; Yang, Y.; He, D.; Zhao, X.; Zhang, W.; Yao, Q.; Liu, H. Characteristics of carbonaceous aerosol in the region of Beijing, Tianjin, and Hebei, China. *Atmos. Environ.* **2013**, *71*, 389–398. [CrossRef]
14. Sheng, L.; Lu, K.; Ma, X.; Hu, J.; Song, Z.; Huang, S.; Zhang, J. The air quality of Beijing-Tianjin-Hebei regions around the Asia-Pacific Economic Cooperation (APEC) meetings. *Atmos. Pollut. Res.* **2015**, *6*, 1066–1072. [CrossRef]
15. Miao, Y.; Liu, S.; Zheng, Y.; Wang, S.; Chen, B.; Zheng, H.; Zhao, J. Numerical study of the effects of local atmospheric circulations on a pollution event over Beijing-Tianjin-Hebei, China. *J. Environ. Sci.* **2015**, *30*, 9–20. [CrossRef]
16. Zhou, Y.; Cheng, S.; Lang, J.; Chen, D.; Zhao, B.; Liu, C.; Xu, R.; Li, T. A comprehensive ammonia emission inventory with high-resolution and its evaluation in the Beijing-Tianjin-Hebei (BTH) region, China. *Atmos. Environ.* **2015**, *106*, 305–317. [CrossRef]
17. Han, F.; Xu, J.; He, Y.; Dang, H.; Yang, X.; Meng, F. Vertical structure of foggy haze over the Beijing-Tianjin-Hebei area in January 2013. *Atmos. Environ.* **2016**, *139*, 192–204. [CrossRef]
18. Guo, X.; Fu, L.; Ji, M.; Lang, J.; Chen, D.; Cheng, S. Scenario analysis to vehicular emission reduction in Beijing-Tianjin-Hebei (BTH) region, China. *Environ. Pollut.* **2016**, *216*, 470–479. [CrossRef] [PubMed]
19. Chen, L.; Shi, M.; Li, S.; Gao, S.; Zhang, H.; Sun, Y.; Mao, J.; Bai, Z.; Wang, Z.; Zhou, J. Quantifying public health benefits of environmental strategy of PM2.5 air quality management in Beijing-Tianjin-Hebei region, China. *J. Environ. Sci.* **2017**, *57*, 33–40. [CrossRef] [PubMed]
20. Zhu, L.; Gan, Q.; Liu, Y.; Yan, Z. The impact of foreign direct investment on SO2 emissions in the Beijing-Tianjin-Hebei region: A spatial econometric analysis. *J. Clean. Prod.* **2017**, *166*, 189–196. [CrossRef]
21. Wang, Y.; Liu, H.; Mao, G.; Zuo, J.; Ma, J. Inter-regional and sectoral linkage analysis of air pollution in Beijing-Tianjin-Hebei (Jing-Jin-Ji) urban agglomeration of China. *J. Clean. Prod.* **2017**, *165*, 1436–1444. [CrossRef]
22. Zhang, X.; Shi, M.; Li, Y.; Pang, R.; Xiang, N. Correlating PM2.5 concentrations with air pollutant emissions: A longitudinal study of the Beijing-Tianjin-Hebei region. *J. Clean. Prod.* **2018**, *179*, 103–113. [CrossRef]
23. Data Center of China's Ministry of Environmental Protection. Concentration of main pollutants in the PRD Region, 2015–2018. Available online: http://datacenter.sepa.gov.cn/ (accessed on 17 July 2018).
24. Wang, S.Y.; Zhu, S.S. On fuzzy portfolio selection problems. *Fuzzy Optim. Decis. Mak.* **2002**, *1*, 361–377. [CrossRef]
25. Xu, W.; Ma, J.; Wang, S.Y.; Hao, G. Vague soft sets and their properties. *Comput. Math. Appl.* **2010**, *59*, 787–794. [CrossRef]
26. China's National Environmental Monitoring Center. The City Air Quality Publishing Platform. Available online: http://106.37.208.233:20035/ (accessed on 17 July 2018).
27. Geng, X.; Dong, X. Concept evaluation approach based on rough information axiom and Cloud Model. *Comput. Integr. Manuf. Syst.* **2017**, *23*, 661–669.
28. Li, K. *Government Work Report at the Fourth Session of the Twelfth National People's Congress (March 5, 2016)*; People's Publishing House: Beijing, China, 2016.
29. Luo, Q. Beijing's investment in atmospheric governance exceeds 30 billion yuan in two years. *Beijing Daily*, 13 December 2017. Available online: http://bjrb.bjd.com.cn/html/2017-12/13/content_202216.htm (accessed on 12 February 2019).
30. Chen, J. *Beijing Municipal Government Work Report at the First Session of the Fifteenth Beijing Municipal People's Congress (January 24, 2018)*; Beijing Daily: Beijing, China, 2018.
31. Beijing Municipal Commission of Housing and Urban-Rural Development. *Detailed Implementation Plan for Beijing of the "Action Plan for Comprehensive Prevention and Control of Atmospheric Pollution in the Beijing-Tianjin-Hebei Region and the Surrounding Areas in the Autumn and Winter 2017–2018"*; Beijing Municipal Commission of Housing and Urban-Rural Development: Beijing, China, 2017.

32. Beijing Municipal People's Government. *Beijing Municipal Air Heavy Pollution Emergency Plan (2016 Edition)*; Beijing Municipal People's Government: Beijing, China, 2016.

33. Yang, X. Cars with National Level I and Level II Emission Standards are forbidden within the 5th Ring Road of Beijing during workdays. *Economic Daily*, 2017 13 February.

34. Beijing Municipal People's Government. *Production Technology and Equipment Upgrade & Retirement List of Heavy Polluting Industries of Beijing (2017 Edition)*; Beijing Municipal People's Government: Beijing, China, 2017.

35. The First Environmental Protection Inspectorate of the Central Government. *2017 Tianjin Environmental Protection Supervision Report*; The First Environmental Protection Inspectorate of the Central Government: Tianjin, China, 2017.

36. Tianjin Municipal People's Government. *The Action Plan for the Comprehensive Prevention and Control of Atmospheric Pollution in the Autumn and Winter of 2017-2018 in Tianjin*; Tianjin Municipal People's Government: Tianjin, China, 2017.

37. SOHU Finance Cangzhou Bohai New Area: Beijing's First Choice for Corporate Relocation. Available online: http://www.sohu.com/a/150424973_232843 (accessed on 21 April 2018).

38. China Economic Net Lingyun Building Materials & Chemical Co.,Ltd., Beijing's first Central-Government-Owned Enterprise Relocated in Handan. Available online: http://www.ce.cn/cysc/yq/dt/201405/16/t20140516_2824568.shtml (accessed on 21 April 2018).

39. Hebei Provincial People's Government. *From Contract to Production: The 18 Months of Victor's Relocation to Hengshui*; Hebei Provincial People's Government: Hengshui, China, 2015.

40. Xinhua News Agency Tangshan: Undertaking the Transfer of Beijing and Tianjin Industry and Promoting Economic Development. Available online: http://www.xinhuanet.com/photo/2018-03/24/c_1122585335.htm (accessed on 21 April 2018).

41. Wang, Q.; Li, B. Enterprises migrated to Caofeidian from Beijing. *People's Daily*. 22 December 2017. Available online: http://paper.people.com.cn/rmrb/html/2017-12/22/nw.D110000renmrb_20171222_3-14.htm (accessed on 12 February 2019).

42. Tangshan Environmental Protection Bureau. *Public Tender Announcement for Accurate Monitoring and Decision Support System for Grid Pollution Control of Air Pollution in Tangshan*; Tangshan Environmental Protection Bureau: Tangshan, China, 2017.

43. Tangshan Municipal People's Government. *The scheme of Staggering-Peak Production for Tangshan City's Steel Industry in 2017–2018 Heating Season*; Tangshan Municipal People's Government: Tangshan, China, 2017.

44. Hebei Provincial Department of Environmental Protection Beijing Evening News: Fighting for the blue sky in Tangshan. Available online: http://www.hebhb.gov.cn/xwzx/mtbb/201712/t20171214_58883.html (accessed on 21 April 2018).

45. Chen, J.; Zhang, Y.; Yang, P.; Qian, W.; Wang, X.; Han, J. Pollution process and optical properties during a dust aerosol event in Shijiazhuang. *China Environ. Sci.* **2016**, *36*, 979–989.

46. Kang, A.; Li, Y.; Zhang, B.; Zhong, H. Reasons and Treatment Measures of Haze Formation in Shijiazhuang City. *J. Hebei Univ. Econ. Bus. Compr. Ed.* **2015**, *15*, 89–90, 100.

47. Zhang, B.; Cao, J.; Du, J.; Sun, L. The Distribution of The Atmospheric Pollutants and Its Weather Background in Qinhuangdao. *J. Environ. Manag. Coll. China* **2016**, *26*, 79–82.

48. Yang, W.; Li, L. Energy Efficiency, Ownership Structure, and Sustainable Development: Evidence from China. *Sustainability* **2017**, *9*, 912. [CrossRef]

49. Yang, W.; Li, L. Analysis of Total Factor Efficiency of Water Resource and Energy in China: A Study Based on DEA-SBM Model. *Sustainability* **2017**, *9*, 1316. [CrossRef]

50. Yang, W.; Li, L. Efficiency Evaluation and Policy Analysis of Industrial Wastewater Control in China. *Energies* **2017**, *10*, 1201. [CrossRef]

51. Li, L.; Yang, W. Total Factor Efficiency Study on China's Industrial Coal Input and Wastewater Control with Dual Target Variables. *Sustainability* **2018**, *10*, 2121. [CrossRef]

52. Veelenturf, L.P.J. *Analysis and Applications of Artificial Neural Networks*; Prentice Hall: London, UK, 1995.

sustainability

MDPI

Article

Does Whistleblowing Work for Air Pollution Control in China? A Study Based on Three-party Evolutionary Game Model under Incomplete Information

Yunpeng Yang [†] and Weixin Yang *,[†]

Business School, University of Shanghai for Science and Technology, Shanghai 200093, China; yang_yunpeng@outlook.com
* Correspondence: iamywx@outlook.com; Tel.: +86-21-5596-0082
† Yunpeng Yang and Weixin Yang are joint first authors. They contributed equally to this paper.

Received: 21 November 2018; Accepted: 7 January 2019; Published: 10 January 2019

Abstract: During China's air pollution campaign, whistleblowing has become an important way for the central government to discover local environmental issues. The three parties involved in whistleblowing are: the central government environmental protection departments, the local government officials, and the whistleblowers. Based on these players, this paper has constructed an Evolutionary Game Model under incomplete information and introduced the expected return as well as replicator dynamics equations of various game agents based on analysis of the game agents, assumptions, and payoff functions of the model in order to study the strategic dynamic trend and stability of the evolutionary game model. Furthermore, this paper has conducted simulation experiments on the evolution of game agents' behaviors by combining the constraints and replicator dynamics equations. The conclusions are: the central environmental protection departments are able to effectively improve the environmental awareness of local government officials by measures such as strengthening punishment on local governments that do not pay attention to pollution issues and lowering the cost of whistleblowing, thus nurturing a good governance and virtuous circle among the central environmental protection departments, local government officials, and whistleblowers. Based on the study above, this paper has provided policy recommendations in the conclusion.

Keywords: whistleblowing; air pollution; evolutionary game; environmental supervision

1. Introduction

The Chinese economy has maintained rapid growth since the reform and opening up in 1978. However, along with rapid growth, environmental issues such as air pollution have become increasingly prominent, seriously threatening the urban environment and quality of life [1–4]. In order to control air pollution, China has taken various measures [5–8], in which whistleblowing is a pollution control measure with Chinese characteristics [9].

On 5 June 2009, China's Ministry of Environmental Protection officially launched the "010-12369" hotline for environment-related whistleblowing, which is open to people all over the country, and accepts complaints about all kinds of environmental issues [10]. Based on that, the Ministry of Environmental Protection launched smartphone APPs such as the "Green Knight" Environmental Snapshot in 2014 [11], and further launched the "12369 Environment Whistleblowing Platform" on WeChat in June 2015, the most widely used smartphone social media app in China [12], and completed the integration of all whistleblowing databases nationwide by the end of 2016 to form China's "12369 Integrated Environmental Whistleblowing Management Platform" [13]. This platform has integrated whistleblowing from various channels including the "12369" hotline, WeChat, and online platforms at four levels (nation-level, province-level, city-level and county-level), and has enabled interconnection and information sharing among all these four levels [14]. This platform has helped whistleblowing to become one important channel for the central environmental protection department to discover local environmental issues [15].

In 2017, China's Ministry of Environmental Protection's national environmental whistleblowing management platform received 618,856 whistleblowing reports in total, of which 409,548 were reported through the "12369" hotline (accounting for 66.18% of the total reports), 129,423 were reported on WeChat (accounting for 20.91% of the total reports), and 79,885 were reported online (accounting for 12.91% of the total reports) [16]. It can be seen that hotlines and WeChat are two major sources of whistleblowing, accounting for 87.09% in total. In terms of the whistleblowing from hotlines, most whistleblowing reports were from the eastern and central provinces and municipalities of China, including Jiangsu, Chongqing, Shanghai, Beijing, Liaoning, and Hainan. In terms of whistleblowing on WeChat, the number of whistleblowing reports increased by 96.4% in 2017 on a year-on-year basis, and the top five provinces in terms of number of whistleblowing reports are: Guangdong, Henan, Shandong, Jiangsu, and Hebei, accounting for around 48% of the total whistleblowing reports nationwide (see Figure 1) [16].

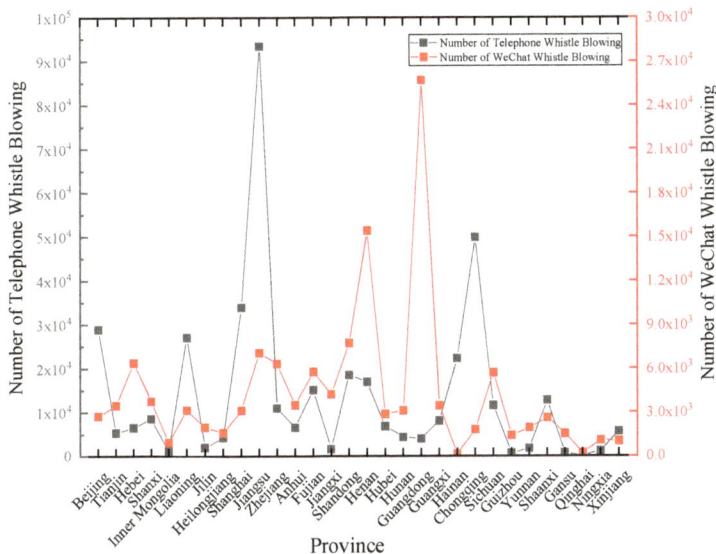

Figure 1. Number of environmental whistleblowing in different provinces of China in 2017, through "12369" Hotline and WeChat.

As seen from the type of pollution in whistleblowing, air pollution is the most prominent, accounting for around 56.7% in all pollution cases [16], which is also the study subject of this paper. In the process of air pollution whistleblowing, there are three main players involved: the central environmental protection department (the Ministry of Environmental Protection), the local government (official), and the whistleblower (including citizens, enterprises, and other organizations). According to the "Environmental Protection Law of the People's Republic of China", the central environmental protection department can directly penalize the local government if the complaint is verified [9]. Whether from an economic perspective or from the evaluation and career development perspective of local officials, the local governments try their best to minimize the number of whistleblowing [17–19]. On the other hand, although the whistleblowers are rewarded after the complaint is verified [20–22], in practice, they are often harassed in real life and their rights and interests are often violated [23,24], which to some extent demotivates the act of whistleblowing. In this case, the central environmental protection department, the local government, and the whistleblower constitute an evolutionary game under incomplete information [25–27].

The academic community has applied the evolutionary game model to environmental protection research and made great progress. For example, Liu et al. (2015) developed an evolutionary game model of two enterprise populations' dynamics and stability for decision-making. By building a system dynamics model, they simulated the aforementioned game and found that the initial choice of strategy was essential to the final result. They concluded that it is important to check the saddle point and external factors [28]. Zhao et al. (2016) established an evolutionary game model to investigate the responses of enterprises to carbon-reduction policies. They examined the individual and combined intervention of carbon-reduction policies through analysis of two scenarios in the simulation based on the game model. They concluded that the combination of these two policies was more efficient than an individual one [29]. Based on the game between the government and enterprises in the context of a complex network, Wu et al. (2017) built an evolutionary model for low-carbon strategies research. By introducing government policy encouragement in the decision-making process of companies, they argued that enterprises' expectations of government encouragement determined whether low-carbon strategies could be diffused. They also concluded that those enterprises quickly adjusting their expectations in the game learned to adopt and follow effective low-carbon strategies [30]. Li et al. (2017) founded a tripartite evolutionary game model based on the relationships among the central and local governments, and land-lost farmers, to analyze China's land expropriation during the rapid urbanization process. They found that serious asymmetry of information between stakeholders led to the low efficiency of the game by simulation analysis. Moreover, they established the reference for the central and local governments to reduce conflicts during land expropriation [31]. Chen (2017) established a three-stage evolutionary game model of the ecological industry chain and achieved the stable strategies by analyzing the equilibrium points of replicator dynamics equations. The author also constructed a multi-agent model to analyze evolutionary paths and trends. Finally, the case of Poyang Lake was used to examine the evolutionary game method [32]. Zhang and Li (2018) built an evolutionary game model of haze cooperative control between governments. They analyzed the dynamic evolution path and stable strategy of this game. Their results showed that the stable state of cooperation cannot be formed between governments spontaneously because of the heterogeneity. Hence the superior government should use administrative penalties to promote the stability of cooperation [33].

However, there are still few studies analyzing environment-related whistleblowing by the evolutionary game method, especially the Three-party Evolutionary Game Model under incomplete information. In order to fill in the blank spaces in game studies on environment-related whistleblowing, this paper has constructed an evolutionary game model under incomplete information that involves the central environmental protection department, the local government, and the whistleblower. In the following parts of this paper, Section 2 introduces the three-party evolutionary game model and conducts detailed analysis on the game agents, assumptions, and payoff functions of the model. Based on the three-party evolutionary game model, Section 3 introduces the expected return and replicator dynamics equations of the three parties in order to perform equilibrium analysis on the game model. Section 4 conducts simulation experiments on the evolution of game agents' behaviors by combining the constraints and replicator dynamics equations and analyzes the impact of changes in parameters on the final evolution result. Section 5 provides conclusions as well as policy recommendations.

2. Methods

2.1. Game Agent Analysis and Assumptions

According to China's current laws and regulations and administrative governance structure, the players, and their relationships in this three-party evolutionary game model, are as follows:

(1) Overall relationship: Currently in China, the local government is responsible for air pollution control and governance in different regions of the country under a centralized environmental regulation [9,34]. The enterprises that release pollution into the air during production and operation are required to declare the types, quantities, concentrations, emission destinations, and emission methods of their pollutants to the local government. The local government is responsible for pollution control and law enforcement [35], while as the highest regulatory authority for air pollution in China, the Ministry of Environmental Protection oversees and manages air pollution control and governance by local governments at all levels and directly accepts air pollution whistleblowing through hotlines, WeChat, and online platforms [36]. According to common practice in academia [37–41], and also because the local governments at all levels in China generally have the same benefits and costs when they are supervised by the central environmental protection department [42–46], this paper has made "the local government" one single player in the three-party evolutionary game model, and does not distinguish local governments by their administrative levels in China (i.e., province-level, city-level, county-level, district-level, etc.).

(2) The Ministry of Environmental Protection as the central supervision department for environmental protection: The Ministry of Environmental Protection is the highest environmental protection regulator in the country [9]. It analyzes and supervises environmental whistleblowing throughout the country. It passes on whistleblowing cases to relevant local governments to verify and solve problems. It also supervises and handles the issues reported in media or by the public, or problems not properly handled by the local governments [9,36]. Due to limitations in monitoring technology and staffing, it is very difficult for the central environmental supervision department to implement strict all-round supervision throughout the country, and the supervision sometimes fails [18,47–49]. Therefore, the central environmental supervision department's behavior strategy space is (Strict Supervision, Loose Supervision).

(3) The local governments: The behavior pattern of the local governments has two sides. On the one hand, due to the need for political achievement and promotion, local governments consider that "excessive" air pollution control will affect the development of local enterprises as well as local GDP growth and their political achievement, and therefore take a negative attitude and a position of inaction towards air pollution [50–52], and even conduct extreme acts, such as retaliation, against air pollution whistleblowers [53]. On the other hand, air pollution has not only received great emphasis from the central government, it has also caused great damage to the local environment. It threatens the local ecological environment and sustainable development in the long term, and causes irreparable damage to the physical and mental health of local residents, which also motivates the local governments to emphasize and control air pollution to a certain extent [54–56]. The local governments' behavioral strategy space is (Put Emphasis on Pollution Control, Neglect Pollution Control).

(4) The whistleblower: The whistleblowers include the public, the media, and non-governmental environmental protection organizations, etc., who are the direct victims of air pollution. From the perspective of air pollution hazards, the whistleblowers are all affected by air pollution in the long term, and with their increasing environmental awareness and legal awareness, the public will become more and more active in participation in supervision. The whistleblowers can report air pollution issues by reporting, petitioning, exposure, litigation, and claims in order to protect their rights and interests [10,12,35]. However, the whistleblowers may also give up on whistleblowing due to the time and effort it costs to collect evidence and proof, the complicated process for supervision and implementation after whistleblowing, the possibilities of retaliation from the local government, etc. [53]. Therefore, the whistleblower's behavioral strategy space is (Blow the Whistle, Not Blow the Whistle).

Based on the analysis above, it can be seen that there is a game relation among the three parties: the central environmental protection department, the local government, and the whistleblower. This paper has made the following assumptions regarding the three-party evolutionary game model:

Assumption 1. *For the central environmental protection department, suppose that the proportion of the central environmental protection department selecting Strict Regulation strategy is x ($0 < x < 1$); then the proportion* selecting Loose Regulation strategy is $1 - x$. The central environmental protection department's pay-off is constant at R_1, whether it pursues strict regulation or not. If it pursues strict regulation, the regulation cost is C_1, and the long-term improved social welfare will be W (including reduced air pollution and better health). However, strict regulation would restrict the local economic development, resulting in a loss of L_1 due to GDP growth slow-down. If the central environmental protection department does not pursue strict regulation, its regulation cost would be C_{11} ($C_1 > C_{11}$), and there will not be any long-term social welfare improvement.

Assumption 2. *For local governments, suppose that the proportion of local governments selecting Put Emphasis on Pollution Control strategy is y ($0 < y < 1$); then the proportion selecting Neglect Pollution strategy is $1 - y$. If they put emphasis on environmental issues and pollution control, their pay-off would be R_2 (such as reputation and public opinion that would help their promotion), while the cost would be C_2 (including time cost, economic cost, and loss in GDP growth). If the local officials neglect environmental protection, their short-term pay-off would be G (including short-term GDP growth), while the cost would be C_{22} ($C_2 > C_{22}$). However, if they do not make efforts on pollution control, they would face a penalty of L_2 if this is discovered by the central environmental protection department, whether through inspections or whistleblowing.*

Assumption 3. *For the whistleblowers, suppose that the proportion of the whistleblowers selecting Blow the Whistle strategy is z ($0 < z < 1$); then the proportion selecting Not Blow the Whistle strategy is $1 - z$. The cost of whistleblowing is C_3. If the central environmental protection department pursues strict regulation, the whistleblower would get a reward of R_3 from the central environmental protection department as well as a compensation of R_{33} from the local government. However, if the central environmental protection department does not pursue strict regulation, the local government would call off the whistleblowing with a bribery of R_{333}. Furthermore, the whistleblower would suffer a loss of L_3 due to the harms of environmental pollution.*

Assumption 4. *According to the experience since China's reform and opening up, the air pollution issue is becoming increasingly serious with the rapid economic development. In order to achieve effective environmental protection, this paper assumes that the three-party game system should adopt a mixed strategy of strict supervision by the central environmental department, emphasis on pollution control by the local governments, and active participation and reporting by the whistleblowers.*

The parameters and variable descriptions of our model are listed in Table 1.

Table 1. Model parameters and variable descriptions.

Parameter	Description
R_1	The normal income of the central environmental protection department (wage income from government finance, which is the same under strict regulation and loose regulation)
C_1	The regulation cost of the central environmental protection department under strict regulation
W	The social welfare improvements achieved by the central environmental protection department's strict regulation
L_1	The impact of strict regulations on local economic growth (losses including decline in GDP growth)
C_{11}	The regulation cost of the central environmental protection department under loose regulation ($C_1 > C_{11}$)
R_2	The reputation and public opinion gain of local government officials if they put the emphasis on environmental issues

Table 1. *Cont.*

Parameter	Description
C_2	Cost of local government officials if they work on environmental issues (including economic cost and time cost of pollution control)
G	The short-term gain of local government officials if they neglect environmental issues (such as short-term growth of local GDP)
C_{22}	The cost to local government officials if they neglect environmental issues ($C_2 > C_{22}$)
L_2	The penalty on local government officials if they neglect environmental issues (if discovered by inspections or due to whistleblowing)
R_3	The rewards to the whistleblower by the central environmental protection department if it pursues strict regulation
C_3	The cost of whistleblowing
R_{33}	The compensation to the whistleblower by the local government if the central environmental protection department pursues strict regulation
R_{333}	The cost to call off whistleblowing paid by the local government (such as "hush money") if the central environmental protection department does not pursue strict regulation and the local government neglects pollution issues
L_3	The loss suffered by the whistleblower due to environmental pollution when the local government officials neglect pollution issues

2.2. The Payoff Function

Since the payoff of each game agent is affected by the strategy of the other two game agents, there are eight strategy combinations for the evolutionary game between the central environmental department (a), local government officials (b), and whistleblowers (c), as below (see Figure 2): (a_1 Strict Regulation, b_1 Put Emphasis on Pollution Control, c_1 Blow the Whistle); (a_2 Strict Regulation, b_2 Neglect Pollution Control, c_2 Blow the Whistle); (a_3 Loose Regulation, b_3 Put Emphasis on Pollution Control, c_3 Blow the Whistle); (a_4 Loose Regulation, b_4 Neglect Pollution Control, c_4 Blow the Whistle); (a_5 Strict Regulation, b_5 Put Emphasis on Pollution Control, c_5 Not Blow the Whistle); (a_6 Strict Regulation, b_6 Neglect Pollution Control, c_6 Not Blow the Whistle); (a_7 Loose Regulation, b_7 Put Emphasis on Pollution Control, c_7 Not Blow the Whistle); (a_8 Loose Regulation, b_8 Neglect Pollution Control, c_8 Not Blow the Whistle).

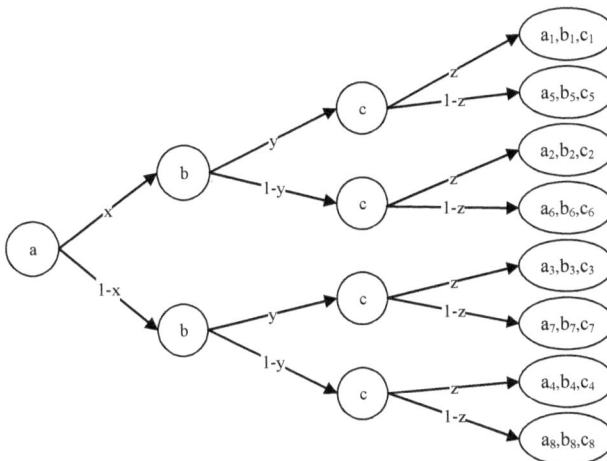

Figure 2. Three-party evolutionary game flow chart.

When the strategy of the central environmental department, local government officials, and whistleblowers is (a_1 Strict Regulation, b_1 Put Emphasis on Pollution Control, c_1 Blow the Whistle), the central environmental protection department's pay-off is constant at R_1, and the regulation cost is C_1, and the long-term improved social welfare will be W. However, strict regulation would restrict local economic development, resulting in a loss of L_1 due to GDP growth slow-down. The local governments' pay-off would be R_2, while the cost would be C_2. The whistleblower would get a reward of R_3 from the central environmental protection department, and the cost of whistleblowing would be C_3. Similarly, the benefits of the central environmental department, local government officials, and whistleblowers under other strategic combinations can be derived. (see Table 2).

Table 2. The Payoff matrix of the three-party evolutionary game.

Central Environmental Protection Department: a		Strict Regulation (x)		Loose Regulation (1 − x)	
Local Government Officials: b		Put Emphasis on Pollution Control (y)	Neglect Pollution Control (1 − y)	Put Emphasis on Pollution Control (y)	Neglect Pollution Control (1 − y)
Whistle Blower: c	Blow the Whistle: (z)	(a_1, b_1, c_1)	(a_2, b_2, c_2)	(a_3, b_3, c_3)	(a_4, b_4, c_4)
	Not Blow the Whistle: (1 − z)	(a_5, b_5, c_5)	(a_6, b_6, c_6)	(a_7, b_7, c_7)	(a_8, b_8, c_8)

Based on assumptions in Section 2.1, the payoffs of the different game agents under different strategy combinations can be written as follows:

$$
\begin{aligned}
a_1 &= W + R_1 - L_1 - C_1 - R_3; \quad b_1 = R_2 - C_2; \quad c_1 = R_3 - C_3; \\
a_2 &= W + R_1 - L_1 - C_1 - R_3 + L_2; \quad b_2 = G - C_{22} - L_2 - R_{33}; \quad c_2 = R_3 - C_3 + R_{33} - L_3; \\
a_3 &= R_1 - C_{11}; \quad b_3 = R_2 - C_2; \quad c_3 = -C_3; \\
a_4 &= R_1 - C_{11} + L_2; \quad b_4 = G - C_{22} - L_2 - R_{333}; \quad c_4 = -C_3 + R_{333} - L_3; \\
a_5 &= W + R_1 - L_1 - C_1; \quad b_5 = R_2 - C_2; \quad c_5 = 0; \\
a_6 &= W + R_1 - L_1 - C_1 + L_2; \quad b_6 = G - C_{22} - L_2 - R_{33}; \quad c_6 = R_3 - L_3; \\
a_7 &= R_1 - C_{11}; \quad b_7 = R_2 - C_2; \quad c_7 = 0; \\
a_8 &= R_1 - C_{11}; \quad b_8 = G - C_{22}; \quad c_8 = -L_3
\end{aligned}
\tag{1}
$$

3. Equilibrium Analysis

3.1. The Expected Payoff and Replicator Dynamics Equation of the Three Game Agents

1. The replicator dynamics equation of the central environmental protection department

Let U_{X1} represent the expected payoff of the central environmental protection department if it pursues strict regulations, and U_{X2} represent the payoff of the central environmental protection department if it does not pursue strict regulations. U_X represents the average expected payoff of the central environmental protection department. U_{X1}, U_{X2}, and U_X can be expressed as:

$$
U_{X1} = y \times z \times a_1 + (1-y) \times z \times a_2 + y \times (1-z) \times a_5 + (1-y) \times (1-z) \times a_6 \tag{2}
$$

$$
U_{X2} = y \times z \times a_3 + (1-y) \times z \times a_4 + y \times (1-z) \times a_7 + (1-y) \times (1-z) \times a_8 \tag{3}
$$

$$
U_X = x \times U_{X1} + (1-x) \times U_{X2} \tag{4}
$$

Therefore, the replicator dynamics equation of the central environmental protection department can be written as:

$$
\begin{aligned}
F(x) &= \tfrac{dx}{dt} = x \times (U_{X1} - U_X) \\
&= -(-1 + x)x(W - C_1 + C_{11} - L_1 + L_2 - yL_2 - zL_2 + yzL_2 - zR_3)
\end{aligned}
\tag{5}
$$

2. The replicator dynamics equation of local government officials

Let U_{Y1} represent the expected payoff of local government officials if they put emphasis on environmental issues, and U_{Y2} represent the payoff of local government officials if they neglect environmental issues. U_Y represents the average expected payoff of local government officials. Then U_{Y1}, U_{Y2}, and U_Y can be expressed as:

$$U_{Y1} = x \times z \times b_1 + x \times (1-z) \times b_5 + (1-x) \times z \times b_3 + (1-x) \times (1-z) \times b_7 \tag{6}$$

$$U_{Y2} = x \times z \times b_2 + x \times (1-z) \times b_6 + (1-x) \times z \times b_4 + (1-x) \times (1-z) \times b_8 \tag{7}$$

$$U_Y = y \times U_{Y1} + (1-y) \times U_{Y2} \tag{8}$$

The replicator dynamics equation of the local government can be written as:

$$F(y) = \tfrac{dy}{dt} = y \times (U_{Y1} - U_Y)$$
$$= (y-1)y(G + C_2 - C_{22} - xL_2 - zL_2 + xzL_2 - R_2 - xR_{33} - zR_{333} + xzR_{333}) \tag{9}$$

3. The replicator dynamics equation of the whistleblower

Let U_{Z1} represent the expected payoff of the whistleblower if he/she decides to blow the whistle, and U_{Z2} represent the expected payoff of the whistleblower if he/she decides not to blow the whistle. U_Z represents the average expected payoff of the whistleblower. Then U_{Z1}, U_{Z2}, and U_Z can be expressed as:

$$U_{Z1} = x \times y \times c_1 + x \times (1-y) \times c_2 + (1-x) \times y \times c_3 + (1-x) \times (1-y) \times c_4 \tag{10}$$

$$U_{Z2} = x \times y \times c_5 + x \times (1-y) \times c_6 + (1-x) \times y \times c_7 + (1-x) \times (1-y) \times c_8 \tag{11}$$

$$U_Z = z \times U_{Z1} + (1-z) \times U_{Z2} \tag{12}$$

The replicator dynamics equation of the whistleblower can be written as:

$$F(z) = \tfrac{dz}{dt} = z \times (U_{Z1} - U_Z)$$
$$= (-1+z)z(C_3 - xyR_3 + (-1+y)(xR_{33} - (-1+x)R_{333})) \tag{13}$$

3.2. Stability Analysis of the Evolutionary Game

By combining these replicator dynamics equations above, we can obtain a three-dimensional dynamic group evolution system with three agents. In this three-dimensional dynamic system, the probability of the three agents' different strategies is affected by time. The solution domain of $F(x)$, $F(y)$, and $F(z)$ is $[0,1] \times [0,1] \times [0,1]$. By letting $F(x) = 0$, $F(y) = 0$, $F(z) = 0$, i.e., letting the rate of strategy change be zero, we can obtain the equilibrium points of this dynamic system, which are $(0,0,0)$, $(0,0,1)$, $(0,1,0)$, $(1,0,0)$, $(1,1,0)$, $(1,0,1)$, $(0,1,1)$, and $(1,1,1)$, respectively. These eight equilibrium points constitute the boundary of the domain of this evolutionary game, and the stability of these equilibrium points in this evolutionary system can be obtained by local stability analysis of the Jacobian matrix [57–59].

By making derivation on these three replicator dynamics equations, we can get:

$$F'(x) = (1-2x)(W - C_1 + C_{11} - L_1 + L_2 - yL_2 - zL_2 + yzL_2 - zR_3) \tag{14}$$

$$F'(y) = -(1-2y)(G + C_2 - C_{22} - xL_2 - zL_2 + xzL_2 - R_2 - xR_{33} - zR_{333} + xzR_{333}) \tag{15}$$

$$F'(z) = -(1-2z)(C_3 - xyR_3 + (-1+y)(xR_{33} - (-1+x)R_{333})) \tag{16}$$

According to the characteristics of the evolutionary game, after substituting these equilibrium points into the formula above, if $F'(x) < 0$, $F'(y) < 0$, $F'(z) < 0$, it means the equilibrium strategies x, y, and z respectively represent the stable strategy adopted by the three agents in the evolution process. Then we will use Wolfram Mathematica for stability analysis.

(1) Analysis of the Asymptotic Stability of the Central Environmental Protection Department

According to Formula (5), if $W - C_1 + C_{11} - L_1 + L_2 - yL_2 - zL_2 + yzL_2 - zR_3 = 0$, then $F(x) \equiv 0$, indicating the boundary of the stable state. Its phase diagram is shown in Figure 3a.

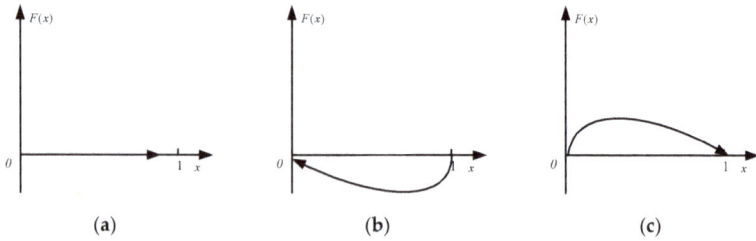

(a)   (b)   (c)

Figure 3. Replicator dynamics phase diagram of the central environmental protection department's strategy: (a) $z = \frac{W-C_1+C_{11}-L_1+L_2-yL_2}{L_2-yL_2+R_3}$; (b) $z < \frac{W-C_1+C_{11}-L_1+L_2-yL_2}{L_2-yL_2+R_3}$; (c) $z > \frac{W-C_1+C_{11}-L_1+L_2-yL_2}{L_2-yL_2+R_3}$.

If $W - C_1 + C_{11} - L_1 + L_2 - yL_2 - zL_2 + yzL_2 - zR_3 \neq 0$, let $F(x) = 0$, and we can get two stable points of $x = 0$ and $x = 1$. If $W - C_1 + C_{11} - L_1 + L_2 - yL_2 - zL_2 + yzL_2 - zR_3 < 0$, i.e., $z < \frac{W-C_1+C_{11}-L_1+L_2-yL_2}{L_2-yL_2+R_3}$, then $F'(x)|x - 0 < 0$ and $F'(x)|x - 1 > 0$. Therefore, $x = 0$ is the stable strategy, and the central environmental protection department will decide not to pursue strict regulation. Its phase diagram is shown in Figure 3b. On the contrary, if $W - C_1 + C_{11} - L_1 + L_2 - yL_2 - zL_2 + yzL_2 - zR_3 > 0$, i.e., $z > \frac{W-C_1+C_{11}-L_1+L_2-yL_2}{L_2-yL_2+R_3}$, then $F'(x)|x - 0 > 0$ and $F'(x)|x - 1 < 0$. Therefore, $x = 1$ is the stable strategy and the central environmental protection department will decide to pursue strict regulation. Its phase diagram is shown in Figure 3c.

By combining the three cases in Figure 3a–c, this paper has obtained the dynamic trend and stability of the central environmental protection department's strategy in three scenarios, as shown in Figure 4.

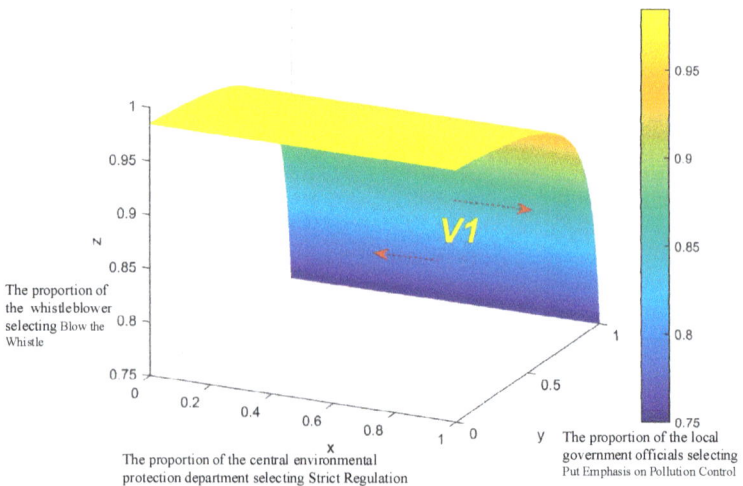

Figure 4. Dynamic trend and stability analysis of the central environmental protection department's strategy.

The eight equilibrium stable points in Figure 4 are (0,0,0), (0,0,1), (0,1,0), (1,0,0), (1,1,0), (1,0,1), (0,1,1), and (1,1,1), respectively. These eight equilibrium points constitute the boundary of the domain of this evolutionary game. Under the constraints of $z < \frac{W-C_1+C_{11}-L_1+L_2-yL_2}{L_2-yL_2+R_3}$, the three-party game hybrid strategy moves to $x = 0$, and we can prove that when the proportion of the whistleblower selecting Blow the Whistle is less than the critical value, the central environmental protection department tends to select Loose Regulation strategy. On the contrary, when the proportion of the whistleblower selecting Blow the Whistle is greater than the critical value, the central environmental protection department tends to select Strict Regulation strategy.

(2) Analysis of the Asymptotic Stability of Local Government Officials

According to the formula (9), if $G + C_2 - C_{22} - xL_2 - zL_2 + xzL_2 - R_2 - xR_{33} - zR_{333} + xzR_{333} = 0$, then $F(y) \equiv 0$, indicating the boundary of the stable state. Its phase diagram is shown in Figure 5a.

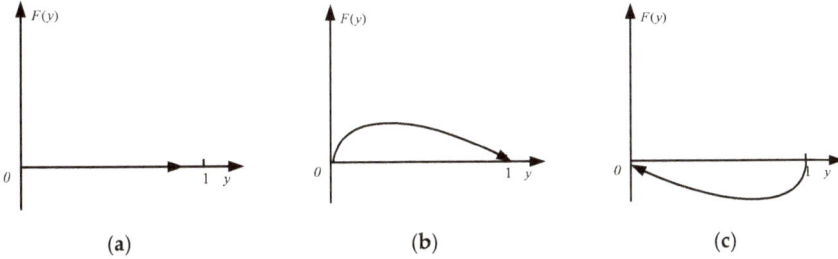

Figure 5. Replicator dynamics phase diagram of local government officials' strategy: **(a)** $z = \frac{G+C_2-C_{22}-xL_2-R_2-xR_{33}}{L_2-xL_2+R_{333}-xR_{333}}$; **(b)** $z > \frac{G+C_2-C_{22}-xL_2-R_2-xR_{33}}{L_2-xL_2+R_{333}-xR_{333}}$; **(c)** $z < \frac{G+C_2-C_{22}-xL_2-R_2-xR_{33}}{L_2-xL_2+R_{333}-xR_{333}}$.

If $G + C_2 - C_{22} - xL_2 - zL_2 + xzL_2 - R_2 - xR_{33} - zR_{333} + xzR_{333} \neq 0$, let $F(y) = 0$, and we can get two stable points of $y = 0$ and $y = 1$. If $-(G + C_2 - C_{22} - xL_2 - zL_2 + xzL_2 - R_2 - xR_{33} - zR_{333} + xzR_{333}) < 0$, i.e., $z > \frac{G+C_2-C_{22}-xL_2-R_2-xR_{33}}{L_2-xL_2+R_{333}-xR_{333}}$, then $F'(y)|y - 0 > 0$ and $F'(y)|y - 1 < 0$. Therefore, $y = 1$ is the stable strategy, and the local government officials will choose to put emphasis on pollution issues. Its phase diagram is shown in Figure 5b. On the contrary, if $G + C_2 - C_{22} - xL_2 - zL_2 + xzL_2 - R_2 - xR_{33} - zR_{333} + xzR_{333} > 0$, i.e., $z < \frac{G+C_2-C_{22}-xL_2-R_2-xR_{33}}{L_2-xL_2+R_{333}-xR_{333}}$, then $F'(y)|y - 0 < 0$ and $F'(y)|y - 1 > 0$. Therefore, $y = 0$ is the stable strategy and the local government officials will choose to neglect pollution issues. Its phase diagram is shown in Figure 5c.

By combining the three cases in Figure 5a–c, this paper has obtained the dynamic trend and stability of local government officials' strategy in three scenarios as shown in Figure 6.

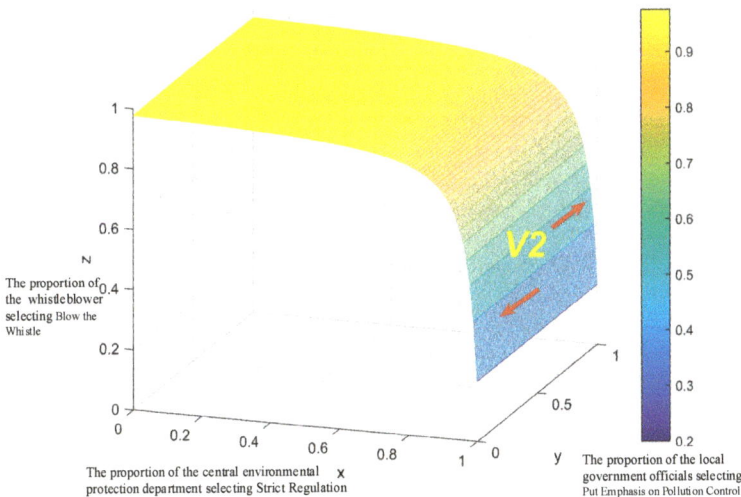

Figure 6. Dynamic trend and stability analysis of local government officials' strategy.

The eight equilibrium stable points in Figure 6 are (0,0,0), (0,0,1), (0,1,0), (1,0,0), (1,1,0), (1,0,1), (0,1,1), and (1,1,1), respectively. These eight equilibrium points constitute the boundary of the domain of this evolutionary game. Under the constraints of $z > \frac{G+C_2-C_{22}-xL_2-R_2-xR_{33}}{L_2-xL_2+R_{333}-xR_{333}}$, the three-party game hybrid strategy moves to $y = 1$, and we can prove that when the proportion of the whistleblower selecting Blow the Whistle is greater than the critical value, the local government officials tend to select Put Emphasis on Pollution Control strategy. On the contrary, when the proportion of the whistleblower selecting Blow the Whistle is less than the critical value, the local government officials tend to select Neglect Pollution Control strategy.

(3) Analysis of the Asymptotic Stability of the Whistleblower

According to the formula (13), if $(C_3 - xyR_3 + (-1+y)(xR_{33} - (-1+x)R_{333})) = 0$, then F(z)$\equiv$0, indicating the boundary of the stable state. Its phase diagram is shown in Figure 7a.

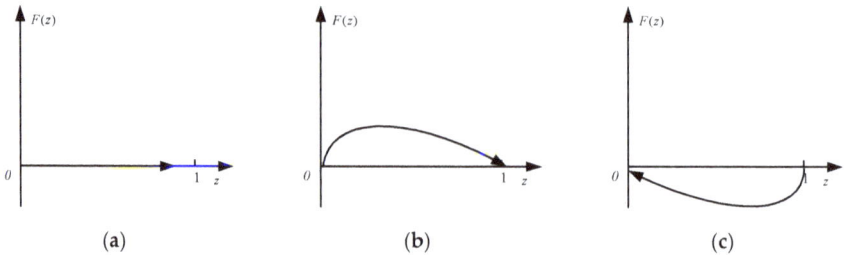

| | | |
| (a) | (b) | (c) |

Figure 7. Replicator dynamics phase diagram of the whistleblower's strategy: (a) $x = \frac{C_3-R_{333}+yR_{333}}{yR_3+R_{33}-yR_{33}-R_{333}+yR_{333}}$; (b) $x > \frac{C_3-R_{333}+yR_{333}}{yR_3+R_{33}-yR_{33}-R_{333}+yR_{333}}$; (c) $x < \frac{C_3-R_{333}+yR_{333}}{yR_3+R_{33}-yR_{33}-R_{333}+yR_{333}}$.

Z is constant. If $(C_3 - xyR_3 + (-1+y)(xR_{33} - (-1+x)R_{333})) \neq 0$, let F(z) = 0, and we can get two stable points of z = 0 and z = 1. If $-(C_3 - xyR_3 + (-1+y)(xR_{33} - (-1+x)R_{333})) < 0$, i.e., $x > \frac{C_3-R_{333}+yR_{333}}{yR_3+R_{33}-yR_{33}-R_{333}+yR_{333}}$, then $F'(z)|z - 0 > 0$ and $F'(z)|z - 1 < 0$. Therefore, z = 1 is the stable strategy, and the whistleblower will decide to blow the whistle. Its phase diagram is shown in Figure 7b. On the contrary, if $-(C_3 - xyR_3 + (-1+y)(xR_{33} - (-1+x)R_{333})) > 0$, i.e., $x < \frac{C_3-R_{333}+yR_{333}}{yR_3+R_{33}-yR_{33}-R_{333}+yR_{333}}$, then $F'(z)|z - 0 < 0$ and $F'(z)|z - 1 > 0$. Therefore, z = 0 is the stable strategy, and the whistleblower will decide not to blow the whistle. Its phase diagram is shown in Figure 7c.

By combining the three cases in Figure 7a–c, this paper has obtained the dynamic trend and stability of the whistleblower's strategy in three scenarios as shown in Figure 8.

The eight equilibrium stable points in Figure 8 are (0,0,0), (0,0,1), (0,1,0), (1,0,0), (1,1,0), (1,0,1), (0,1,1), and (1,1,1), respectively. These eight equilibrium points constitute the boundary of the domain of this evolutionary game. Under the constraints of $x > \frac{C_3-R_{333}+yR_{333}}{yR_3+R_{33}-yR_{33}-R_{333}+yR_{333}}$, the three-party game hybrid strategy moves to z = 1, and we can prove that when the proportion of the central environmental protection department selecting Strict Regulation is greater than the critical value, the whistleblowers tend to select Blow the Whistle strategy. On the contrary, when the proportion of the central environmental protection department selecting Strict Regulation is less than the critical value, the whistleblowers tend to select Not Blow the Whistle strategy.

In summary, when the proportion of the whistleblower selecting Blow the Whistle is greater than the critical value, the central environmental protection department tends to select Strict Regulation strategy(see Figure 4); when the proportion of the whistleblower selecting Blow the Whistle is greater than the critical value, the local government officials tend to select Put Emphasis on Pollution Control strategy (see Figure 6); when the proportion of the central environmental protection department selecting Strict Regulation is greater than the critical value, the whistleblowers tend to select Blow the Whistle strategy (see Figure 8).

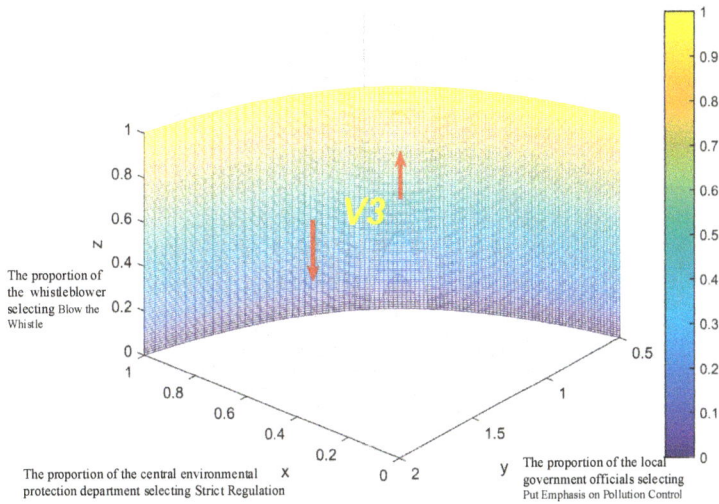

Figure 8. Dynamic trend and stability analysis of the whistleblower's strategy.

4. Simulations and Discussions

Based on the stability analysis of the evolutionary game, it can be seen that the evolutionary equilibrium of the central environmental protection department changes with y (the probability that the local government officials put emphasis on pollution control) and z (the probability that the whistleblower blows the whistle); the evolutionary equilibrium of local government officials changes with x (the probability that the central environmental protection department chooses strict regulation) and z (the probability that the whistleblower blows the whistle); and the evolutionary equilibrium of the whistleblower changes with x (the probability that the central environmental protection department chooses strict regulation) and y (the probability that the local government officials put emphasis on pollution control).

Since the values of x, y, and z change from time to time during the evolution process, and since the equilibrium of this evolutionary game is not robust against small changes in the values of x, y, and z, we cannot make the three-party game evolve towards the expected stable equilibrium merely by adjusting the initial conditions. This paper is committed to promoting the evolution of this three-party game towards the social rational model, that is, strict regulation by the central environmental protection department, great emphasis on pollution control by local government officials, and whistleblowing by the citizens (i.e., $x = 1$, $y = 1$, $z = 1$). Therefore, we could guide the behavior of different agents towards the desired direction by controlling or adjusting related variables. More specifically, this paper has conducted numerical experiments on the evolution process of the behaviors of the three parties by combining the constraints and replicator dynamic Equations (5), (9), and (13) in order to analyze the impact of changes in parameters on the evolution result. The study adopts MATLAB R2015b to simulate the evolution process for the behavior strategies.

(1) The Dynamic Evolution of Central Supervision Authorities, Local Government Officials, and Whistleblowers in the Initial State

x_0, y_0, z_0 respectively indicate the initial proportion or probability of the central environmental department choosing the "strict supervision" strategy, the local governments choosing the "emphasis on pollution control" strategy, and the whistleblowers choosing the "participation" strategy, with the initial time of 0 and evolution end time of 5. The values of the parameters are: $R_1 = 20$, $C_1 = 10$, $W = 26.5$, $L_1 = 20$, $C_{11} = 5$, $R_2 = 22.5$, $C_2 = 20$, $G = 50$, $C_{22} = 10$, $L_2 = 30$, $R_3 = 2$, $R_3 = 2$, $R_{33} = 3$, $R_{333} = 3.1$, and $C_3 = 1$. Let $x_0 = 0.1, y_0 = 0.1, z_0 = 0.1$, and the initial system simulation results are shown in Figure 9.

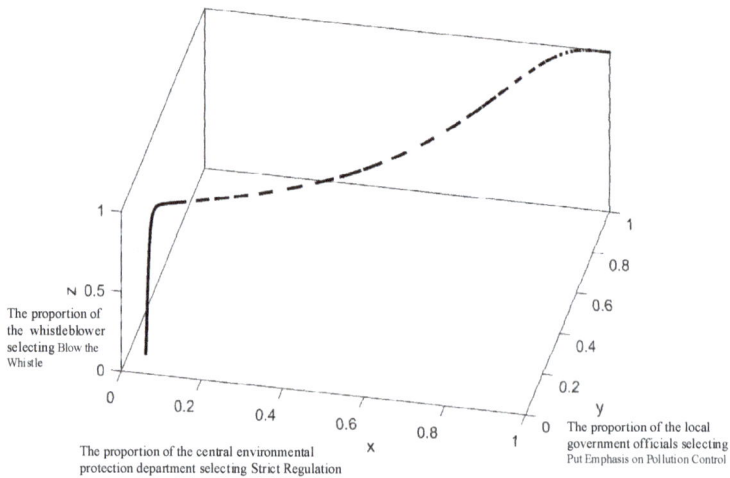

Figure 9. System evolution results in the initial state ($x_0 = 0.1$, $y_0 = 0.1$, $z_0 = 0.1$).

The above simulation results have verified the stability of the above equilibrium points as well as the derivation of the system evolution results. The system finally reaches a stable equilibrium state (1,1,1), which satisfies Assumption 4, i.e., the three-party game system adopts the mixed strategy of strict supervision by the central environmental department, emphasis on pollution control by the local governments, and active participation and reporting by the whistleblowers.

Furthermore, through the simulation experiments with different changes to the initial state, this paper has obtained the impacts of initial state change on the evolution result when all other parameters are unchanged. The results of the simulation experiments with initial state change are shown in Figure 9.

As shown in the simulation results comparison, the initial state of the three-party game agents has significant impact on the evolution of the game system towards the stable equilibrium state. The higher the initial x, y, z are, the shorter time it takes for the system to reach the stable equilibrium state. When (x, y, z), (i.e., strict regulation, emphasis on pollution control, participation in whistleblowing) take the strategy proportions of (0.2, 0.3, 0.4), the game system will reach the stable equilibrium state at around $t = 3$. When the proportions increase to (0.5, 0.6, 0.7), the game system will reach the stable equilibrium state at around $t = 2.5$. When the proportions increase to (0.8, 0.8, 0.8), the game system will reach the stable equilibrium state at around $t = 1.5$. This indicates that the increase in initial state proportions would help shorten the time it takes for the game system to evolve to the stable equilibrium state, that is, increasing the proportion of strategy choices of (strict regulation, emphasis on pollution control, participation in whistleblowing) by the game agents would help facilitate pollution control.

Furthermore, it is shown by Figure 9 that as long as the central supervision department increases the probability of strict supervision, it can indirectly increase the incentive probability, thus greatly enhancing the enthusiasm of whistleblowers and quickly reaching the equilibrium state of the whistleblowers. However, when the initial state is comparatively low ($x_0 = 0.1$, $y_0 = 0.1$, $z_0 = 0.1$), the local governments do not put much emphasis on pollution control, and the increase in whistleblowing probability cannot enhance local governments' emphasis on the environment. With the increase in the probability of strict supervision by the central supervision department, the local government officials would greatly raise their emphasis on environmental pollution ($y > 0.8008$) when there is a high probability of strict supervision by the central supervision department ($x > 0.8017$), and would evolve towards the equilibrium state (0.9934, 0.9999, 0.9997). The initial stage of environmental pollution control in China is a relatively low initial state. First, the central supervision department needs to increase the probability of strict supervision, and then local governments need to raise their emphasis on pollution control, and the whistleblowers should also be encouraged to participate in environmental whistleblowing. Only when the three parties increase their environmental awareness ($x > x_0$, $y > y_0$, $z > z_0$), can the environmental pollution issues be efficiently managed and controlled.

(2) The Dynamic Evolution of Central Supervision Authorities, Local Government Officials, and Whistleblowers under Different Parameters

Based on the stability analysis of the evolutionary game, it can be seen that a number of influencing factors also have a great impact on the evolution of the system.

When $z > \frac{W - C_1 + C_{11} - L_1 + L_2 - yL_2}{L_2 - yL_2 + R_3}$, $x \to 1$, which means the central environmental protection department would eventually choose strict regulation. Therefore, by increasing the penalty on local government officials who neglect pollution control (L_2) and increasing the rewards to whistleblowers (R_3), we could encourage the central environmental protection department to evolve towards the strategy of strict regulations.

When $z > \frac{G + C_2 - C_{22} - xL_2 - R_2 - xR_{33}}{L_2 - xL_2 + R_{333} - xR_{333}}$, $y \to 1$, which means local government officials would eventually put an emphasis on pollution control. Therefore, by raising the compensation to whistleblowers by local government officials who neglect pollution control (R_{33}), lowering the pollution control cost of local governments (C_2), enhancing the role of reputation and public opinion in political career or official promotion (R_2), increasing the penalty on local government officials who neglect pollution control (L_2) as well as the cost on local government officials if they neglect pollution control (C_{22}), and to some extent making it more difficult for local government officials to use measures such as "hush money" to call off whistleblowing (i.e., lowering R_{333}), we could also motivate local government officials to evolve towards the strategy of emphasizing pollution control and environmental issues.

When $y > \frac{-C_3 + xR_{33} + R_{333} - xR_{333}}{-xR_3 + xR_{33} + R_{333} - xR_{333}} = 1 - \frac{xR_3 - C_3}{(R_3 - R_{33} + R_{333})x - R_{333}}$, $z \to 1$, which means the whistleblower would eventually decide to blow the whistle. Therefore, by increasing the rewards to whistleblowers by the central environmental protection department (R_3), lowering the cost of whistleblowing (C_3), raising the compensation to whistleblowers by local government officials who neglect pollution control (R_{33}), and to some extent making it more difficult for local government officials to use measures such as "hush money" to call off whistleblowing (i.e., lowering R_{333}), we could encourage whistleblowers to evolve towards the strategy of blowing the whistle.

Therefore, we could select multiple factors to analyze the impact of parameter change on the evolution result. While keeping all other parameters and the low-level initial state of ($x_0 = 0.1$, $y_0 = 0.1$, $z_0 = 0.1$) unchanged, the results of the simulation experiments where the central environmental protection department strengthens punishment on local governments that neglect air pollution issues and lowers the cost of whistleblowing are shown in Figure 10.

The simulation results indicate that by strengthening the punishment on local governments that neglect air pollution issues and lowering the cost of whistleblowing, the central environmental protection department could significantly enhance the local government officials' attention to environmental pollution and facilitate the local governments' strategy to evolve towards the stable equilibrium. Moreover, in the long term, this could also effectively reduce central supervision departments' regulatory cost on local governments and achieve the strategy of emphasis on pollution control by the local governments and active participation by the whistleblowers with a comparatively lower probability of strict supervision.

In the same way, the impact of other parameter changes that are conducive to pollution control, such as increasing the weight of public opinion in local government officials' performance evaluation and promotion, reducing the pollution control costs of local governments, improving the rewards to whistleblowers, and enhancing the compensation from the government to whistleblowers, could also be experimented with the above method. Due to space limitations, this paper will not elaborate further. There are two important factors that affect the speed at which the three-party game system reaches the evolutionary stability strategy (ESS). One is the initial probability. The larger the initial probability, the faster the system reaches the ESS (see Figure 10). The other is the magnitude of penalties, incentives, and environmental governance costs. It is necessary to reduce the cost of pollution control by local government and the cost of whistleblowers, based on increasing the magnitude of penalties to local governments' negative response to environmental protection and the effective reports from whistleblowers (see Figure 11).

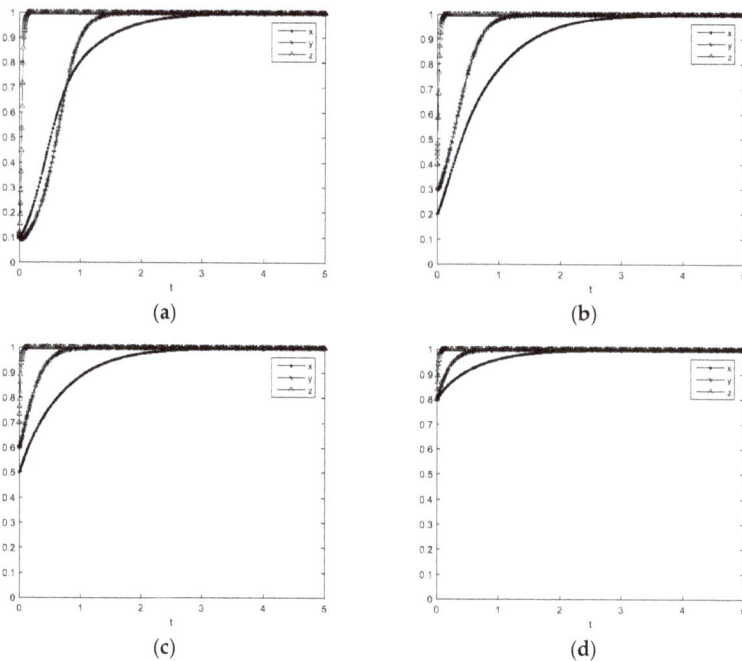

Figure 10. The effect of state changes on system evolution results: (a) $x_0 = 0.1$, $y_0 = 0.1$, $z_0 = 0.1$; (b) $x_0 = 0.2$, $y_0 = 0.3$, $z_0 = 0.5$; (c) $x_0 = 0.5$, $y_0 = 0.6$, $z_0 = 0.7$; (d) $x_0 = 0.8$, $y_0 = 0.8$, $z_0 = 0.8$.

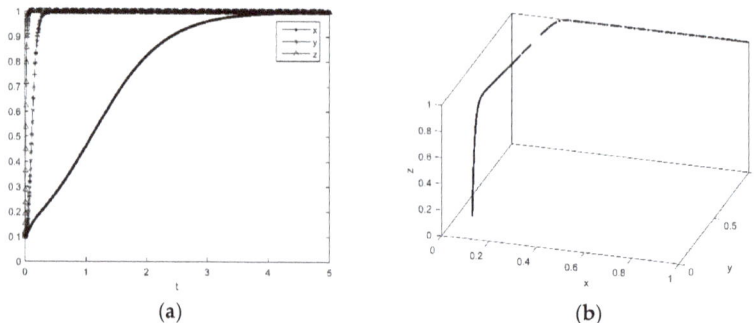

Figure 11. Effect of parameter changes on system evolution results from the initial state ($x_0 = 0.1$, $y_0 = 0.1$, $z_0 = 0.1$): (a) two-dimensional diagram of the result; (b) three-dimensional diagram of the result.

It is worth noting that as indicated by the simulation results, the reward to whistleblowers can urge local governments to raise their emphasis on environmental pollution control and management to some extent; however, from the sustainability perspective, when the emphasis on environmental pollution is generally low, the supervision and punishment of the central supervision department plays a crucial role, and reducing the pollution control costs of the local governments can achieve the same effect as punishment, such as increasing the central government's subsidies to environmental pollution control, setting the environmental quality as one performance indicator in local governments' performance evaluation, etc.

5. Conclusions

This paper has focused on the three main parties involved in environmental pollution control—the central supervision department, local governments, and whistleblowers—and constructed a three-party evolutionary game model based on evolutionary game theory. This paper has also analyzed the equilibrium points as well as the evolutionary stable strategy in the three-dimensional dynamic system using the replicator dynamics equation. This paper has then conducted numerical simulations to demonstrate the impact of different values on the evolution result of environmental pollution control. The results of this study indicate that the positive feedback among behaviors of the central supervision department, local government officials, and the whistleblowers can facilitate the improvement in environmental awareness of the entire society, enhance environmental pollution control, and promote the harmonious coexistence of human society and nature. Based on the above conclusions, this paper has provided the following recommendations for further improvement of China's environmental pollution control:

(1) Public participation is an indispensable part of environmental pollution control and one of the fundamental ways to solve environmental problems [60,61]. It is suggested that whistleblowing be encouraged by reducing the whistleblowing cost and increasing the reward to a certain level [53]. The traditional whistleblowing channels (such as petitioning and hotlines) have high time and economic costs and low efficiency, while the rapid development of the Internet and smart phones has offered new methods with comparatively low economic costs. Therefore, we should make full use of low cost online channels, such as WeChat, Weibo, and other online platforms to encourage whistleblowing. The convenient online channels could effectively enhance the interactions between whistleblowers and the Ministry of Environmental Protection, and indirectly increase the rewards for whistleblowing while saving time and economic costs.

(2) It is necessary to further increase the punishment on local government officials that neglect pollution control and use subsidies to lessen the concern of local government officials on pollution control costs. It is recommended that "promotion" be used as an incentive for local government officials, and to comprehensively evaluate government officials' political achievements from multiple aspects including local economic indicators and environmental indicators. We should not only supervise local government officials' emphasis on local environmental pollution control, but also conduct spot checks on the achievements of pollution control with reference to whistleblowers' feedback. Compared with developed countries, currently the local air pollution control performance in China does not have much weight in the promotion of local government officials [62,63]. In future policy design, it is recommended that bonus points be given to local government officials that work hard on environment improvement and air pollution control in promotion evaluation, or lower the requirements on economic indicators in performance assessment. At the same time, the central environmental protection department should provide proper and reasonable subsidies based on the local policy and measures targeting air pollution according to the actual situation.

(3) Establish and improve the disclosure system for air pollution information and supervision progress. The public disclosure of air pollution information is a new environment management method different from the "administrative control measures" and "market economy measures" [64,65]. It could put pressure on the polluters and environmental departments by disclosing relevant environmental information and utilizing public opinion, thus urging them to change their behavior with help of public supervision in order to eventually achieve the goal of environmental protection. The central

supervision department should make efforts to properly collect and disclose the whistleblowing data from different regions related to environmental pollution. It should not only publicly disclose relevant information such as the number of whistleblowing cases in different regions, but also urge various local governments to establish and improve air pollution information disclosure and supervision progress disclosure systems as soon as possible, in order to form a positive trend in which various parties actively and efficiently participate in air pollution control.

Author Contributions: Y.Y. and W.Y. are joint first authors. They contributed equally to this paper. Conceptualization, W.Y.; Methodology, Y.Y. and W.Y.; Resources, Y.Y. and W.Y.; Software, Y.Y.; Validation, W.Y.; Formal Analysis, Y.Y. and W.Y.; Data Curation, Y.Y. and W.Y.; Writing—Original Draft Preparation, Y.Y. and W.Y.; Writing—Review and Editing, Y.Y. and W.Y.

Funding: This research was funded by the Humanities and Social Sciences Research Fund of the University of Shanghai for Science and Technology, and the Decision-making Consultation Research Project of Shanghai Municipal Government. We gratefully acknowledge the above financial supports.

Conflicts of Interest: The authors declare no conflicts of interest.

References

1. Yang, W.X.; Li, L.G. Energy Efficiency, Ownership Structure, and Sustainable Development: Evidence from China. *Sustainability* **2017**, *9*, 912. [CrossRef]
2. Yang, W.; Li, L. Analysis of Total Factor Efficiency of Water Resource and Energy in China: A Study Based on DEA-SBM Model. *Sustainability* **2017**, *9*, 1316. [CrossRef]
3. Zhai, B.; Chen, J.; Yin, W.; Huang, Z. Relevance Analysis on the Variety Characteristics of PM2.5 Concentrations in Beijing, China. *Sustainability* **2018**, *10*, 3228. [CrossRef]
4. Li, L.G.; Yang, W.X. Total Factor Efficiency Study on China's Industrial Coal Input and Wastewater Control with Dual Target Variables. *Sustainability* **2018**, *10*, 2121. [CrossRef]
5. Huang, W.; Wang, H.; Wei, Y. Endogenous or Exogenous? Examining Trans-Boundary Air Pollution by Using the Air Quality Index (AQI): A Case Study of 30 Provinces and Autonomous Regions in China. *Sustainability* **2018**, *10*, 4220. [CrossRef]
6. Liu, R.; Liu, X.; Pan, B.; Zhu, H.; Yuan, Z.; Lu, Y. Willingness to Pay for Improved Air Quality and Influencing Factors among Manufacturing Workers in Nanchang, China. *Sustainability* **2018**, *10*, 1613. [CrossRef]
7. Wang, F.; Wang, K. Assessing the Effect of Eco-City Practices on Urban Sustainability Using an Extended Ecological Footprint Model: A Case Study in Xi'an, China. *Sustainability* **2017**, *9*, 1591. [CrossRef]
8. Yang, W.; Li, L. Efficiency evaluation of industrial waste gas control in China: A study based on data envelopment analysis (DEA) model. *J. Clean. Prod.* **2018**, *179*, 1–11. [CrossRef]
9. Standing Committee of the Twelfth National people's Congress of the people's Republic of China. *The Environmental Protection Law of the People's Republic of China*; The Law Press of China: Beijing, China, 2014.
10. Ministry of Environmental Protection of the People's Republic of China. The Ministry of Environmental Protection Will Open the "010-12369" Environmental Whistleblowing Hotline on the World Environment Day (5 June). Available online: http://www.mee.gov.cn/gkml/sthjbgw/qt/200910/t20091023_179602.htm (accessed on 17 November 2018).
11. Ministry of Environmental Protection of the People's Republic of China. The "Green Knight" Environmental Snapshot APP for Smartphone Is Officially Released. Available online: http://www.mee.gov.cn/xxgk/hjyw/201406/t20140623_277270.shtml (accessed on 17 November 2018).
12. Ministry of Environmental Protection of the People's Republic of China. The "12369 Environment Whistleblowing Platform" on WeChat Has Been Working Well Since Its Opening in 2015. Available online: http://www.mee.gov.cn/gkml/sthjbgw/qt/201603/t20160322_334063.htm (accessed on 17 November 2018).
13. Ministry of Environmental Protection of the People's Republic of China. Report on the "12369" Environment Whistleblowing in 2016. Available online: http://www.mee.gov.cn/gkml/hbb/bgth/201705/t20170512_414013.htm (accessed on 17 November 2018).
14. China Environment News. Strictly Supervising and Promoting the Coordinated Development of China's Environment and Economy. Available online: http://www.cenews.com.cn/opinion/hjsp/201710/t20171013_854057.html (accessed on 17 November 2018).

15. Environmental Emergency and Accident Investigation Center of the Ministry of Environmental Protection. 12369: Promoting Environmental Information Disclosure and Public Participation. *World Environ.* **2017**, *4*, 76–77.
16. Ministry of Environmental Protection of the People's Republic of China. Report on the "12369" Environment Whistleblowing in 2017. Available online: http://www.mee.gov.cn/gkml/sthjbgw/qt/201801/t20180123_430188.htm (accessed on 17 November 2018).
17. Wei, J.; Lu, S. Investigation and penalty on major industrial accidents in China: The influence of environmental pressures. *Saf. Sci.* **2015**, *76*, 32–41. [CrossRef]
18. Zhang, B.; Chen, X.; Guo, H. Does central supervision enhance local environmental enforcement? Quasi-experimental evidence from China. *J. Public Econ.* **2018**, *164*, 70–90. [CrossRef]
19. Chen, H.; Hao, Y.; Li, J.; Song, X. The impact of environmental regulation, shadow economy, and corruption on environmental quality: Theory and empirical evidence from China. *J. Clean. Prod.* **2018**, *195*, 200–214. [CrossRef]
20. Xinhua Net. 300,000 Yuan! A Domestic Reward Record Has Been Created for Environmental Whistleblowing in Jingjiang, Jiangsu Province. Available online: http://www.xinhuanet.com//legal/2017-11/18/c_1121974431.htm (accessed on 17 November 2018).
21. Zhang, Y. Rewarding Pollution Whistleblowing Highlights Environmental Protection. *Environ. Prot. Circ. Econ.* **2017**, *37*, 1.
22. Huang, G. A Number of Major Environmental Protection Cases were Uncovered by Reward for Whistleblowing. *Pestic. Mark. News* **2018**, *11*, 16–17.
23. Du, T. "Environmental Guardian" Was Arrested Twice—Suspected of Retaliation for Whistleblowing. Available online: http://www.mzyfz.com/index.php/cms/item-view-id-1199432 (accessed on 17 November 2018).
24. The Paper News. The Polluted Enterprises Retaliated the Resident after His Whistleblowing to the Central Environmental Protection Inspectorate. Available online: https://www.thepaper.cn/newsDetail_forward_1512811 (accessed on 17 November 2018).
25. Jin, J.; Zhuang, J.; Zhao, Q. Supervision after Certification: An Evolutionary Game Analysis for Chinese Environmental Labeled Enterprises. *Sustainability* **2018**, *10*, 1494. [CrossRef]
26. Zhao, X.; Zhang, Y. The System Dynamics (SD) Analysis of the Government and Power Producers' Evolutionary Game Strategies Based on Carbon Trading (CT) Mechanism: A Case of China. *Sustainability* **2018**, *10*, 1150. [CrossRef]
27. Gao, L.; Zhao, Z.-Y. System Dynamics Analysis of Evolutionary Game Strategies between the Government and Investors Based on New Energy Power Construction Public-Private-Partnership (PPP) Project. *Sustainability* **2018**, *10*, 2533. [CrossRef]
28. Liu, L.; Feng, C.; Zhang, H.; Zhang, X. Game Analysis and Simulation of the River Basin Sustainable Development Strategy Integrating Water Emission Trading. *Sustainability* **2015**, *7*, 4952–4972. [CrossRef]
29. Zhao, R.; Zhou, X.; Han, J.; Liu, C. For the sustainable performance of the carbon reduction labeling policies under an evolutionary game simulation. *Technol. Forecast. Soc. Chang.* **2016**, *112*, 262–274. [CrossRef]
30. Wu, B.; Liu, P.; Xu, X. An evolutionary analysis of low-carbon strategies based on the government–enterprise game in the complex network context. *J. Clean. Prod.* **2017**, *141*, 168–179. [CrossRef]
31. Li, Q.; Bao, H.; Peng, Y.; Wang, H.; Zhang, X. The Collective Strategies of Major Stakeholders in Land Expropriation: A Tripartite Game Analysis of Central Government, Local Governments, and Land-Lost Farmers. *Sustainability* **2017**, *9*, 648. [CrossRef]
32. Chen, S. An Evolutionary Game Study of an Ecological Industry Chain Based on Multi-Agent Simulation: A Case Study of the Poyang Lake Eco-Economic Zone. *Sustainability* **2017**, *9*, 1165. [CrossRef]
33. Zhang, M.; Li, H. New evolutionary game model of the regional governance of haze pollution in China. *Appl. Math. Model.* **2018**, *63*, 577–590. [CrossRef]
34. Standing Committee of the Twelfth National people's Congress of the people's Republic of China. *Law of the People's Republic of China on Air Pollution Prevention and Control (Latest Revision)*; The Law Press of China: Beijing, China, 2018.
35. State Council of the People's Republic of China. Notice of the State Council on Printing and Dispatching the Air Pollution Prevention and Control Action Plan. Available online: http://www.gov.cn/zhengce/content/2013-09/13/content_4561.htm (accessed on 17 November 2018).

36. Ministry of Ecology and Environment of the People's Republic of China. The Responsibility of the Ministry of Ecology and Environment of the People's Republic of China. Available online: http://www.mee.gov.cn/zjhb/zyzz/ (accessed on 17 November2018).
37. Shen, X.; Wang, L.; Wu, C.; Lv, T.; Lu, Z.; Luo, W.; Li, G. Local interests or centralized targets? How China's local government implements the farmland policy of Requisition–Compensation Balance. *Land Use Policy* **2017**, *67*, 716–724. [CrossRef]
38. Yang, W.; Li, L. Efficiency Evaluation and Policy Analysis of Industrial Wastewater Control in China. *Energies* **2017**, *10*, 1201. [CrossRef]
39. Wang, D.; Chen, C.; Richards, D. A prioritization-based analysis of local open government data portals: A case study of Chinese province-level governments. *Gov. Inf. Q.* **2018**, *35*, 644–656. [CrossRef]
40. Cao, R.; Zhang, A.; Wen, L. Trans-regional compensation mechanism under imbalanced land development: From the local government economic welfare perspective. *Habitat Int.* **2018**, *77*, 56–63. [CrossRef]
41. Zhang, L.; Wu, B. Farmer innovation system and government intervention: An empirical study of straw utilisation technology development and diffusion in China. *J. Clean. Prod.* **2018**, *188*, 698–707. [CrossRef]
42. Zhou, Y.; Zhu, S.; He, C. How do environmental regulations affect industrial dynamics? Evidence from China's pollution-intensive industries. *Habitat Int.* **2017**, *60*, 10–18. [CrossRef]
43. Cull, R.; Xu, L.C.; Yang, X.; Zhou, L.-A.; Zhu, T. Market facilitation by local government and firm efficiency: Evidence from China. *J. Corp. Financ.* **2017**, *42*, 460–480. [CrossRef]
44. Wu, J.; Xu, M.; Zhang, P. The impacts of governmental performance assessment policy and citizen participation on improving environmental performance across Chinese provinces. *J. Clean. Prod.* **2018**, *184*, 227–238. [CrossRef]
45. Geall, S.; Shen, W. Gongbuzeren Solar energy for poverty alleviation in China: State ambitions, bureaucratic interests, and local realities. *Energy Res. Soc. Sci.* **2018**, *41*, 238–248. [CrossRef]
46. Yang, Y.; Yu, G. The analysis of social resource mobilization on new media: A case study of Chinese environmental protection documentary Under the Dome. *Telematics Inform.* **2018**. [CrossRef]
47. Shen, L.; Wang, Y. Supervision mechanism for pollution behavior of Chinese enterprises based on haze governance. *J. Clean. Prod.* **2018**, *197*, 571–582. [CrossRef]
48. Guttman, D.; Young, O.; Jing, Y.; Bramble, B.; Bu, M.; Chen, C.; Furst, K.; Hu, T.; Li, Y.; Logan, K.; et al. Environmental governance in China: Interactions between the state and "nonstate actors". *J. Environ. Manag.* **2018**, *220*, 126–135. [CrossRef] [PubMed]
49. Deng, J. The National Supervision Commission: A New Anti-corruption Model in China. *Int. J. Law Crime Justice* **2018**, *52*, 58–73. [CrossRef]
50. Liu, L.; Wu, T.; Li, S.; de Jong, M.; Sun, Y. The drivers of local environmental policy in China: An analysis of Shenzhen's environmental performance management system, 2007–2015. *J. Clean. Prod.* **2017**, *165*, 656–666. [CrossRef]
51. Zhang, L.; An, Y. The government capacity on industrial pollution management in Shanxi province: A response impulse analysis. *J. Environ. Manag.* **2018**, *223*, 1037–1046. [CrossRef]
52. Jia, S.; Liu, X.; Yan, G. Effect of APCF policy on the haze pollution in China: A system dynamics approach. *Energy Policy* **2019**, *125*, 33–44. [CrossRef]
53. Wang, L. Research on Environmental Right of Reporting and Its Protection in China. Master's Thesis, Soochow University, Suzhou, China, 2017.
54. Zhang, H.; Xiong, L.; Qiu, Y.; Zhou, D. How Have Political Incentives for Local Officials Reduced the Environmental Pollution of Resource-depleted Cities? *Energy Procedia* **2017**, *143*, 873–879. [CrossRef]
55. Yang, J.; Zhang, B. Air pollution and healthcare expenditure: Implication for the benefit of air pollution control in China. *Environ. Int.* **2018**, *120*, 443–455. [CrossRef]
56. Xia, Q.; Jin, M.; Wu, H.; Yang, C. A DEA-based decision framework to determine the subsidy rate of emission reduction for local government. *J. Clean. Prod.* **2018**, *202*, 846–852. [CrossRef]
57. Zhao, X.-G.; Ren, L.Z.; Zhang, Y.Z.; Wan, G. Evolutionary game analysis on the behavior strategies of power producers in renewable portfolio standard. *Energy* **2018**, *162*, 505–516. [CrossRef]
58. Chen, Y.; Ding, S.; Zheng, H.; Zhang, Y.; Yang, S. Exploring diffusion strategies for mHealth promotion using evolutionary game model. *Appl. Math. Comput.* **2018**, *336*, 148–161. [CrossRef]

59. Vigliassi, M.P.; Massignan, J.A.D.; Delbem, A.C.B.; London, J.B.A. Multi-objective evolutionary algorithm in tables for placement of SCADA and PMU considering the concept of Pareto Frontier. *Int. J. Electr. Power Energy Syst.* **2019**, *106*, 373–382. [CrossRef]

60. Rollason, E.; Bracken, L.J.; Hardy, R.J.; Large, A.R.G. Evaluating the success of public participation in integrated catchment management. *J. Environ. Manag.* **2018**, *228*, 267–278. [CrossRef]

61. Brombal, D.; Moriggi, A.; Marcomini, A. Evaluating public participation in Chinese EIA. An integrated Public Participation Index and its application to the case of the New Beijing Airport. *Environ. Impact Assess. Rev.* **2017**, *62*, 49–60. [CrossRef]

62. Pu, Z.; Fu, J. Economic growth, environmental sustainability and China mayors' promotion. *J. Clean. Prod.* **2018**, *172*, 454–465. [CrossRef]

63. Taylor, C.M.; Gallagher, E.A.; Pollard, S.J.T.; Rocks, S.A.; Smith, H.M.; Leinster, P.; Angus, A.J. Environmental regulation in transition: Policy officials' views of regulatory instruments and their mapping to environmental risks. *Sci. Total Environ.* **2019**, *646*, 811–820. [CrossRef]

64. Tian, X.-L.; Guo, Q.-G.; Han, C.; Ahmad, N. Different extent of environmental information disclosure across chinese cities: Contributing factors and correlation with local pollution. *Glob. Environ. Chang.* **2016**, *39*, 244–257. [CrossRef]

65. Kasim, M.T. Evaluating the effectiveness of an environmental disclosure policy: An application to New South Wales. *Resour. Energy Econ.* **2017**, *49*, 113–131. [CrossRef]

sustainability

MDPI

Article

Relevance Analysis on the Variety Characteristics of PM$_{2.5}$ Concentrations in Beijing, China

Binxu Zhai [1,2], Jianguo Chen [1,2,*], Wenwen Yin [1,2] and Zhongliang Huang [1,2]

[1] Department of Engineering Physics, Tsinghua University, Beijing 100084, China;
     dbx15@mails.tsinghua.edu.cn (B.Z.); yww17@mails.tsinghua.edu.cn (W.Y.); hzlthu@foxmail.com (Z.H.)
[2] Beijing Key Laboratory of City Integrated Emergency Response Science, Tsinghua University,
     Beijing 100084, China
*   Correspondence: chenjianguo@mail.tsinghua.edu.cn

Received: 9 August 2018; Accepted: 7 September 2018; Published: 10 September 2018

Abstract: Air pollution has become one of the most serious environmental problems in the world. Considering Beijing and six surrounding cities as main research areas, this study takes the daily average pollutant concentrations and meteorological factors from 2 December 2013 to 30 June 2017 into account and studies the spatial and temporal distribution characteristics and the relevant relationship of particulate matter smaller than 2.5 μm (PM$_{2.5}$) concentrations in Beijing. Based on correlation analysis and geo-statistics techniques, the inter-annual, seasonal, and diurnal variation trends and temporal spatial distribution characteristics of PM$_{2.5}$ concentration in Beijing are studied. The study results demonstrate that the pollutant concentrations in Beijing exhibit obvious seasonal and cyclical fluctuation patterns. Air pollution is more serious in winter and spring and slightly better in summer and autumn, with the spatial distribution of pollutants fluctuating dramatically in different seasons. The pollution in southern Beijing areas is more serious and the air quality in northern areas is better in general. The diurnal variation of air quality shows a typical seasonal difference and the daily variation of PM$_{2.5}$ concentrations present a "W" type of mode with twin peaks. Besides emission and accumulation of local pollutants, air quality is easily affected by the transport effect from the southwest. The PM$_{2.5}$ and PM$_{10}$ concentrations measured from the city of Langfang are taken as the most important factors of surrounding pollution factors to PM$_{2.5}$ in Beijing. The concentrations of PM$_{10}$ and carbon monoxide (CO) concentrations in Beijing are the most significant local influencing factors to PM$_{2.5}$ in Beijing. Extreme wind speeds and maximal wind speeds are considered to be the most significant meteorological factors affecting the transport of pollutants across the region. When the wind direction is weak southwest wind, the probability of air pollution is greater and when the wind direction is north, the air quality is generally better.

Keywords: relevance analysis; spatial and temporal distribution characteristics; PM$_{2.5}$; Beijing

---

1. Introduction

Ambient fine particulate matter smaller than 2.5 μm (PM$_{2.5}$) is a major environmental problem and is harmful to human health [1,2]. Numerous studies have documented that short-term and long-term exposure to PM$_{2.5}$ can increase the risks of allergies, respiratory system diseases, and cardiovascular diseases [3–6]. Meanwhile, the haze caused by PM$_{2.5}$ reduces visibility [7,8] and affects transportation, causing huge economic losses [9]. Chinese cities suffer heavily from ambient air pollution [10], particularly the capital Beijing [11]. A Global Burden of Disease (GBD) study ranks ambient particulate matter pollution (PM$_{2.5}$) as the 5th leading risk factor for early death and disability in China [12]. Thus, it is necessary to carry out research on PM$_{2.5}$ in China. Current research mainly focuses on the physical and chemical properties of pollutants [13–15], although some studies focus on social [16] and natural factors [17]. Previous studies have shown that meteorological factors, including those of wind speed, wind direction, precipitation, relative humidity, and atmospheric pressure, have a very significant impact on air quality. For example, Guo et al. found that adverse weather conditions, including those of low wind speed and high relative humidity would promote the concentration of pollutants in the study area and lead to the rapid deterioration of air quality in the short term [18]. Li et al. [19] identified ideal meteorological regions, according to the quantified spatial relationships between PM and meteorological elements. Lv et al. [20] found that wind speed and wind direction have more complex effects on air pollution. The prevailing wind direction or wind speed is beneficial to the dilution and diffusion of pollutants in the region and then reduce the concentrations of pollutants. The effect of non-prevailing wind (not long-lasting) or of low wind speed is just the opposite. The specific effect of wind direction is related to the distribution of surrounding pollution sources. In addition, it has also been found that the periodic changes in the haze weather in Beijing are closely related to the specific geographical location of Beijing and the growth of high pressure cyclones in the upper reaches of Siberia [21]. In addition, the impact of the cross-regional transportation of particulates on local air quality has been gradually emphasized in recent research [22–26]. Wang et al. [27] used an integrated Fifth-Generation NCAR/Penn State Mesoscale Model-Community Multiscale Air Quality (MM5-CMAQ) modeling system to analyze the backward trajectory of pollutants in Beijing and found that the southwest is the most influential transport channel. Ma et al. [28] demonstrated that regional transport from southern Beijing is a leading influencing factor that spurs initial $PM_{2.5}$ increases.

$PM_{2.5}$ concentration is affected by local pollutants, surrounding pollutants factors, and meteorological factors and it has temporal and spatial variability. However, in most studies on the temporal and spatial distribution characteristics and the relevant relationship of $PM_{2.5}$ concentrations, only local pollutants and meteorological factors were considered [17,19]. Few studies have considered comprehensive factors. In this study, we consider Beijing and six surrounding cities as main research areas, taking the daily average pollutant concentrations and meteorological elements from 2 December 2013 to 13 October 2017 into account and study the spatial and temporal distribution characteristics and the relevant relationship of $PM_{2.5}$ concentrations in Beijing. Investigating the temporal and spatial distribution characteristics and the relevant relationship of $PM_{2.5}$ is important for understanding the mechanisms underlying $PM_{2.5}$ pollution and for preventing haze. Therefore, this study has great practical value, it can elucidate the factors contributing to this air pollution and provide scientific reference for joint control measures in future.

Based on correlation analysis and geo-statistics techniques, this paper studies the inter-annual, seasonal, diurnal variation trends, and temporal spatial distribution characteristics of $PM_{2.5}$ concentration in Beijing. The relevant relationships between $PM_{2.5}$ and major local pollutants, surrounding pollutants, and meteorological factors are also analyzed.

2. Study Area and the Data

2.1. Study Area

Beijing, ranging from 39.4° N to 41.6° N and 115.7° E to 117.4° E and located in the north of the North China Plain, is surrounded by Hebei Province, along with Tianjin. The terrain is generally characterized by high altitude in the west and low altitude in the east. The western mountains belong to the Taihang Mountains and the northern mountains belong to the Yanshan Mountains. The central and southeastern parts between the two mountains are plain areas, with large mountainous areas and a total elevation between 20–2300 m. It has 16 districts, 6 of which are located in the downtown area and the remaining 10 are located in the suburbs, covering an area of 16,410.54 square kilometers. The seasonal distribution of precipitation is fairly inhomogeneous. 80% of the annual precipitation occurs in the three months of summer (June, July and August). The sunshine duration is the longest in spring, followed by autumn. In summer, the sunshine duration is slightly shorter due to the plentiful precipitation and the sunshine duration is the shortest in winter. Taking into account the effects of transport, this study focuses on Beijing and its surrounding cities, including Baoding, Chengde, Langfang, Tianjin, Zhangjiakou, and Tangshan. The study area is illustrated in Figure 1.

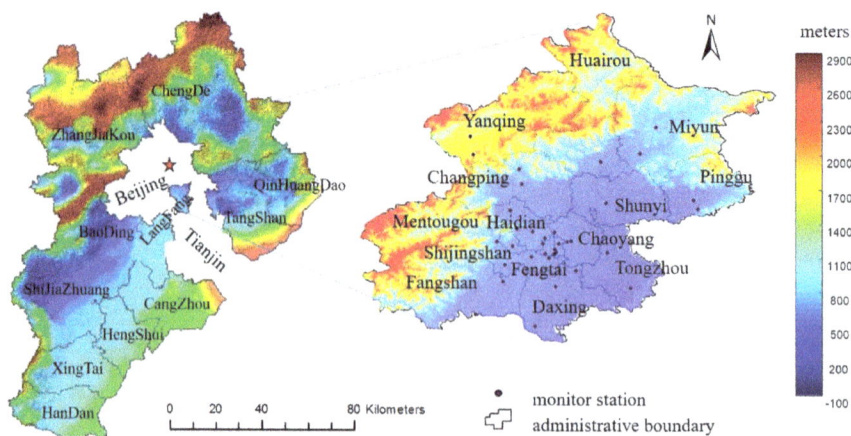

Figure 1. The physical geography of the study areas with colors denoting altitudes above sea level.

2.2. Data Collection

Historical site records of major pollutant concentrations and meteorological data from 2 December 2013 to 30 June 2017 were collected from open sources, spanning a total of 1307 days. Ground-measured hourly pollutant concentrations, including those of PM_{10}, $PM_{2.5}$, carbon monoxide (CO), nitrogen dioxide (NO_2), sulfur dioxide (SO_2), ozone (O_3) and the Air Quality Index (AQI), were collected from the Beijing Municipal Environment Monitoring Center (BJMEMC, http://zx.bjmemc.com.cn/) and then calculated as daily mean values. The locations of the 35 monitoring stations around the city are shown as dots in Figure 1. Original meteorological data on air temperature (TEM), daily sunshine duration (SSD), wind direction (WD), and wind speed (WIN), etc. were acquired from the website of the National Meteorological Information Center (NMIC, http://data.cma.cn/). Pollutant concentrations measured in surrounding cities were obtained from the Ministry of Environmental Protection of China data center (MEP, http://www.mep.gov.cn/). The raw parameters selected from open sources are listed in Table 1.

For meteorological data, NMIC provides the Air Pollution Index (API) interface for data acquisition. After authentication, it can be downloaded directly and the original data can be obtained through analysis. There are no historical archived pollutant concentration data and the BJMEMC official website only provides real-time online display. This paper applies a crawler based on the Scrapy framework to scrawl pollutant concentration records.

The data flow in Scrapy is controlled by the central engine. The engine opens a website for a crawler, requests the URL address for the site and then schedules it in the scheduler. The engine sends the URL to the downloader by the download middleware and the download middleware then disguises itself as a normal client in response to the anti-scrawling strategy and generates feedback from the parsing page and returns to the engine. Once received by the engine, the parsed page is sent to the crawler by the crawler middleware, then the content resolved by the crawler is sent back to the engine. The collected data are saved to a database through the pipeline and item middleware and then returned to the scheduler. After that, the procedures above are repeated until all requests are processed. The engine closes the website and gets all the data.

Table 1. Description of raw data parameters.

Parameter		Type *	Unit	Description
Pollutant factors	$PM_{2.5}$-BJ	N	$\mu g/m^3$	Daily averaged concentration of $PM_{2.5}$ in Beijing
	PM_{10}-BJ	N	$\mu g/m^3$	Daily averaged concentration of PM_{10} in Beijing
	SO_2-BJ	N	$\mu g/m^3$	Daily averaged concentration of SO_2 in Beijing
	CO-BJ	N	mg/m^3	Daily averaged concentration of CO in Beijing
	NO_2-BJ	N	$\mu g/m^3$	Daily averaged concentration of NO_2 in Beijing
	O_3-BJ	N	$\mu g/m^3$	Daily averaged concentration of O_3 in Beijing
	AQI-BJ	N	-	Daily air quality index of Beijing
	Class-BJ	C	1–5	Daily grade of air quality in Beijing
Meteorological factors	EVP	N	0.1 mm	Daily evaporation capacity of Beijing
	GST-mean	N	0.1 °C	Daily averaged ground surface temperature
	GST-max	N	0.1 °C	Daily maximal ground surface temperature
	GST-min	N	0.1 °C	Daily minimal ground surface temperature
	PRE-208	N	0.1 mm	Precipitation from 20:00 p.m. to 8:00 a.m.
	PRE-820	N	0.1 mm	Precipitation from 8:00 a.m. to 20:00 p.m.
	PRE-2020	N	0.1 mm	Precipitation from 20:00 p.m. to 20:00 p.m.
	PRS-mean	N	0.1 hPa	Daily averaged barometric pressure
	PRS-max	N	0.1 hPa	Daily maximal barometric pressure
	PRS-min	N	0.1 hPa	Daily minimal barometric pressure
	RHU-mean	N	1%	Daily averaged relative humidity
	RHU-min	N	1%	Daily minimal relative humidity
	SSD	N	0.1 h	Daily duration of sunshine
	TEM-mean	N	0.1 °C	Daily averaged air temperature
	TEM-max	N	0.1 °C	Daily maximal air temperature
	TEM-min	N	0.1 °C	Daily minimal air temperature
	WIN-mean	N	0.1 m/s	Daily averaged wind speed
	WIN-max	N	0.1 m/s	Daily maximal wind speed
	WD-max	C	1–16	Wind direction of maximal wind speed in category
	WIN-ext	N	0.1 m/s	Extreme wind speed
	WD-ext	C	1–16	Wind direction of extreme wind speed in category

* The C in column Type indicates category variables, while the N indicates numerical ones.

3. Methodology

The main data analysis methods adopted in this study include statistical analysis, spatial analysis, and visualization technology. The statistical analysis technique principally included variance analysis, correlation analysis, regression analysis, factor analysis, and so on, analyzing the complicated relationship between $PM_{2.5}$ concentrations, meteorological factors, and surrounding factors. Spatial analysis, including Spatial Center Statistics (SCS) and Exploratory Spatial Data Analysis (ESDA) were adopted to reveal the temporal and spatial characteristics of pollutant diffusion. The Spatial Center Statistics focused on depicting spatial distribution, which was mainly realized by calculating the basic parameters of the distribution, while the Exploratory Spatial Data Analysis emphasized the description of data, the identification of data statistical characteristics, and the preliminary judgment of the structure of the data through relevant assumptions, aimed at revealing spatial data characteristics, identifying outliers or regions, exploring spatial association patterns, recognizing accumulate or hotspot areas, implementing spatial zoning, and discovering spatial heterogeneity through geographical visualization. The data visualization methods used in this paper mainly included scatter plot, wind rose chart, and violin diagram, so as to intuitively illustrate the atmospheric phenomena varying with time and space behind data and help to find out the potential development pattern.

4. Results and Discussion

4.1. General Statistical Characteristics of $PM_{2.5}$ Concentrations and Exploratory Data Analysis

Statistical descriptions of the main indicators measured for the observation period are presented in Table 2. It can be seen from the table that the air quality situation in Beijing is certainly not optimistic, considering how the average 24-h value of $PM_{2.5}$ concentrations reached 77.40 $\mu g/m^3$. This is three times more than the WHO guidance value (25 $\mu g/m^3$) and the 24-h average of the PM_{10} concentrations reached 104.90 $\mu g/m^3$, which is two times more than that of the WHO guidance value (50 $\mu g/m^3$). The 24-h maximum concentrations of $PM_{2.5}$ and PM_{10} reached 477 $\mu g/m^3$ and 820 $\mu g/m^3$, respectively. In addition, the variance of $PM_{2.5}$ and PM_{10} concentrations were also closed to the mean levels respectively, indicating the volatility of the major pollutant concentrations and the instability of the regional air quality. The annual mean value of other gaseous pollutants has not exceeded the national standard at present but the peak value was higher in different degrees than the national level-2 standard in the same period. It reveals that all of the pollutant concentrations in the heavily polluted days have reached reasonably high levels and the long-term exposure to such an environment is very harmful to the human body and corresponding protection measures should be taken as precautions.

Table 2. The statistical description for main indicators.

	$PM_{2.5}$-bj	PM_{10}-bj	NO_2-bj	CO-bj	SO_2-bj	O_3-bj
Unit	$\mu g/m^3$	$\mu g/m^3$	$\mu g/m^3$	mg/m^3	$\mu g/m^3$	$\mu g/m^3$
Mean	77.40	104.90	50.41	1.23	14.52	97.93
Standard Deviation	68.43	80.86	24.38	1.01	17.01	64.70
25%	29	48	33	0.60	4	50
50%	58	88	44	0.90	8	85
75%	103	135	61	1.40	18	138
Range	(5, 477)	(0, 820)	(8, 155)	(0.2, 8.0)	(2, 133)	(2, 294)

Figure 2 shows the fluctuation of the $PM_{2.5}$ concentrations in Beijing during the study period. The solid line represents the variety of $PM_{2.5}$ concentrations, the points are the daily mean values of $PM_{2.5}$ concentration in six different colors according to the Individual Air Quality Index (IAQI) level respectively. The 24-h mean value of $PM_{2.5}$ concentrations in the study period and the corresponding WHO guidance value are marked by a dotted line. It can be seen that about 40% of Beijing's air quality exceeded the national standard in one year and the air quality exceeded the WHO recommended standard about 80% of the time.

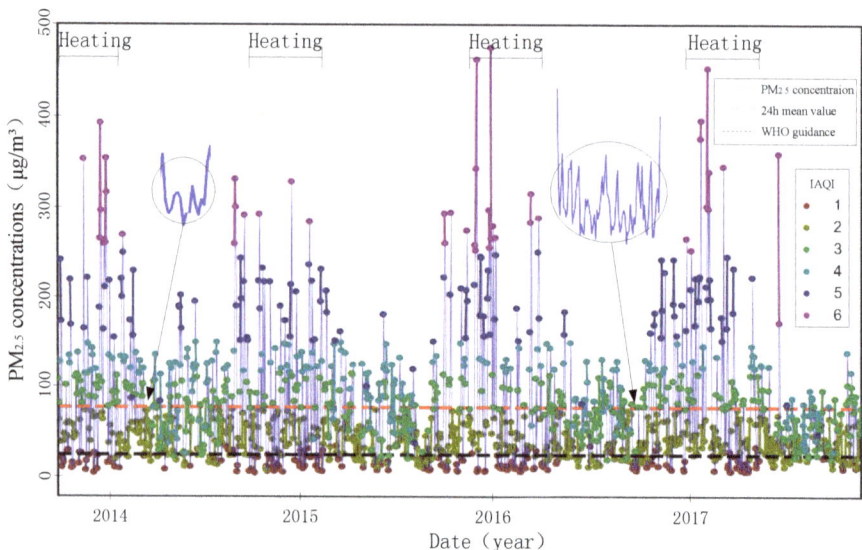

Figure 2. The fluctuation of the $PM_{2.5}$ concentrations in Beijing during the study period.

It can be drawn from Figure 2 that the air pollution caused by fine particles exhibits remarkable volatility and irregularity, at the same time, it shows a certain seasonal and periodic fluctuation in the overall variation tendency. The peak values of pollutant concentrations generally concentrate on heating periods, indicating that the energy structure of central heating has had an obvious influence on the ambient air quality in Beijing. According to the management methods for central heating in Beijing, the statutory heating period in Beijing is generally from 15 November to 15 March of the next year and fluctuates slightly in accordance to the current situation. The annual variation of $PM_{2.5}$ concentrations were relatively stable, with the annual mean values of year 2014 to 2017 were 84.83 μg/m^3, 80.25 μg/m^3, 73.01 μg/m^3, and 57.83 μg/m^3 respectively. The annual mean values spread over a decreasing tendency, which indicates that the current treatment measures for air pollution have achieved initial success. In terms of the periodical tendency, the short cycle of the pollutant concentration fluctuations was about one week (the left magnified curve in Figure 2) to one month (the right magnified curve in Figure 2), the long period is one year, and the peak of the pollutant concentrations occurred alternately throughout the year.

4.2. Seasonal Variation of $PM_{2.5}$ Concentrations in Beijing

Figure 3 shows the temporal and spatial distribution of seasonal variation of $PM_{2.5}$ concentrations in Beijing in the year 2017. According to climatological classification, the spring in Beijing is regarded as months March, April, and May, the summer from June to August, the autumn from September to November, and the winter will be regarded as the time from December to February of the next year. This distribution situation diagram uses the monitoring data of 35 monitoring stations in the city, taking into account the anisotropy, autocorrelation, and the trend of data distribution, which is drawn by the Kriging interpolation method in geo-statistics [29,30].

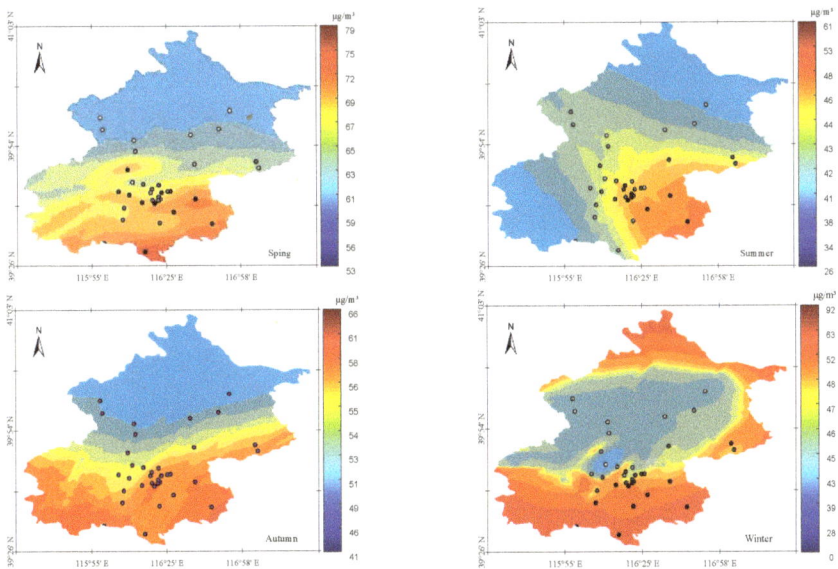

Figure 3. Temporal and spatial distribution of seasonal variation of $PM_{2.5}$ concentrations in Beijing in 2017.

It can be drawn from Figure 3 that the air pollution in Beijing is more serious in spring and winter, and slightly better in summer and autumn. The distribution of pollutants varied dramatically in different seasons. The most seriously polluted regions in spring were the southern and central Beijing. In summer, the overall air pollution situation was better, only the southeast part of the city was more polluted. The main polluted areas in autumn were also concentrated in the southeast and southwest regions. In winter, the pollution situation was further aggravated, and some regions in the north also registered by high level pollution in PM$_{2.5}$ concentrations. In general, the spatial distributional characteristics of regional air quality were quite different in the four seasons, the pollution levels in the southern regions were more serious, and the concentrations of pollutants gradually reduced from the southwest to the northeast. The seasonal variation of pollutant spatial and temporal distribution may have been caused by different meteorological conditions and the distribution of pollution sources [31,32]. For example, the meteorological conditions formed by the combination of dry climate and strong wind in spring are conducive to the formation and development of sandstorms. The humid and hot environment and the increase of irradiation intensity in summer are beneficial to the formation of photochemical reactions, resulting in secondary pollution. The spatial distribution of pollutants in autumn was mainly due to the regional transport caused by unfavorable weather conditions, which was the main cause of air pollution in this period. In winter, the air quality was inseparable from the biomass burning [33] and coal combustion [34]. With the weakening of the wind and the decrease of the atmospheric height, the diffusion and convection in the horizontal and vertical direction were gradually restricted. The accumulation effect of local pollutants aggravated the outbreak of serious pollution events in winter.

4.3. Diurnal Variation Characteristics of PM$_{2.5}$ Concentrations in Beijing

Figure 4 shows the diurnal variation of PM$_{2.5}$ concentrations in different seasons of the year 2017 in Beijing. The curves in different colors represent the variation of the seasonal average PM$_{2.5}$ concentrations at different times in one day of the corresponding season. The value of each time is the average measured values of 35 monitoring stations around the city. The dashed lines represent the annual mean value of the PM$_{2.5}$ concentrations in the year 2017 (red line) and the guiding value given by WHO (black line).

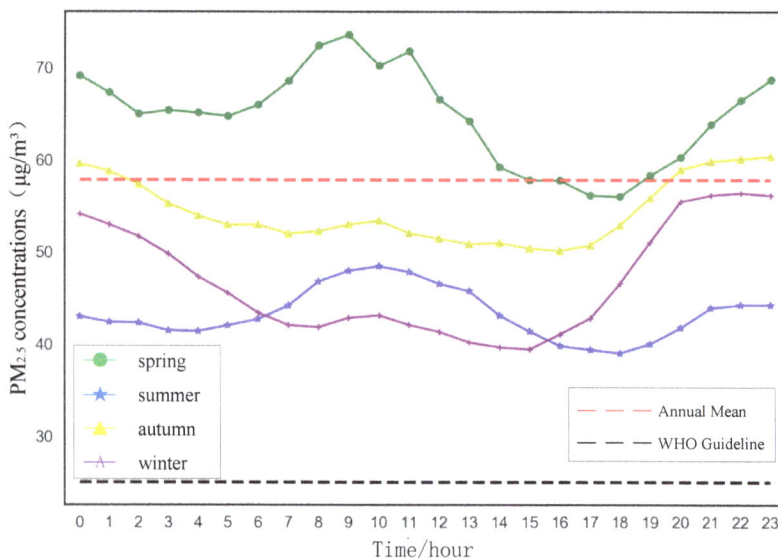

Figure 4. Diurnal variation of PM$_{2.5}$ concentrations in different seasons in Beijing.

It can be observed from Figure 4 that the diurnal variation of air quality presented a certain seasonal difference and there was a certain fluctuation in diurnal concentrations. In the seasons of summer and autumn, the diurnal variation was small, with the daily fluctuation lingering around

10 μg/m^3, while the diurnal variations in winter and spring was large, with the daytime fluctuation reaching about 20 μg/m^3. The diurnal variation of PM$_{2.5}$ concentrations was by and large characterized by a "W" type double wave. The peak value in the daytime occurred between 08:00 and 11:00 in the morning, and then continued to decrease to a trough. The peak in the night appeared after 19:00 and then gradually decreased in the early hours of the morning. The occurrence of the peak pollutant concentrations in the daytime could be related to the increase of human activity during the early peak period. With the increase of temperature at noon, the pollutant concentrations gradually decreased, aided by the weather conditions. Subsequently, with the approach of evening peak, the increase of restaurant emissions, and the reduction of the height of the planetary boundary layer, the concentration of pollutants increased further [34]. In the seasons of spring and summer, the average concentrations of pollutants were higher during the daytime and reduced at night, which contrasts with the situations in the autumn and winter. This difference was mainly due to the diverse sources of pollution and their distinct formation mechanisms in different seasons. The air quality in spring and summer was more affected by human activities. With the advent of night and the decrease of human activities, the concentration of pollutants dropped to a lower level in these two quarters. The main influencing factors of outdoor air quality in autumn and winter were the transport and diffusion effect of external pollution sources. The impact of human activity was relatively small and superseded by meteorological conditions. Therefore, during the night time, the lower atmosphere and stagnant wind conditions aggravated the accumulation of pollutants and increased the PM$_{2.5}$ concentrations [20].

To further reveal the diurnal variation of the pollutant concentrations in Beijing, Figure 5 shows the temporal and spatial variation of the PM$_{2.5}$ concentrations in Beijing on 25 December 2015. On 25 December 2015, a serious particulate matter pollution incident occurred in Beijing. The concentration of PM$_{2.5}$ in some areas reached over 700 μg/m^3, causing widespread international and social concerns.

As can be drawn from Figure 5, the particulate matters in Beijing were mainly concentrated on the southeast and central areas at 00:00 in the early morning, and the air quality in the northern and western mountainous areas was better than in other regions. At 04:00 and 08:00, the pollution bound expanded to the northern and western regions and the pollution levels in the southern part of the region were also aggravated, however the northern and western parts of the region still maintained high levels of air quality. By noon, the concentration of particulate matters in the city reached a peak and the pollution range was further expanded. From the southwest to the northeast, almost the whole city was immersed in serious pollutions of middle and above-recommended levels and the concentration of PM$_{2.5}$ in some areas reached more than 705 μg/m^3, creating the record of the highest concentration in a single day in the year. At this time, there were still some regions in the northern mountainous areas that were unaffected. At 16:00, the pollution range expanded once again. The core pollution areas were concentrated in the Fengtai, Chaoyang, and Haidian districts in the center of the city and the northern regions were also thereby affected. After nightfall, the concentration of pollutants gradually decreased but the pollution areas did not shrink. The average concentration of pollutants in the city dropped to the levels of 08:00 in the morning. In general, the heavily polluted areas in this serious pollution event were still concentrated in the southern and central areas, and were obviously affected by the transport effect from southwest directions. At the early stage of the development of this air pollution event (before 12:00 a.m.), the air quality level was mainly affected by the local emission and accumulation effects. Influenced by meteorological conditions, the transport effect of the surrounding pollution sources became the leading factor for the overall air quality levels in the city, which aggravated the severity of the air pollution and promoted the outbreak of a serious air pollution event.

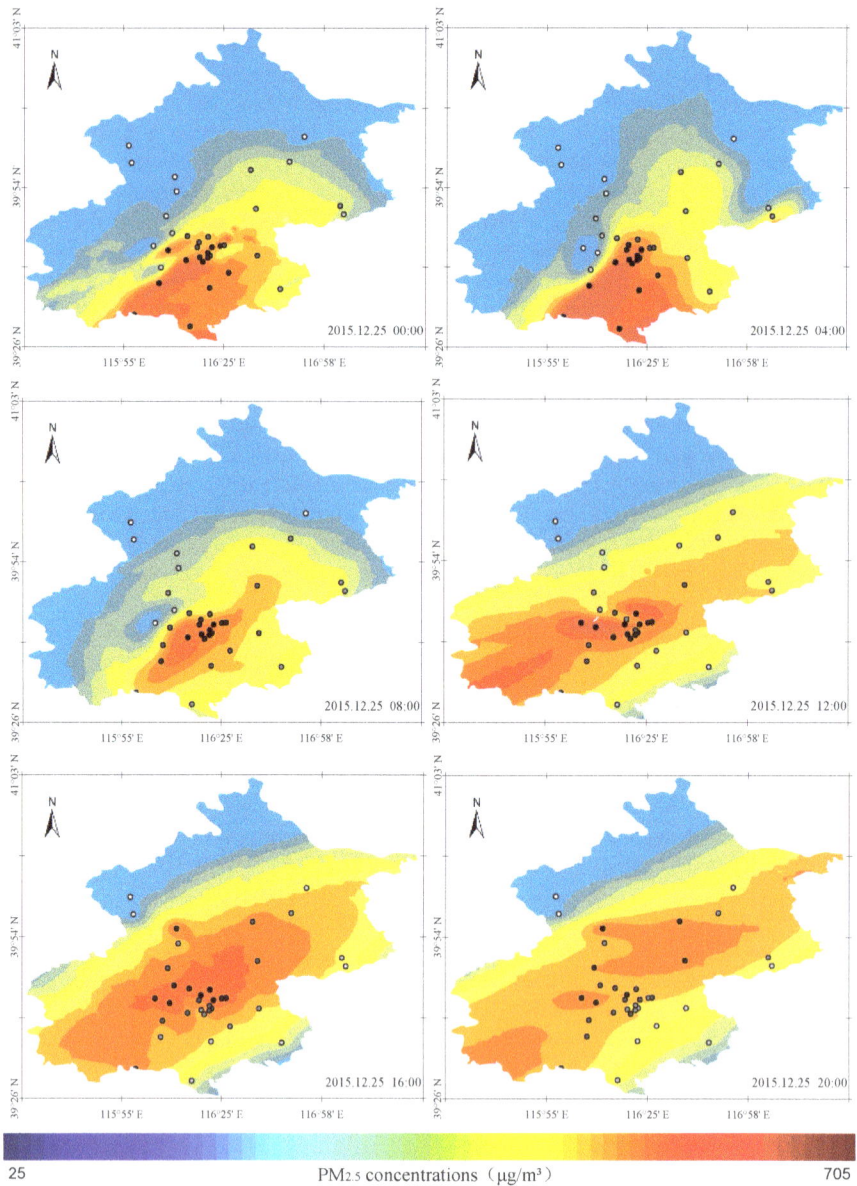

Figure 5. The temporal and spatial variation of the PM$_{2.5}$ concentrations in Beijing on 25 December 2015.

4.4. Relevance Analysis between PM$_{2.5}$ and Major Pollutants

The raw data can be divided into two types, the numerical variables and categorical ones. For numerical variables, the Pearson coefficient analysis showed that the top five linearly related raw parameters with PM$_{2.5}$ concentrations in Beijing were the PM$_{2.5}$ concentrations in Langfang (0.85), PM$_{10}$ concentrations in Beijing (0.85), PM$_{10}$ concentrations in Langfang (0.83), CO concentrations in Beijing (0.83), and PM$_{2.5}$ concentrations in Chengde (0.83), as shown in the upper right and lower left corners of the correlation matrix in Figure 6. At the same time, there was a linear correlation between the variables, as shown in the middle of the correlation matrix. Figure 6 shows a linear correlation matrix of the top 15 variables, with a high linear correlation with $PM_{2.5}$ concentration. A significance test was performed on these correlation coefficients and found that sig = 0.000, indicating that the significance level p value was less than 0.001, further indicating that the correlation does exist. It can be seen from the figure that there was a strong linear correlation between $PM_{2.5}$ concentration in Beijing and pollutants in surrounding cities, such as: Langfang, Chengde, Baoding, etc.

Moreover, obvious nonlinear relationships between several independent variables and dependent variables could be found during the exploratory data analysis, as depicted in Figure 7. It can be seen from the figure that the distribution of most numerical variables exhibited different degrees of skewness (the diagonal part of the figure), and there was a significant nonlinear relationship between the independent variable and dependent variable (the upper right and the lower left corner). Meanwhile, the linear relationship between these parameters indicated the risk of multi-collinearity (lower right).

For example, the O_3 concentrations, evaporation capacity, and extreme wind speed exhibited an apparent exponential relationship with $PM_{2.5}$, while the PM_{10} and $PM_{2.5}$ concentrations of Langfang presented potential logarithm relevance.

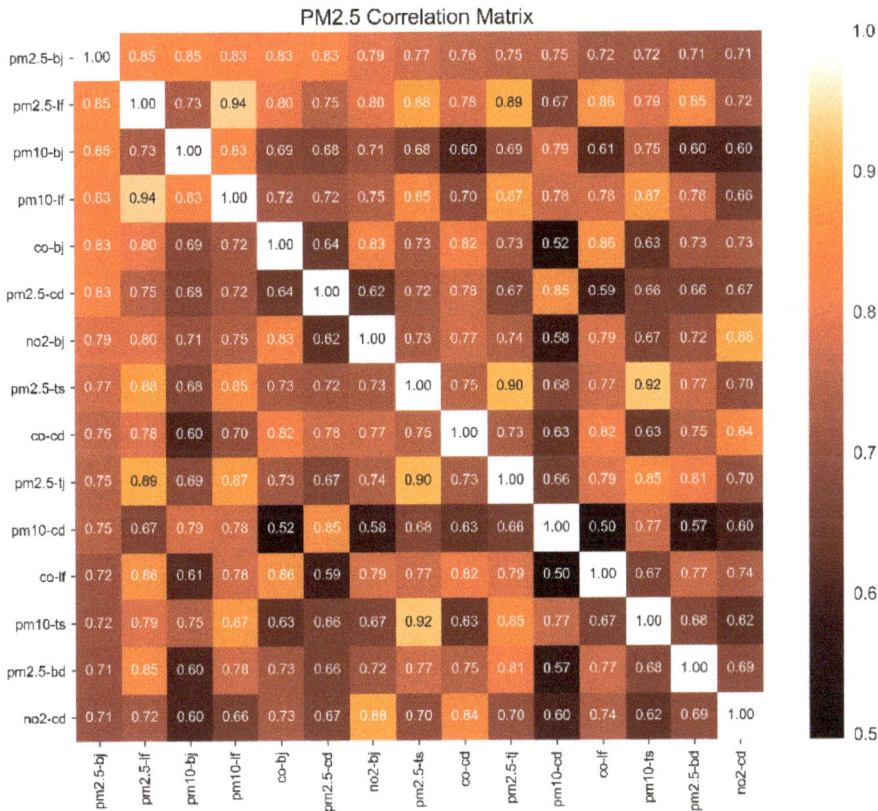

Figure 6. $PM_{2.5}$ concentration linear correlation matrix.

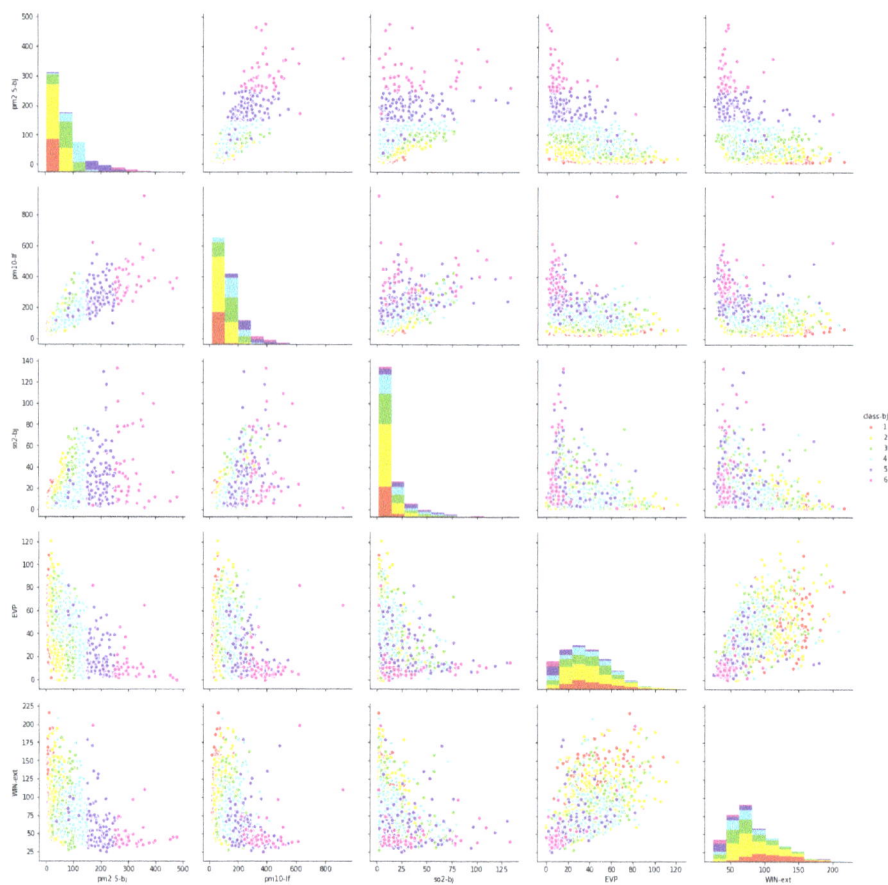

Figure 7. Typical nonlinear relationship between features and PM$_{2.5}$.

For categorical variables, there was a linear correlation between the variables and the target values. Most of the variables exhibited typical periodic variation and fluctuation characteristics, as shown in Figure 8. Figure 8a refers to the violin plot of PM$_{2.5}$ monthly concentration. It can be seen that the change in PM$_{2.5}$ concentration showed typical seasonal fluctuations and that the PM$_{2.5}$ concentration was smaller in the summer and autumn from June to September. The rest of the months fluctuated greatly and the lowest PM$_{2.5}$ concentration appeared around August. This is consistent with the conclusions of previous studies. Guo et al. found the lowest and highest monthly mean PM$_{2.5}$ concentrations appeared in August and January, respectively [35]. We encoded the wind direction from 1 to 16 clockwise and 1 represents a north wind direction. Figure 8b refers to the scatter plot of wind direction of extreme wind speed and PM$_{2.5}$ concentration. It shows that the PM$_{2.5}$ concentration exhibited periodic rhythm with the change of extreme wind speed direction and the highest concentration of pollutants occurred when the extreme wind speed direction was northeast and southwest (wind direction code is 3 and 11). When the wind direction was west and northwest (wind direction codes 13 and 15), air quality conditions were generally good. Figure 8c,d further reveal this phenomenon through wind rose for wind direction of maximal and extreme wind speed against PM$_{2.5}$ concentrations. The radius refers to the frequency of specific wind direction and the intensity refers to the value of PM$_{2.5}$ concentrations. The prevailing wind direction of Beijing's maximum wind speed and daily maximum wind speed is northeast-southwest, where the daily maximum wind speed is slightly east. When the wind direction is weak southwest wind, the probability of air pollution is greater, and when the wind direction is north, the air quality is generally better. This phenomenon may be related to the topographical features of the three sides mountains of Beijing and the distribution of southern industrial areas [36,37].

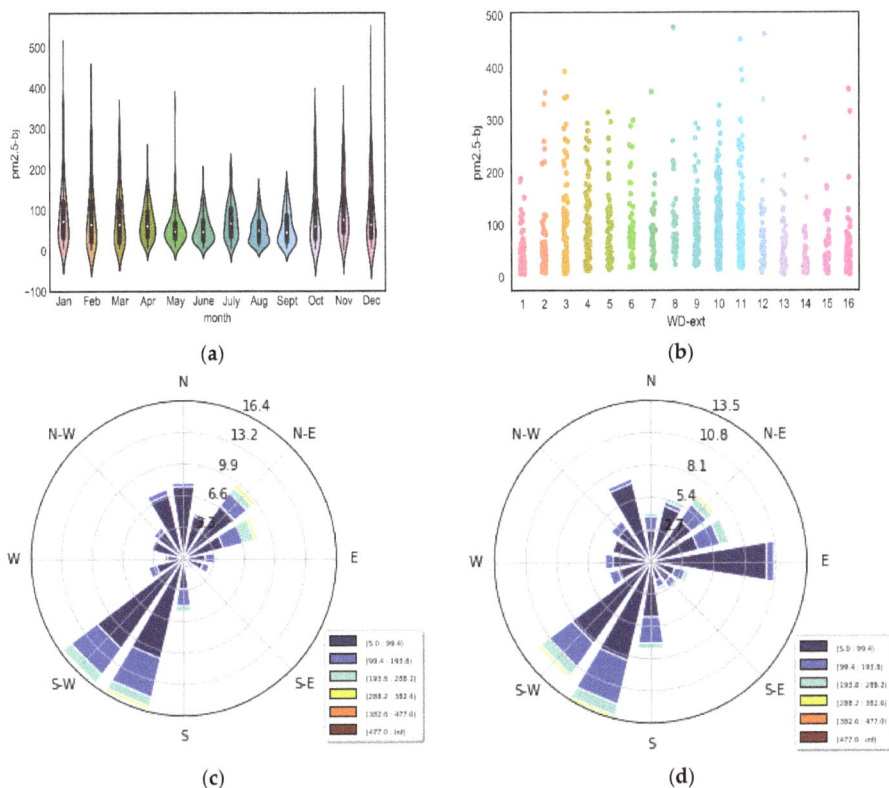

Figure 8. The relationship between categorical variables and PM$_{2.5}$. (a: violin plot of PM$_{2.5}$ monthly concentration; b: scatter plot of wind direction of extreme wind speed and PM$_{2.5}$ concentration; c: wind rose for wind direction of maximal wind speed against PM$_{2.5}$ concentrations; d: wind rose for wind direction of extreme wind speed against PM$_{2.5}$ concentrations).

Combined with the analysis of the correlation between surrounding pollutants, meteorological factors, and PM$_{2.5}$ in Beijing, it can further explain the reason why the air quality in southern Beijing is generally better than that in the north. The surrounding pollutants have a strong influence on Beijing's air quality and Beijing's prevailing winds are mostly southerly, so the areas in the south of Beijing have a greater impact and Langfang is closer than Baoding in geographical distance. Therefore, in the correlation analysis, the pollutants in Langfang have a greater impact on Beijing than Baoding.

5. Conclusions

Today, air pollution has become one of the most serious environmental problems in the world. Fine particulate matters (PM$_{2.5}$) are harmful to ambient air quality, economic development and human health. Considering Beijing and six surrounding cities as main research areas, this study took the daily average pollutant concentrations and meteorological elements from 2 December 2013 to 13 October 2017 into account and studied the spatial and temporal distribution characteristics, the primary influencing factors, and the forecasting method of PM$_{2.5}$ concentrations in Beijing in order to provide guidance for coping with extreme meteorological disasters and to provide references for improving municipal crisis response and emergency planning.

In this paper, the inter-annual, seasonal and diurnal variation trends, and temporal spatial distribution characteristics of $PM_{2.5}$ concentration in Beijing were studied by correlation analysis and geo-statistics. The main conclusions are as follows:

(1) The pollutant concentrations in Beijing exhibit obvious seasonal and cyclical fluctuation patterns. Air pollution is more serious in winter and spring and slightly better in summer and autumn, with the spatial distribution of pollutants fluctuating dramatically in different seasons. The pollution in southern Beijing areas are more grievous and the air quality in northern areas are better in general. The diurnal variation of air quality shows a typical seasonal difference and the daily variation of $PM_{2.5}$ concentrations by and large presented a "W" type of mode with twin peaks. Except for the emissions and accumulation of local pollutants, air quality is susceptible to the transport effect from southwest.

(2) A feature importance analysis reveals that PM_{10} and $PM_{2.5}$ concentrations measured from the city of Langfang should be taken as the most important elements of surrounding pollution factors to $PM_{2.5}$ in Beijing. These concentrations of PM_{10} and CO are the most significant local factors to $PM_{2.5}$ in Beijing. Extreme wind speeds and maximal wind speeds are considered to extend most effects of meteorological factors to the cross-regional transportation of contaminants. Pollutants found in the cities of Langfang have a stronger impact on air quality in Beijing than other surrounding factors. Each element affects the air quality of the study areas in a different way.

This study elaborated the spatial and temporal distribution characteristics of $PM_{2.5}$ concentrations in Beijing and the influencing modes of various factors on $PM_{2.5}$ concentrations in Beijing. It helps to thoroughly recognize and understand the formation mechanisms of serious haze events.

Author Contributions: B.Z. and J.C. conceived the key ideas and the system architecture; B.Z. conducted the research; W.Y. and Z.H. analyzed the data; B.Z. and J.C. wrote the paper; and W.Y. reviewed the process.

Funding: This research was funded by National Science Foundation of China grant number 71790613.

Acknowledgments: This research was supported through the National Science Foundation of China under Grant No. 71790613 and by the Beijing Key Laboratory of City Integrated Emergency Response Science.

Conflicts of Interest: The authors declare no conflict of interest.

References

1. Apte, J.S.; Marshall, J.D.; Cohen, A.J.; Brauer, M. Addressing global mortality from ambient $PM_{2.5}$. *Environ. Sci. Technol.* **2015**, *49*, 8057–8066. [CrossRef] [PubMed]
2. Kim, K.H.; Kabir, K.; Kabir, S. A review on the human health impact of airborne particulate matter. *Environ. Int.* **2015**, *74*, 136–143. [CrossRef] [PubMed]
3. Kim, K.H.; Jahan, S.A.; Kabir, E. A review on human health perspective of air pollution with respect to allergies and asthma. *Environ. Int.* **2013**, *59*, 41–52. [CrossRef] [PubMed]
4. Lelieveld, J.; Evans, J.S.; Fnais, M.; Giannadaki, D.; Pozzer, A. The contribution of outdoor air pollution sources to premature mortality on a global scale. *Nature* **2015**, *525*, 367–371. [CrossRef] [PubMed]
5. Maji, K.J.; Dikshit, A.K.; Arora, M.; Deshpande, A. Estimating premature mortality attributable to $PM_{2.5}$ exposure and benefit of air pollution control policies in China for 2020. *Sci. Total Environ.* **2017**, *612*, 683–693. [CrossRef] [PubMed]
6. West, J.J.; Cohen, A.; Dentener, F.; Brunekreef, B.; Zhu, T.; Armstrong, B.; Bell, M.L.; Brauer, M.; Carmichael, G.; Costa, D.L.; et al. What we breathe impacts our health: Improving understanding of the link between air pollution and health. *Environ. Sci. Technol.* **2016**, *50*, 4895–4904. [CrossRef] [PubMed]
7. Liu, X.; Zhang, Y.; Cheng, Y.; Hu, M.; Han, T. Aerosol hygroscopicity and its impact on atmospheric visibility and radiative forcing in Guangzhou during the 2006 PRIDE-PRD campaign. *Atmos. Environ.* **2012**, *60*, 59–67. [CrossRef]
8. Zhang, Q.; Quan, J.; Tie, X.; Li, X.; Liu, Q.; Gao, Y.; Zhao, D. Effects of meteorology and secondary particle formation on visibility during heavy haze events in Beijing, China. *Sci. Total Environ.* **2015**, *502*, 578–584. [CrossRef] [PubMed]

9. Mu, Q.; Zhang, S.Q. An evaluation of the economic loss due to the heavy haze during January 2013 in China. *China Environ. Sci.* **2013**, *33*, 2087–2094.

10. Wang, L.T.; Wei, Z.; Yang, J.; Zhang, Y.; Zhang, F.F.; Su, J.; Meng, C.C.; Zhang, Q. The 2013 severe haze over southern Hebei, China: Model evaluation, source apportionment, and policy implications. *Atmos. Chem. Phys.* **2014**, *14*, 3151–3173. [CrossRef]

11. Cai, S.; Wang, Y.; Zhao, B.; Wang, S.; Chang, X.; Hao, J. The impact of the "air pollution prevention and control action plan" on $PM_{2.5}$ concentrations in Jing-Jin-Ji region during 2012–2020. *Sci. Total Environ.* **2017**, *580*, 197–209. [CrossRef] [PubMed]

12. Forouzanfar, M.H.; Afshin, A.; Alexander, L.T.; Anderson, H.R.; Bhutta, Z.A.; Biryukov, S.; Brauer, M.; Burnett, R.; Cercy, K.; Charlson, F.J.; et al. Global, regional, and national comparative risk assessment of 79 behavioural, environmental and occupational, and metabolic risks or clusters of risks, 1990–2015: A systematic analysis for the Global Burden of Disease Study 2015. *Lancet* **2016**, *388*, 1659–1724. [CrossRef]

13. Xu, W.; Chen, H.; Li, D.; Zhao, F.; Yang, Y. A case study of aerosol characteristics during a haze episode over Beijing. *Procedia Environ. Sci.* **2013**, *18*, 404–411. [CrossRef]

14. Zhang, X.; Huang, Y.; Zhu, W.; Rao, R. Aerosol characteristics during summer haze episodes from different source regions over the coast city of North China Plain. *J. Quant. Spectrosc. Radiat. Transf.* **2013**, *122*, 180–193. [CrossRef]

15. Jansen, R.C.; Shi, Y.; Chen, J.; Hu, Y.; Xu, C.; Hong, S.; Li, J.; Zhang, M. Using hourly measurements to explore the role of secondary inorganic aerosol in $PM_{2.5}$ during haze and fog in Hangzhou, China. *Adv. Atmos. Sci.* **2014**, *31*, 1427–1434. [CrossRef]

16. Wu, J.; Zhang, P.; Yi, H.; Qin, Z. What Causes Haze Pollution? An Empirical Study of $PM_{2.5}$ Concentrations in Chinese Cities. *Sustainability* **2016**, *8*, 132. [CrossRef]

17. Zhang, Z.; Zhang, X.; Gong, D.; Quan, W.; Zhao, X.; Ma, Z.; Kim, S.J. Evolution of surface O_3 and $PM_{2.5}$ concentrations and their relationships with meteorological conditions over the last decade in Beijing. *Atmos. Environ.* **2015**, *108*, 67–75. [CrossRef]

18. Guo, S.; Hu, M.; Zamora, M.L.; Peng, J.; Shang, D.; Zheng, J.; Du, Z.; Wu, Z.; Shao, M.; Zeng, L.; et al. Elucidating severe urban haze formation in China. *Proc. Natl. Acad. Sci. USA* **2014**, *111*, 17373–17378. [CrossRef] [PubMed]

19. Li, X.; Chen, X.; Yuan, X.; Zeng, G.; León, T.; Liang, J.; Chen, G.; Yuan, X. Characteristics of Particulate Pollution ($PM_{2.5}$ and PM_{10}) and Their Spacescale-Dependent Relationships with Meteorological Elements in China. *Sustainability* **2017**, *9*, 2330. [CrossRef]

20. Lv, B.; Cai, J.; Xu, B.; Bai, Y. Understanding the rising phase of the $PM_{2.5}$ concentration evolution in Large China cities. *Sci. Rep.* **2017**, *7*, 46456. [CrossRef] [PubMed]

21. Zheng, G.J.; Duan, F.K.; Su, H.; Ma, Y.L.; Cheng, Y.; Zheng, B.; Zhang, Q.; Huang, T.; Kimoto, T.; Chang, D.; et al. Exploring the severe winter haze in Beijing: The impact of synoptic weather, regional transport and heterogeneous reactions. *Atmos. Chem. Phys.* **2015**, *15*, 2969–2983. [CrossRef]

22. Chen, D.; Liu, X.; Lang, J.; Zhou, Y.; Wei, L.; Wang, X.; Guo, X. Estimating the contribution of regional transport to $PM_{2.5}$ air pollution in a rural area on the North China Plain. *Sci. Total Environ.* **2017**, *583*, 280–291. [CrossRef] [PubMed]

23. Gao, J.; Wang, K.; Wang, Y.; Liu, S.; Zhu, C.; Hao, J.; Liu, H.; Hua, S.; Tian, H. Temporal-spatial characteristics and source apportionment of $PM_{2.5}$ as well as its associated chemical species in the Beijing-Tianjin-Hebei region of China. *Environ. Pollut.* **2018**, *233*, 714–724. [CrossRef] [PubMed]

24. Li, P.; Yan, R.; Yu, S.; Wang, S.; Liu, W.; Bao, H. Reinstate regional transport of $PM_{2.5}$ as a major cause of severe haze in Beijing. *Proc. Natl. Acad. Sci. USA* **2015**, *112*, E2739–E2740. [CrossRef] [PubMed]

25. Wang, H.; Xu, J.; Zhang, M.; Yang, Y.; Shen, X.; Wang, Y.; Chen, D.; Guo, J. A study of the meteorological causes of a prolonged and severe haze episode in January 2013 over central-eastern China. *Atmos. Environ.* **2014**, *98*, 146–157. [CrossRef]

26. Wang, Y.; Zhang, Y.; Schauer, J.J.; de Foy, B.; Guo, B.; Zhang, Y. Relative impact of emissions controls and meteorology on air pollution mitigation associated with the Asia-Pacific Economic Cooperation (APEC) conference in Beijing, China. *Sci. Total Environ.* **2016**, *571*, 1467–1476. [CrossRef] [PubMed]

27. Wang, F.; Chen, D.S.; Cheng, S.Y.; Li, J.B.; Li, M.J.; Ren, Z.H. Identification of regional atmospheric PM_{10} transport pathways using HYSPLIT, MM5-CMAQ and synoptic pressure pattern analysis. *Environ. Model. Softw.* **2010**, *25*, 927–934. [CrossRef]

28. Ma, Q.; Wu, Y.; Zhang, D.; Wang, X.; Xia, Y.; Liu, X.; Tian, P.; Han, Z.; Xia, X.; Wang, Y.; et al. Roles of regional transport and heterogeneous reactions in the $PM_{2.5}$ increase during winter haze episodes in Beijing. *Sci. Total Environ.* **2017**, *599*, 246–253. [CrossRef] [PubMed]

29. Hoek, G.; Beelen, R.; De Hoogh, K.; Vienneau, D.; Gulliver, J.; Fischer, P.; Briggs, D. A review of land-use regression models to assess spatial variation of outdoor air pollution. *Atmos. Environ.* **2008**, *42*, 7561–7578. [CrossRef]

30. Oliver, M.A.; Webster, R. Kriging: A method of interpolation for geographical information systems. *Int. J. Geogr. Inf. Syst.* **1990**, *4*, 313–332. [CrossRef]

31. Hu, J.; Wang, Y.; Ying, Q.; Zhang, H. Spatial and temporal variability of $PM_{2.5}$ and PM_{10} over the North China Plain and the Yangtze River Delta, China. *Atmos. Environ.* **2014**, *95*, 598–609. [CrossRef]

32. Liu, Z.; Hu, B.; Wang, L.; Wu, F.; Gao, W.; Wang, Y. Seasonal and diurnal variation in particulate matter (PM_{10} and $PM_{2.5}$) at an urban site of Beijing: Analyses from a 9-year study. *Environ. Sci. Pollut. Res.* **2015**, *22*, 627–642. [CrossRef] [PubMed]

33. Ricciardelli, I.; Bacco, D.; Rinaldi, M. A three-year investigation of daily $PM_{2.5}$ main chemical components in four sites: The routine measurement program of the Supersito Project (Po Valley, Italy). *Atmos. Environ.* **2017**, *152*, 418–430. [CrossRef]

34. Li, R.; Li, Z.; Gao, W.; Ding, W.; Xu, Q.; Song, X. Diurnal, seasonal, and spatial variation of $PM_{2.5}$ in Beijing. *Sci. Bull.* **2015**, *60*, 387–395. [CrossRef]

35. Guo, H.; Cheng, T.; Gu, X.; Wang, Y.; Chen, H.; Bao, F.; Shi, S.; Xu, B.; Wang, W.; Zuo, X.; et al. Assessment of $PM_{2.5}$ concentrations and exposure throughout China using ground observations. *Sci. Total Environ.* **2017**, *601*, 1024–1030. [CrossRef] [PubMed]

36. Shen, R.; Schaefer, K.; Schnelle-Kreis, J.; Shao, L.; Norra, S.; Kramar, U.; Michalke, B.; Abbaszade, G.; Streibel, T.; Fricker, M.; et al. Characteristics and sources of PM in seasonal perspective—A case study from one year continuously sampling in Beijing. *Atmos. Pollut. Res.* **2016**, *7*, 235–248. [CrossRef]

37. Yang, F.; Tan, J.; Zhao, Q.; Du, Z.; He, K.; Ma, Y.; Duan, F.; Chen, G. Characteristics of $PM_{2.5}$ speciation in representative megacities and across China. *Atmos. Chem. Phys.* **2011**, *11*, 5207–5219. [CrossRef]

sustainability

MDPI

Article

Performance-Based or Politic-Related Decomposition of Environmental Targets: A Multilevel Analysis in China

Pan Zhang [1,2] and Jiannan Wu [1,3,*]

1   School of International and Public Affairs, Shanghai Jiao Tong University, Shanghai 200030, China;
     zhang_pan@sjtu.edu.cn
2   Center for Chinese Local Governance Innovations, Xi'an Jiaotong University, Xi'an 710049, Shaanxi, China
3   China Institute for Urban Governance, Shanghai Jiao Tong University, Shanghai 200030, China
*   Correspondence: jnwu@sjtu.edu.cn; Tel.: +86-136-0198-5852

Received: 18 August 2018; Accepted: 21 September 2018; Published: 25 September 2018

Abstract: China relies on the total pollutant emission control and environmental target responsibility system to curb environmental pollution and improve energy conversation. How the central government breaks down environmental targets among provincial governments lies at the core, but little research has been done to explore the determinants of environmental target-setting empirically. This work models the decomposition process of environmental targets by focusing on the roles of historical performance and provinces' political status. With the method of hierarchical linear model, data on five kinds of environmental obligatory targets (energy consumption per unit GDP and other four kinds of pollutants) during China's "12th Five-year Plan" period is used to test the hypotheses. The results show that provincial historical structural performance is negatively significantly correlated with their environmental target levels, while the effects of historical scale performance and intensity performance are not significant. Besides, provinces with higher political rankings tend to be allocated higher targets, which is in accordance with the model effect hypothesis rather than the bargaining effect hypothesis.

Keywords: environmental target-setting; performance; hierarchical linear model; environmental governance; China

---

1. Introduction

Echoing management by objectives, many countries in the world have already adopted various kinds of result-oriented goal management reforms [1–4]. In these result-oriented reforms, goal-setting lies in the core and plays the "baton" role in directing the behaviors of persons for whom the goals are set. As an important management instrument, the positive effects of goal-setting on performance improvement have been empirically confirmed by dozens of studies in the past few decades [5–7]. Goals can usually be elaborated through goal dimensions and goal aspiration levels: goal dimensions are used to illustrate what is considered important (concentrating on the goal priorities or goal preferences), while goal aspiration levels are used to explain which performance levels are expected to be achieved on certain goal dimensions [8–10]. Nowadays, to examine determinants of goal aspiration levels and goal priorities has become an important channel to explore the behavioral logic of decision-makers.

Actually, the top-down target responsibility system in China provides an ideal practical situation for conducting research on determinants of government performance goal-setting [10]. In the Chinese target responsibility system, performance goals are divided into two forms, namely, anticipated targets and obligatory targets. Anticipated targets refer to the development goals that the central government and local governments expect to achieve and set by themselves (e.g., GDP), while obligatory targets refer to the task requirements for local governments that are allocated or set by their superior governments in public service and other policy domains involving public interest (e.g., pollutant emission control). Previous research on goal-setting concentrated only on anticipated targets [10–12] and failed to explore determinants of the setting of obligatory targets. There is a great difference between the setting of anticipated targets and obligatory targets in China, because anticipated targets are set by local governments independently and obligatory targets are allocated by their superior governments, although local governments can also express their own opinions or adjust the targets in the decomposition of obligatory targets to some extent [13].

Due to the goal-setting theory being developed mainly based on anticipated targets, it is of great importance to explore determinants and features of the setting of obligatory targets in order to understand the target-setting logic in China more thoroughly. Along with more and more severe environmental pollution [14], China relies on the total pollutant emission control and environmental target responsibility system to curb environmental pollution, the core of which is the disaggregation of obligatory environmental targets [13,15,16]. Thus, this work models the decomposition process of environmental targets and focuses on the effects of performance feedback and political features by controlling a set of factors advocated in previous goal-setting research. This work contributes to current literature by first dividing historical performance into three dimensions and then extending previous frameworks by including the politic-related factors based on the feature of the setting of obligatory targets. It also additionally extends previous literature that mainly focused on the context of setting anticipated targets to the context of the setting of obligatory targets. By taking the setting of environmental targets as a case, this work uses the data of Chinese provinces in the "12th Five-year Plan" period to examine this framework empirically.

The remainder of this work is organized as follows. The second section briefly reviews current research related to organizational goal-setting and the third section proposes the theoretical framework and two clusters of hypotheses. The fourth section illustrates the methodology, including samples, measures, and data sources, and the analytical approach used, followed by the fifth section which presents the empirical results in detail. Finally, the theoretical and practical implications of these findings are discussed, followed by limitations and research avenues for future research.

2. Literature Review

Research on organizational target-setting or goal-setting can be tracked back to the behavioral theory of the firm developed by Cyert and March, which holds that the prior performance, the previous goal, and comparable peers' previous performance of one organization are important predictors of its goal-setting [8]. Following this research stream, some studies have examined how one organization's previous goal and previous goal attainment discrepancy (the gap between previous performance and the previous goal) influence its decision-making in the private sector [17,18]. Recently, the behavioral theory of the firm has become one of the most influential theories in organizational studies and is advocated by many followers [9,19]. However, current evidence mainly focuses on the goal-setting process in the private sector and whether it can be generalized to the decision-making process in the public sector, which is usually faced with multiple tasks and characterized by a hierarchical structure, is still an outstanding question [10].

Similarly, public administration scholars have also developed the Bayesian theory of public organizations' decision-making to explain the decision-making process in the public sector [20]. It argues that the historical performance in the last period, performance gaps, peers' performance, superiors' preferences can figure as benchmarks that can be referred to in organizations' decision-making in the public sector [20]. Recently, some public administration scholars have examined impacts of these factors on the setting of goal priorities empirically based on survey data in the public education system [11,12]. Moreover, Ma also empirically explored determinants of the setting of Gross Domestic Product (GDP) goal aspiration levels by Chinese provincial governments and found that it followed prior historical GDP aspiration levels and was related to the average performance aspiration levels of peers and performance gaps when compared with peers horizontally [10]. However, all of these studies focus on the setting of anticipated targets rather than obligatory targets in the public sector.

After reviewing research on setting of environmental obligatory targets in detail, we find that English articles about environmental target-setting published in international journals are really scarce. In this aspect, Zhao and Wu interpreted the evolution of China's energy-saving target allocation system from the policy learning perspective and described the target allocation process [13]. Some studies also either discussed principles which should be followed in decomposing environmental targets (e.g., effect, efficiency, equity, transparency, feasibility, continuity, consistency, responsibility, and capability) [21,22] or used some economic factors to model the allocation of environmental targets among local governments, including resource distribution, energy consumption, fiscal revenue, GDP per capita, and industry development features [23,24]. Though these studies have provided some clues to understand the behavioral logic of environmental target allocation, no research was found that used archive data to examine determinants of environmental target-setting in China empirically.

Overall, due to the importance of target-setting, it becomes an emerging hot topic to explore determinants of organizational target-setting in recent years. Though current studies have helped us understand organizational decision-making logic and behaviors, there are still several significant research gaps. First, many studies have explored what determines the setting of goal priorities and goal aspiration levels of individuals [6] and organizations in the private sector [8,17–19], while research on determinants of organizational goal priorities and goal aspiration levels in the public sector is still much less than what we need [10,11,20]. Second, current studies mainly concentrate on the setting of anticipated targets rather than obligatory targets [10], which makes it difficult to generalize these findings in different situations. Third, current research mainly focuses the roles of historical performance in target-setting from the perspective of rational decision-making, while target-setting in the public sector is also a political process and no research examines the political features of target-setting.

3. Theoretical Framework and Hypotheses

3.1. Theoretical Framework: Performance-Based or Politic-Related

Both the behavioral theory of the firm and the Bayesian theory of public organizations' decision-making put organizations' historical performance at the center of organizational target-setting [8,20]. They argue that organizations tend to set their own targets using their own prior performance as a key reference point from the perspective of rational decision-making. Actually, target-setting is an important policy instrument to improve performance [7,25,26]. Organizational prior performance records can influence organizations' expectations about their performance in the next period. As far as China's environmental pollution control concerned, previous pollutant emission and energy consumption performance is also related to local governments' potentials to reduce their pollutant emissions and energy consumption [16]. Thus, it is reasonable that the central government in China should decompose national targets down to provincial governments according to provinces' historical performance, which implies environmental target-setting in China is expected to be performance-based.

In addition, the decomposition and allocation of environmental targets in China is a top-down process in which the central government first releases the national environmental targets, provinces propose their own possible targets, the central government further compiles and reviews the proposed targets, and then, the central government negotiates with provinces and finalizes target assignments [13,15]. Actually, some studies argue that local governments strive to stand out in policy areas prioritized by their superior governments [27,28], while environmental target-setting research also holds that the decomposition of China's environmental targets is a political process full of intensive central–local bargaining [13]. This means provincial governments may be stimulated by political factors to submit targets higher or lower than the national environmental targets, and the primary difference between the initial proposed targets and the finalized targets of the provinces in the "11th Five-year Plan" period provides partial evidence about the political features of this process [13]. Hence, in order to simulate the decomposition process of environmental targets in China, this work integrates the performance-based and politic-related decomposition perspectives together to establish a theoretical framework (Figure 1). In the following section, it proposes detailed hypotheses from these two perspectives.

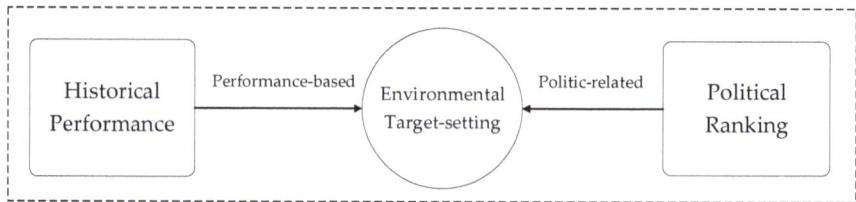

Figure 1. Theoretical Framework.

3.2. Hypotheses Denoting Performance-Based Decomposition

Historical performance in the last period can figure as the simplest benchmark that can be referred to in decision-making [8,20]. On one hand, historical performance is related to goal attainment discrepancies or performance gaps and, thus, can influence goal aspiration level adaptation [18,20]. On the other hand, managers tend to lower the weight of this performance dimension when the performance on this dimension has improved in this period; otherwise, the manager tend to increase the weight of the performance dimension [12]. When one organization had a better performance record in the last period, it would be more confident in setting higher-level performance goals; on the contrary, when the performance record in the last period was worse, the organization would tend to lower its goal aspiration level in order to attain its performance goal in this period [10].

The effect of historical performance is expected to be different in the setting of anticipatory targets and in the decomposition of environmental obligatory targets. As far as the setting of anticipatory targets, historical performance tends to be positively related to the goal aspiration level in the current period, because a better historical performance record can act as an anchor and indicate that better performance is possible, thereby strengthening organizational confidence in setting a higher performance goal in the next period [10,18]. However, the setting of environmental obligatory targets is a top-to-bottom decomposition process in China, and historical performance can negatively influence the setting of environmental obligatory targets because it reflects local governments' potentials to reduce pollutant emissions in the current period [16]. Specifically, when the historical performance of one province is low (that means it emitted more pollutants), it is expected to have a larger potential to improve performance on this dimension in the next period.

Previous research focuses on the effect of one-dimension performance on governmental goal-setting, such as the GDP growth rate [10]. In the research field of carbon emissions, some scholars typically decompose the carbon emissions into production, structural, and efficiency effects [29,30], which implies that we can improve environmental performance through multiple channels, such as reducing the production scale, adjusting the structure, and improving production efficiency. Similarly, environmental performance can also be measured from these three perspectives with scale performance, structural performance, and intensity performance, which reflects the total gross of pollutant emissions or energy consumption, the structure of pollutant emissions or energy consumption, and the pollutant emission efficiency or energy consumption efficiency, respectively. Thus, the environmental targets are hypothesized to be decomposed mainly based on historical environmental performance records and thus we propose:

Hypothesis H1. *With other conditions being equal, the allocated target of one province is negatively related to its historical energy consumption and pollutant emission performance.*

Hypothesis H1a. *With other conditions being equal, the allocated target of one province is negatively related to its historical energy consumption and pollutant emission scale performance.*

Hypothesis H1b. *With other conditions being equal, the allocated target of one province is negatively related to its historical energy consumption and pollutant emission structural performance.*

Hypothesis H1c. *With other conditions being equal, the allocated target of one province is negatively related to its historical energy consumption and pollutant emission intensity performance.*

3.3. Hypotheses Denoting Politic-Related Decomposition

Local governments enjoy considerable autonomy in setting anticipatory targets in China (such as GDP) [10], while pollutant emission reduction targets, as a kind of obligatory target, are usually decomposed and allocated by the central government in China [31]. Rational local government leaders would not always completely obey performance management activities initiated by their superior governments and would instead usually take some response strategies [32]. In the policy domain of pollutant emission reduction, in order to minimize the pressure of attaining targets set by their superior governments, it is a rational choice for local governments to strive for lower pollutant emission reduction targets by bargaining with their superior governments [13]. On the one hand, the nomenklatura personnel management system of China divides government officials into multiple hierarchical levels, and the central Politburo is located at the top layer [27,33,34]. The higher historical likelihood that leaders (party secretary and governor) of one province are promoted to the Politburo means the province has a higher political ranking in Chinese political system [28,31], which not only makes it easier for the province to get information about the central policy direction in advance, but also gives the province more political capital to bargain with the central government in the setting of environmental targets. In this case, provinces with higher political rankings are expected to be allocated lower environmental targets through bargaining.

Because environmental protection has been one of the top policy priorities of the central government in China since 2007 [35], provinces with higher political rankings are also likely to be driven by higher political promotion incentives to accept higher environmental targets. Some scholars argue that provinces whose leaders have a higher historical likelihood to be promoted to the Politburo are keener on following the policy priority of the central government to control pollutant emissions [28]. Meanwhile, some research also provides empirical evidence that the historical likelihood of one province's leaders to be promoted to the Politburo is not significantly but positively related to the attainment of environmental targets [31], which implies that provinces with higher political rankings indeed are perhaps more aggressive in achieving environmental targets set by the central government. Therefore, a model effect may exist for provinces with higher political rankings in setting environmental targets, which means provinces with higher political rankings, driven by political promotion incentives, are more likely to accept higher environmental targets allocated by the central government. Overall, two competing hypotheses are proposed:

Hypothesis H2a. *With all other conditions being equal, provinces with higher political rankings have more power to bargain with the central government and tend to be allocated lower environmental obligatory targets, defined as the bargaining effect hypothesis.*

Hypothesis H2b. *With all other conditions being equal, provinces with higher political rankings tend to play a leading role and be allocated higher environmental targets, defined as the model effect hypothesis.*

4. Materials and Methods

4.1. Samples

This work uses data from 29 provinces in China (Hong Kong, Taiwan, Tibet, Xinjiang, and Macao are excluded) during the "12th Five-year Plan (2011–2015)" period to test our hypothesis. The samples are province–target combination units. During the "12th Five-year Plan" period, the central government of China decomposed China's national environmental targets down to its provincial governments, providing an appropriate institutional context to explore determinants of the setting of environmental obligatory targets. The reasons why the allocation of environmental targets from the central government to Chinese provinces is chosen lie in the following two aspects: on the one hand, it is the first and most important step to break national targets down to Chinese provinces in the Chinese top-down target assignment decomposition system, which is completely independent from target decomposition and allocation in the following stages; on the other hand, previous empirical research on determinants of goal aspiration levels was conducted at Chinese provincial level [10], and it is better to compare those findings with the findings of this work.

4.2. Measures and Data Sources

4.2.1. Dependent Variable

The dependent variable is the expected reduction rate for the investigated environmental obligatory indicator of the investigated province during the "12th Five-year Plan" period. For example, the dependent variable for the observed sample of "Henan province-sulfur dioxide" is measured with the mandatory reduction percentage of the total amount of sulfur dioxide emissions in Henan province during the "12th Five-year Plan" period; for another example, the dependent variable of the observed sample for "Shandong province-energy consumption per unit GDP" is the mandatory reduction percentage of the energy consumption per unit GDP in Shandong province during the "12th Five-year Plan" period. The data are from the Comprehensive Work Scheme of Energy Conservation and Emission Reduction during the "12th Five-year Plan" Period. The larger the value of the variable is, the stronger the pressure is to improve environmental performance for the focus province.

4.2.2. Independent Variables

As discussed in the third section, historical performance is used to examine the performance-based decomposition feature in China's environmental target allocation system, including historical scale performance, historical structural performance, and historical intensity performance. Historical scale performance represents the relative gross of energy consumption or the relative gross of pollutant emission for the corresponding investigated obligatory indicator of the investigated province in the year before the "12th Five-year Plan" period. For example, the variable of the observed sample of "Henan province-sulfur dioxide" is measured with the ratio between the total sulfur dioxide emission of Henan province and the national emission gross of sulfur dioxide in 2010; for another example, the variable of the observed sample of "Shandong province energy consumption per unit GDP" is measured with the ratio between the total energy consumption of Henan province and the national energy consumption gross in 2010. This measurement can make different pollutant emissions and energy consumption comparable. The data are from the Chinese Statistical Yearbook 2011, Chinese Energy Statistical Yearbook 2011, and Chinese Environment Yearbook 2011. The larger the value of this variable is, the lower the scale performance is.

Historical structural performance is measured with the secondary industry proportion of the investigated province in the year before the "12th Five-year Plan" period, as industrial consumption of energy or industrial pollutant emissions have stronger control elasticity compared to daily life energy consumption or daily life pollutant emissions. The data are from the Chinese Statistical Yearbook 2011. The larger the value of this variable is, the lower the structural performance is.

Historical intensity performance is measured with the ratio between the investigated province's energy consumption per unit GDP and the national energy consumption per unit GDP or the ratio between the investigated province's pollutant emission per unit GDP and the national total pollutant emission per unit GDP for the investigated environmental indicator in the year before the "12th Five-year Plan" period, which makes it comparable among these five kinds of environmental indicators. The data are from the Chinese Statistical Yearbook 2011, Chinese Energy Statistical Yearbook 2011, and Chinese Environment Yearbook 2011. The larger the value of this variable is, the lower the energy utilization efficiency or pollutant control efficiency of the investigated province is.

Besides, the third section also elaborates that provinces' political status measured with political rankings can be used to examine the politic-related decomposition feature in China's environmental target allocation system. Political ranking is measured as the total number of the members of the Politburo who had worked in the investigated province as the party secretary or governor from 1997 to 2011. Members of the Politburo lie at the top in Chinese nomenklatura system; this measurement has been adopted in some other studies [27,28]. These data are coded by the author based on the resumes of the members of the Politburo in the 15th, 16th, and 17th sessions. The larger the value of this variable is, the higher the political ranking is.

4.2.3. Control Variables

A set of variables are controlled in this work. First, previous research holds that organizational goal-setting historically follows its past goal aspiration level, horizontally aligns with the average goal aspiration level of its comparable peers, and vertically complies with the goal aspiration level of its superior [10,20]. Hence, the prior environmental target of the investigated province, the average environmental target of its comparable peers, and the national environmental target are controlled.

Specifically, the variable of the prior target is measured with the target level of the investigated province for the investigated environmental obligatory indicator in the "11th Five-year Plan" period. The data are from the National Control Plan for the Total Emissions of Major Pollutants during the "11th Five-year Plan" Period and the Decomposition Plan for the National Energy Consumption per Unit GDP Reduction Target among Chinese Subnational Regions during the "11th Five-year Plan" Period, which are approved and issued by the State Council in 2006.

The variable of the comparable peers' average environmental target is measured by the average reduction rate for the investigated environmental obligatory indicator of all the comparable provinces of the investigated province during the "12th Five-year Plan" period. Neighboring provinces are treated as proxies of comparable provinces, which are usually used in research on policy diffusion and goal-setting [10,27,36]. Thus, this variable is calculated by the authors according to the Chinese administrative territory and the data are from the Comprehensive Work Scheme of Energy Conservation and Emission Reduction during the "12th Five-year Plan" period.

The variable of the national environmental target is measured with the national reduction rate set by the central government for the investigated environmental obligatory indicator during the "12th Five-year Plan" period. The data are from the Comprehensive Work Scheme of Energy Consumption and Emission Reduction in the "12th Five-year Plan" period. The higher the value of this variable is, the stronger the ambition is for the central government to control pollutant emissions.

Second, unemployment can reflect economic health, and provinces with more unemployed labors should bear less pollutant emission reduction burden in order to guarantee employment. Hence, we also control the unemployment rate of each province, which is measured with the registered urban unemployment rate of the investigated province in the year before the "12th Five-year Plan" period. The data is from the Chinese Statistical Yearbook 2011. Besides, previous studies confirmed that economic development conditions were related to the setting of $PM_{2.5}$ concentration control targets and GDP goal aspiration levels [7,10]. Thus, economic development is also controlled. It is measured by the GDP per capita of the investigated province in the year before the "12th Five-year

Plan" period, and the data are from the Chinese Statistical Yearbook 2011. Details of the variables, measures, and data sources are presented in Table 1.

Table 1. Variables, measures, and sources.

Variables	Measures	Sources
Target	The target level of the investigated province for the investigated environmental obligatory indicator in the "12th Five-year Plan" Period (%)	FYP
E_scale	The ratio between the investigated province's total energy consumption and the national total energy consumption or the ratio of the investigated province's total pollutant emission and the national total pollutant emission (one-year lagged)	CSY CESY CEY
E_structure	The proportion of the secondary industrial output of the investigated province (one-year lagged, %)	CSY
E_intensity	The ratio between the investigated province's energy consumption per unit GDP and the national energy consumption per unit GDP or the ratio between the investigated province's pollutant emission per unit GDP and the national total pollutant emission per unit GDP (one-year lagged)	CSY CESY CEY
P_ranking	Number of secretaries/governors of the investigated province who had served as members of the Politburo from 1997 to 2011	Author
P_target	The target level of the investigated province for the investigated environmental obligatory indicator in the "11th Five-year Plan" period (%)	FYP
Neighbor	The average target level of the investigated province's neighboring provinces for the investigated environmental indicator (%)	Author
Central	The target level of the central government for the investigated environmental indicator (%)	FYP
Deprivation	Unemployment rate of the investigated province (one-year lagged, %)	CSY
GDP	GDP per capita of the investigated province (one-year lagged, log)	CSY

Notes: FYP indicates Five-year Plans; CSY indicates the China Statistical Yearbook; CNSY indicates the China Energy Statistical Yearbook; CEY indicates the China Environment Yearbook.

4.3. Analytical Approach

Given that the data used are cross-sectional data of five kinds of environmental targets (sulfur dioxide, chemical oxygen demand, energy consumption per unit GDP, nitrogen oxide, ammonia nitrogen) of 29 provinces in mainland China (excluding Hong Kong, Taiwan, Tibet, Xinjiang, and Macao) during the "12th Five-year Plan" period, and some variables (e.g., structural performance, political ranking, unemployment rate, and GDP) are shared by samples within the same province, the data present a nested structure. The assumptions for the ordinary least square regression are violated in data with a nested structure and hence, we follow some scholars' recommendations to adopt a two-layered hierarchical linear model (HLM) to analyze the data [37]. There are two kinds of HLM, namely, the random intercept and fixed slope model and the random intercept and random slope model [38]. This work uses the random intercept and fixed slope model for estimation because we focus on the universal effects of the independent variables across layer-2 units. Specifically, the two-layered HLM can be presented as follows.

$$\text{Layer 1: } \text{Target}_{i,p} = \beta_{0i} + \beta_{1i}\text{E\_scale}_{i,p} + \beta_{2i}\text{E\_intensity}_{i,p} + \beta_{3i}\text{Control\_1}_{i,p} + \varepsilon_{i,p} \qquad (1)$$

$$\text{Layer 2: } \beta_{0i} = \gamma_{00} + \gamma_{01}\text{E\_structure}_i + \gamma_{02}\text{P\_ranking}_i + \gamma_{03}\text{Control\_2}_i + \mu_i \qquad (2)$$

In our data, environmental indicator units (layer 1) are nested in province units (layer 2). As described in Equation (1), the dependent variable of Target$_{i,p}$ denotes the mandatory reduction rates for major environmental obligatory indicators of 29 provinces in mainland China for the "12th Five-year

Plan" period, where i denotes the codes of provinces and p denotes the codes of environmental obligatory indicators. According to Equation (1), the target level in province i for environmental indicator j is the sum of the following three parts: average outcome in the province unit i (β_{0i}), outcome predicted by indicator-specific factors (including E_scale$_{i,p}$ denoting scale performance in province i for environmental indicator j, E_intensity$_{i,p}$ denoting intensity performance in province i for environmental indicator j, Control_1$_{i,p}$ denoting other indicator-level control variables, such as the prior target in province i for environmental indicator j, the average target for environmental indicator j of the neighboring peers of province i, and the national target for environmental indicator j), and indicator-level errors ($\varepsilon_{i,p}$).

Besides, according to the intercept Equation (2), the average outcome in province unit i (β_{0i}) is also composed of three parts: the average outcome for all samples (γ_{00}), outcome predicted by province-level factors (including E_structure$_i$ denoting structural performance measured with the proportion of the secondary industrial output in province i, P_ranking$_i$ denoting the political ranking of province i, Control_2$_i$ denoting other province-level control variables, such as the unemployment rate in province i, and the GDP per capita in province i), and province-level random effect (μ_i). Because u_i has different values across provinces, the method can make the intercepts vary among different provinces. Following previous research, we center layer-1 variables around their mean within each layer-2 group and center layer-2 variables around their grand mean of all samples [37]. The software used to run HLM is Stata 14.

5. Results

Descriptive analysis results of the variables are shown in Table 2, which mainly reports the sample size, mean, standard deviation, minimum value, and maximum value. The sample size is 145, comprising 145 combinations of five kinds of environmental indicators (energy consumption per unit GDP, sulfur dioxide, nitrogen oxide, chemical oxygen demand, ammonia nitrogen) and 29 provinces in mainland China. The minimum value of the dependent variable is −34.9, the maximum value is 18, the mean value is 9.75, and the standard deviation is 7.17, indicating great variance of the dependent variable, which provides great space to explore determinants of environmental target-setting. It is noteworthy that observations for some independent variables are 29 (such as E_structure, P_ranking, Unemployment, and GDP) because of the nested data structure. Only three kinds of environmental obligatory targets (energy consumption per unit GDP, sulfur dioxide, and chemical oxygen demand) were set at the national level and decomposed down to the provincial level during the "11th Five-year Plan" period; hence, the number of observations for the variable of the P_target is 87 (87 combinations of 29 provinces and three environmental indicators).

Table 2. Descriptive analysis results.

Variables	N	Mean	S.D.	Min	Max
Target	145	9.75	7.17	−34.90	18
E_scale	145	3.36	2	0.13	8.94
E_structure	29	49.18	7.72	24	57.3
E_intensity	145	1.25	0.76	0.16	4.99
P_ranking	29	2.41	2.69	0	11
P_target	87	13.29	6.76	0	30
Neighbor	145	10.61	3.74	1.48	18
Central	145	10.40	2.95	8	16
Unemployment	29	3.61	0.61	1.37	4.35
GDP	29	0.49	0.2	0.12	0.87

The results of bivariable correlation analysis are shown in Table 3. E_scale, E_structure, and P_ranking are significantly positively related to the dependent variable of Target, which preliminarily supports the corresponding hypotheses. However, the correlation coefficient between

E_intensity and Target is negatively significant, which does not provide evidence to support the corresponding hypothesis. Furthermore, the bivariable correlation coefficients between P_target, Neighbor, and Central are approximately 0.7 and are very significant ($p < 0.01$). Therefore, if we put these three variables into the same regression model, severe multicollinearity problems may exist due to the high correlations among them. The correlation coefficients between the independent variables and other control variables are all smaller than 0.65, indicating low correlations.

Table 3. Correlation analysis results.

Variables	1	2	3	4	5	6	7	8	9
1. Target	1								
2. E_scale	0.38 ***	1							
3. E_structure	0.26 ***	0.41 ***	1						
4. E_intensity	−0.17 **	0.035	0.14 *	1					
5. P_ranking	0.28 ***	0.08	−0.19 **	−0.48 ***	1				
6. P_target	0.70 ***	0.22 **	0.06	−0.16	0.26 **	1			
7. Neighbor	0.59 ***	0.09	−0.05	−0.32 ***	0.30 ***	0.76 ***	1		
8. Central	0.45 ***	0.004	0	−0.065	0	0.69 ***	0.78 ***	1	
9. Unemployment	−0.07	−0.03	0.45 ***	0.34 ***	−0.28 ***	−0.10	−0.13	0	1
10. GDP	0.29 ***	0.119	−0.07	−0.48 ***	0.62 ***	0.31 ***	0.31 ***	0	−0.33 ***

Notes: *** $p < 0.01$, ** $p < 0.05$, * $p < 0.1$.

The results of three regression models based on HLM are shown in Table 4. Due to the strong correlations between P_target, Neighbor, and Central, we first put these three variables into regression models separately in order to control multicollinearity (see Model 1, Model 2, and Model 3). The variance inflation factors of all variables in the three models are all smaller than the critical value 10, and thus, there is no severe multicollinearity [39]. In Model 1, E_structure and Target are significantly positively related ($p < 0.05$), and the coefficients remain significantly positive in Models 2 and 3, strongly supporting the Hypothesis H1b. This finding indicates that provinces with higher proportions of the secondary industrial outputs were allocated higher environmental targets by the central government. In Models 1–3, the coefficients of E_scale are negative and the coefficients of E_intensity are positive, but they are all not significant. Therefore, hypotheses H1a and H1c are not supported, indicating that there was no significant trend of being allocated with higher environmental targets for provinces with lower scale performance and lower intensity performance in energy consumption and pollutant emissions. In all the three models, the coefficients of the P_ranking are all significantly positive ($p < 0.05$), fully supporting Hypothesis H2b. Therefore, provinces with higher probabilities of provincial leaders being promoted to Politburo members were more likely to be allocated higher environmental targets.

Some control variables are also significant in these models. For example, the coefficient of P_target is positive and significant ($p < 0.01$) in Model 1, which means provinces which were allocated higher environmental targets during the "11th Five-year Plan" period also tended to be allocated higher environmental targets during the "12th Five-year Plan" period. In Model 2, the coefficient of Neighbor is positive and significant ($p < 0.01$), and hence, it concludes that provinces whose neighboring provinces were allocated higher environmental targets also tended to be allocated higher environmental targets. In Model 3, the coefficient of Central is positive and significant ($p < 0.01$), indicating that provinces were allocated higher environmental targets for this corresponding indicator when the central government set a higher environmental target for one environmental indicator. Moreover, the variable of Unemployment is significantly negatively related to the dependent variable in Models 2 and 3, which implies provinces with higher unemployment rates were allocated lower environmental targets. The Chi-square values of Models 1–3 are very significant ($p < 0.01$), indicating that they fit the data well. Moreover, when we re-estimate these models with the ordinary least square regression, the results are similar, indicating that the findings are robust.

Table 4. Regression results of hierarchical linear models (HLM).

	Model 1	Model 2	Model 3
Constant	9.988 ***	9.748 ***	9.748 ***
	(0.622)	(0.644)	(0.644)
Layer-2 variables			
P_ranking	0.545 **	0.640 **	0.640 **
	(0.250)	(0.297)	(0.297)
E_structure	0.323 **	0.349 **	0.349 **
	(0.162)	(0.173)	(0.173)
Unemployment	−1.304	−1.472 *	−1.472 *
	(0.860)	(0.806)	(0.806)
GDP	5.094	4.792	4.792
	(3.598)	(3.433)	(3.433)
Layer-1 variables			
E_scale	−1.013	−1.438	−1.186
	(1.153)	(0.885)	(0.907)
E_intensity	3.302	3.911	3.346
	(3.438)	(2.616)	(2.613)
P_target	0.753 ***		
	(0.0949)		
Neighbor		1.188 ***	
		(0.153)	
Central			1.152 ***
			(0.137)
Model statistics			
N (layer 1)	87	145	145
N (layer 2)	29	29	29
Variance (layer 1)	30.43	18.51	22.15
Variance (layer 2)	7.04×10^{-11}	7.92	7.20
Intra-class Correlation Coefficient	2.3135×10^{-12}	0.2997	0.2453
Log pseudolikelihood	−272.023	−433.928	−444.317
Wald Chi2	166.72 ***	175.78 ***	210.86 ***

Notes: *** $p < 0.01$, ** $p < 0.05$, * $p < 0.1$. Robust standard errors are presented in parentheses. Regression results based on the ordinary least square estimation are similar.

Robustness Check

We then re-measure E_scale, E_structure, and E_intensity with the difference between the investigated province's scale performance and the average scale performance of its neighboring provinces for the investigated environmental indicator, the difference between the investigated province's structural performance and the average structural performance of its neighboring provinces for the investigated environmental indicator, and the difference between the investigated province's intensity performance and the average intensity performance of its neighboring provinces for the investigated environmental indicator, which are coded as R_scale, R_structure, and R_intensity, respectively. We then put these three new variables into models to do a robustness check. Results of the three new HLM models are shown in Table 5. The three variables of P_target, Neighbor, and Central are also put into regression models separately in order to control multicollinearity (see Models 4–6). The variance inflation factors of all variables in Models 4–6 are all smaller than the critical value 10, and hence, there is no severe multicollinearity [39].

Table 5. Regression results of HLM.

	Model 4	Model 5	Model 6
Constant	10.06 ***	9.748 ***	9.7483 ***
	(0.608)	(0.668)	(0.668)
Layer-2 variables			
P_ranking	0.569 **	0.672 **	0.672 **
	(0.246)	(0.305)	(0.305)
R_structure	0.270 *	0.302 *	0.302 *
	(0.153)	(0.166)	(0.166)
Unemployment	−0.880	−1.289 *	−1.289 *
	(0.898)	(0.747)	(0.747)
GDP	4.704	4.476	4.476
	(3.362)	(3.420)	(3.420)
Layer-1 variables			
R_scale	−2.006	−1.686 *	−1.354
	(1.289)	(0.894)	(0.956)
R_intensity	4.885	3.967	3.789
	(3.590)	(2.483)	(2.644)
P_target	0.720 ***		
	(0.075)		
Neighbor		1.131 ***	
		(0.121)	
Central			1.097 ***
			(0.108)
Model statistics			
N (layer 1)	87	145	145
N (layer 2)	29	29	29
Variance (layer 1)	28.594	18.200	21.655
Variance (layer 2)	1.478	8.871	8.180
Intraclass Correlation Coefficient	0.0491	0.3277	0.2742
Log pseudolikelihood	−271.403	−434.001	−444.081
Wald Chi2	145.21 ***	191.62 ***	192.23 ***

Notes: *** $p < 0.01$, ** $p < 0.05$, * $p < 0.1$; robust standard error in parentheses. Regression results based on the ordinary least square estimation are similar.

In Models 4–6, R_structure and Target are all significantly positively related ($p < 0.1$), and hence, Hypothesis H1b is still supported; in the Model 5, the coefficient of R_scale is significantly negative ($p < 0.1$), and it remains negative in Models 4 and 6, still failing to support Hypotheses H1a. The coefficients of R_intensity are positive but not significant in Models 4–6, and hence, Hypothesis H1c is still not supported. Besides, the coefficients of P_ranking are all significantly positive ($p < 0.05$) in Models 4–6, thus supporting hypothesis H2b. Overall, these results of Models 4–6 are consistent with those of Models 1–3, implying our findings are highly robust. The results conclude that the decomposition of environmental targets mainly depends on provinces' historical structural performance and their political rankings. This means the decomposition of environmental targets in China is both performance-based and politic-related.

6. Discussion and Conclusions

This work explores the roles of historical performance and provinces' political status in the decomposition of environmental targets by controlling a set of factors advocated in previous goal-setting research. Data from five environmental indicators (sulfur dioxide, chemical oxygen demand, energy consumption for per unit GDP, nitrogen oxide, and ammonia nitrogen) of 29 provinces in mainland China during the "12th Five-year Plan" period are used to test these hypotheses. Considering the nested data structure, HLM is used for estimation. The findings show that the structural performance of energy consumption and pollutant emissions of provinces is significantly negatively related to environmental target levels, while the scale performance and intensity performance of provinces' energy consumption and pollutant emissions are not significantly related to provinces' environmental target levels. Provinces with higher probabilities of provincial leaders being promoted to politburo members tend to be allocated higher environmental targets, though the wealth is controlled. This finding means that the decomposition of environmental targets in China is partially performance-based but strongly politic-related.

The historical structural performance of energy consumption and pollutant emissions is significantly correlated with the setting of environmental target aspiration levels, while the effects of historical scale performance and intensity performance are not significant. This finding confirms the limited and partial role of historical performance in China's environmental target-setting and implies that the central government mainly considered the pollutant emission and energy consumption reduction potentials of Chinese provinces through the adjustment of economic development structures. Similar to the "scale economies effect" in economics, there is still a scale effect in pollutant emission and energy consumption reduction. Thus, it is easier for the province to reduce pollutant emissions and energy consumption when the scale of its previous pollutant emissions and energy consumption is large. However, the nonsignificant effect of the historical scale performance implies that the central government ignored the scale effect when evaluating the pollutant emission reduction potentials. Intensity performance is also not significant, which means the central government also underestimated the reduction potentials by using technological advances to improve pollutant emission efficiency in environmental target-setting. Some scholars hold that carbon intensity convergence can be used to make carbon reduction target allocation reasonable, which means provinces with relatively higher CO_2 intensities of GDP should be assigned higher CO_2 intensity reduction targets and vice versa [40]. Their ideas are in line with our arguments to a great extent.

This work also reveals some clues about the different effects of historical performance in the setting of obligatory targets and the setting of anticipated targets in China. Generally, historical performance is positively related to the setting of anticipatory targets because better historical performance can enhance the organization's confidence in setting higher anticipatory target aspiration levels [10,18]. This observation reveals that the setting of anticipated targets follows the law of "addition", which means organizations set their anticipated targets through increasing certain percentages of performance based on their historical performance. However, historical performance is negatively related to environmental obligatory targets, mainly because environmental obligatory targets are allocated by the superior government, and lower historical pollutant emission performance means greater pollutant emission reduction potentials. This finding reveals that the setting of environmental obligatory targets follows the "law of subtraction", which means organizations set their obligatory targets through decreasing certain percentages of pollutant emissions or energy consumption based on their historical performance.

Chinese provincial governments set their anticipated target aspiration levels themselves, while Chinese provincial governments' obligatory targets are allocated by the central government [13,16]. During the allocation process, local governments interact with the central government; this work finds that provinces with different political rankings set different environmental obligatory target aspiration levels. Based on two competing hypotheses of the "bargaining effect" and "model effect", the empirical results support the "model effect hypothesis". This indicates that provinces with higher political rankings do not take advantage of their own political capital to bargain with the central government to strive for lower environmental obligatory target levels. Instead, these provinces tend to play the role of "leading models" and accept higher environmental obligatory target aspiration levels allocated by the central government. One possible explanation is that these provinces are motivated by political promotion incentives: leaders (party secretary and governor) in the provinces with higher political rankings have higher likelihoods of being promoted to the Politburo in China; in this case, these provinces are more likely to follow the policy preference of the central government and strive to keep their advantages by performing well in the performance dimension and subsequently being favored by the central government [28].

The findings provide some important practical policy implications. First, the national environmental targets should be decomposed based on provinces' historical scale performance, structural performance, and intensity performance in pollutant emissions and energy consumption comprehensively. Specifically, the higher the proportion of the secondary industrial output of one province is, the higher target the province should be allocated; the higher the scale of pollutant emissions and energy consumption of one province is, the higher target the province should be allocated; the higher the pollutant emissions and energy consumption per unit GDP of one province is, the higher target the province should be allocated. Second, provinces with higher probabilities of provincial leaders being promoted to politburo members are incentivized by potential political promotions to comply with national environmental targets, and hence, other kinds of incentives (financial incentives, moral incentives, etc.) can also be provided for provincial governments in order to make them comply with environment-related central directives actively [35].

Though this work takes the first step to explore the hidden logic in the decomposition of environmental targets in China empirically, its limitations also provide some future research avenues. On the one hand, this work focuses on the disaggregation of environmental targets from the central government to Chinese provincial governments and more evidence is needed about the antecedents of the decomposition of environmental targets from Chinese provincial governments to city governments. On the other hand, this work explores determinants of the decomposition of environmental targets in China during the "12th Five-Year Plan" period. Future studies can explore features and determinants of environmental target-setting during the "11th Five-Year Plan" period and the "13th Five-Year Plan" period, which can help reveal the evolution features of environmental target-setting across various "Five-year Plan" periods through comparison.

Author Contributions: P.Z. designed the research, compiled the data, performed initial experiments, and produced early drafts; J.W. and P.Z. reviewed the drafts, did revisions, and approved the final version together.

Funding: This research was funded by the National Social Science Fund of China (grant number [17CGL069]) and the Key Philosophy and Social Science Research Project of the Ministry of Education of China (grant number [17JZD025]).

Conflicts of Interest: The authors declare no conflict of interest.

References

1. Moynihan, D.P. Managing for results in state government: Evaluating a decade of reform. *Public Adm. Rev.* **2006**, *66*, 77–89. [CrossRef]
2. Latham, G.P.; Borgogni, L.; Petitta, L. Goal setting and performance management in the public sector. *Int. Public Manag. J.* **2008**, *11*, 385–403. [CrossRef]
3. Gao, J. Governing by goals and numbers: A case study in the use of performance measurement to build state capacity in China. *Public Adm. Dev.* **2009**, *29*, 21–31. [CrossRef]
4. Mu, R. Bounded rationality in the developmental trajectory of environmental target policy in China, 1972–2016. *Sustainability* **2018**, *10*, 199. [CrossRef]
5. Locke, E.A.; Shaw, K.N.; Saari, L.M.; Latham, G.P. Goal setting and task performance: 1969–1980. *Psychol. Bull.* **1981**, *90*, 125–152. [CrossRef]
6. Locke, E.A.; Latham, G.P. Building a practically useful theory of goal setting and task motivation: A 35-year odyssey. *Am. Psychol.* **2002**, *57*, 705–717. [CrossRef] [PubMed]
7. Zhang, P.; Wu, J.N. Impact of mandatory targets on PM$_{2.5}$ concentration control in Chinese cities. *J. Clean. Prod.* **2018**, *197*, 323–331. [CrossRef]
8. Cyert, R.M.; March, J.G. *A Behavioral Theory of the Firm*, 2nd ed.; Blackwell Business: Malden, MA, USA, 1992.

9. Greve, H.R. *Organization Learning from Performance Feedback: A Behavioral Perspective on Innovation and Change*; Cambridge University Press: Cambridge, UK, 2003.
10. Ma, L. Performance feedback, government goal-setting and aspiration level adaptation: Evidence from Chinese provinces. *Public Adm.* **2016**, *94*, 452–471. [CrossRef]
11. Rutherford, A.; Meier, K.J. Managerial goals in a performance-driven system: Theory and empirical tests in higher education. *Public Adm.* **2015**, *93*, 17–33. [CrossRef]
12. Nielsen, P.A. Learning from performance feedback: Performance information, aspiration levels, and managerial priorities. *Public Adm.* **2014**, *92*, 142–160. [CrossRef]
13. Zhao, X.F.; Wu, L. Interpreting the evolution of the energy-saving target allocation system in China (2006–13): A view of policy learning. *World Dev.* **2016**, *82*, 83–94. [CrossRef]
14. Wu, J.N.; Zhang, P.; Yi, H.T.; Qin, Z. What causes haze pollution? An empirical study of PM2.5 concentrations in Chinese cities. *Sustainability* **2016**, *8*, 132. [CrossRef]
15. Li, H.M.; Zhao, X.F.; Yu, Y.Q.; Wu, T.; Qi, Y. China's numerical management system for reducing national energy intensity. *Energy Policy* **2016**, *94*, 64–76. [CrossRef]
16. Xu, Y. The use of a goal for SO_2 mitigation planning and management in China's 11th Five-Year Plan. *J. Environ. Plan. Manag.* **2011**, *54*, 769–783. [CrossRef]
17. Lant, T.K. Aspiration level adaptation: An empirical exploration. *Manag. Sci.* **1992**, *38*, 623–644. [CrossRef]
18. Mezias, S.J.; Chen, Y.R.; Murphy, P.R. Aspiration-level adaptation in an American financial services organization: A field study. *Manag. Sci.* **2002**, *48*, 1285–1300. [CrossRef]
19. Argote, L.; Greve, H.R. A behavioral theory of the firm—40 years and counting: Introduction and impact. *Organ. Sci.* **2007**, *18*, 337–349. [CrossRef]
20. Meier, K.J.; Favero, N.; Zhu, L. Performance gaps and managerial decisions: A Bayesian decision theory of managerial action. *J. Public Adm. Res. Theory* **2015**, *25*, 1221–1246. [CrossRef]
21. Zhou, L.; Zhang, X.L.; Qi, T.Y.; He, J.K.; Luo, X.H. Regional disaggregation of China's national carbon intensity reduction target by reduction pathway analysis. *Energy Sustain. Dev.* **2014**, *23*, 25–31. [CrossRef]
22. Kanie, N.; Nishimoto, H.; Hijioka, Y.; Kameyama, Y. Allocation and architecture in climate governance deyond Kyoto: Lessons from interdisciplinary research on target setting. *Int. Environ. Agreem. Polit. Law Econ.* **2010**, *10*, 299–315. [CrossRef]
23. Liu, Z.; Shi, Y.R.; Yan, J.M.; Qu, X.M.; Lieu, J. Research on the decomposition model for China's National Renewable Energy total target. *Energy Policy* **2012**, *51*, 110–120. [CrossRef]
24. Zhang, L.X.; Feng, Y.Y.; Zhao, B.H. Disaggregation of energy-saving targets for China's provinces: Modeling results and real choices. *J. Clean. Prod.* **2015**, *103*, 837–846. [CrossRef]
25. Tang, X.; Liu, Z.W.; Yi, H.T. Mandatory targets and environmental performance: An analysis based on regression discontinuity design. *Sustainability* **2016**, *8*, 931. [CrossRef]
26. Wu, J.N.; Xu, M.M.; Zhang, P. The impacts of governmental performance assessment policy and citizen participation on improving environmental performance across Chinese provinces. *J. Clean. Prod.* **2018**, *184*, 227–238. [CrossRef]
27. Wu, J.N.; Zhang, P. Local government innovation diffusion in China: An event history analysis of a performance-based reform programme. *Int. Rev. Adm. Sci.* **2018**, *84*, 63–81. [CrossRef]
28. Liang, J.Q.; Langbein, L. Performance management, high-powered incentives, and environmental policies in China. *Int. Public Manag. J.* **2015**, *18*, 346–385. [CrossRef]
29. Zhao, X.L.; Ma, C.B.; Hong, D.Y. Why did China's energy intensity increase during 1998–2006: Decomposition and policy analysis. *Energy Policy* **2010**, *38*, 1379–1388. [CrossRef]
30. Hammond, G.P.; Norman, J.B. Decomposition analysis of energy-related carbon emissions from UK manufacturing. *Energy* **2012**, *41*, 220–227. [CrossRef]
31. Liang, J.Q. Who maximizes (or satisfices) in performance management? An empirical study of the effects of motivation-related institutional contexts on energy efficiency policy in China. *Public Perform. Manag. Rev.* **2015**, *38*, 284–315. [CrossRef]
32. Li, J.Y. The paradox of performance regimes: Strategic responses to target regimes in Chinese local government. *Public Adm.* **2015**, *93*, 1152–1167. [CrossRef]
33. Manion, M. The cadre management system, post-Mao: The appointment, promotion, transfer and removal of party and state leaders. *China Q.* **1985**, *102*, 203–233. [CrossRef]

34. Chan, H.S. Cadre personnel management in China: The nomenklatura system, 1990–1998. *China Q.* **2004**, *179*, 703–734. [CrossRef]
35. Ran, R. Perverse incentive structure and policy implementation gap in China's local environmental politics. *J. Environ. Policy Plan.* **2013**, *15*, 17–39. [CrossRef]
36. Berry, F.S.; Berry, W.D. State lottery adoptions as policy innovations: An event history analysis. *Am. Polit. Sci. Rev.* **1990**, *84*, 395–415. [CrossRef]
37. Ma, L. Performance management and citizen satisfaction with the government: Evidence from Chinese municipalities. *Public Adm.* **2017**, *95*, 39–59. [CrossRef]
38. Aguinis, H.; Gottfredson, R.K.; Culpepper, S.A. Best-practice recommendations for estimating cross-level interaction effects using multilevel modeling. *J. Manag.* **2013**, *39*, 1490–1528. [CrossRef]
39. Kutner, M.H.; Nachtsheim, C.J.; Neter, J.; Li, W. *Applied Linear Statistical Models*; McGraw-Hill Irwin: New York, NY, USA, 2005.
40. Hao, Y.; Liao, H.; Wei, Y.M. Is China's carbon reduction target allocation reasonable? An Analysis based on carbon intensity convergence. *Appl. Energy* **2015**, *142*, 229–239. [CrossRef]

sustainability

MDPI

Article

A New Study on Air Quality Standards: Air Quality Measurement and Evaluation for Jiangsu Province Based on Six Major Air Pollutants

Xueyan Liu [1] and Xiaolong Gao [2,*]

[1]   School of Economics, Fudan University, Shanghai 200433, China; liuxueyan_fd@126.com
[2]   School of Business, East China University of Science and Technology, Shanghai 200237, China
*   Correspondence: cnxiaolong@126.com; Tel.: +86-21-6425-3583

Received: 27 August 2018; Accepted: 30 September 2018; Published: 5 October 2018

Abstract: China's current Air Quality Index (AQI) system only considers one air pollutant which has the highest concentration value. In order to comprehensively evaluate the urban air quality of Jiangsu Province, this paper has studied the air quality of 13 cities in that province from April 2015 to March 2018 based on an expanded AQI system that includes six major air pollutants. After expanding the existing air quality evaluation standards of China, this paper has calculated the air quality evaluation scores of cities in Jiangsu Province based on the six major air pollutants by using the improved Fuzzy Comprehensive Evaluation Model. This paper has further analyzed the effectiveness of air pollution control policies in Jiangsu Province and its different cities during the study period. The findings are as follows: there are distinct differences in air quality for different cities in Jiangsu Province; except for coastal cities such as Nantong, Yancheng and Lianyungang, the southern cities of Jiangsu generally have better air quality than the northern cities. The causes of these differences include not only natural factors such as geographical location and wind direction, but also economic factors and energy structure. In addition, air pollution control policies have achieved significant results in Nantong, Changzhou, Wuxi, Yangzhou, Suzhou, Yancheng, Zhenjiang, Tai'an and Lianyungang. Among them, Nantong has seen the biggest improvement, 20.28%; Changzhou and Wuxi have improved their air quality by more than 10%, while Yangzhou, Suzhou, and Yancheng have improved their air quality by more than 5%. However, the air quality of Nanjing, Huai'an, Xuzhou, and Suqian has worsened by different degrees compared that of the last period within the beginning period, during which Suqian's air quality has declined by 20.07% and Xuzhou's by 16.32%.

Keywords: air quality; air quality evaluation standards; AQI; Jiangsu province; fuzzy comprehensive evaluation

1. Introduction

Since the reform and opening-up of China, with the rapid expansion in the size of its economy, its ecological environment, especially air quality, has been facing serious threats [1–6]. Increasing energy consumption, the energy structure over-relying on primary energy sources (such as coal) with low conversion efficiency and high pollution emissions, as well as the lack of environmental awareness, have all contributed to the severe deterioration in China's air quality [7–12]. In recent years, China has experienced frequent heavy pollution weathers, especially in the eastern coastal region. The devastating, long-lasting, and wide-ranging smog and haze phenomenon is typical proof of the deterioration in air quality. Apart from its negative impact on traffic, the growing problem of air pollution has also caused great loss to normal people's daily life, health [13,14], and the operation and production of enterprises, which has drawn great attention from the academic community [15–19].

As the most economically developed province in the eastern coastal areas of China, Jiangsu Province has also been seriously plagued by air pollution problems in recent years. According to the "Jiangsu Province Environmental Bulletin" published by the Jiangsu Provincial Environmental Protection Department every year, although the air quality of Jiangsu has improved since 2013, none of its 13 cities has reached the Level II Air Quality Standards stated in China's "Ambient Air Quality Standards (GB3095-2012)" on air pollutants' annual average concentration limits since 2013 (see Table 1 below) [20]. Moreover, the annual average concentration of NO_2 and O_3 rebounded in 2017, and the province's air quality compliance rate decreased by 2.2% in that year [21].

Table 1. Annual Average Concentration of Major Air Pollutants and Air Quality Compliance Rate of Jiangsu Province (2013–2017).

	2013	2014	2015	2016	2017
$PM_{2.5}$ ($\mu g/m^3$)	73	66	58	51	49
PM_{10} ($\mu g/m^3$)	115	106	96	86	81
SO_2 ($\mu g/m^3$)	35	29	25	21	16
NO_2 ($\mu g/m^3$)	41	39	37	37	39
CO (mg/m^3)	2.1	1.7	1.7	1.7	1.5
O_3 ($\mu g/m^3$)	139	154	167	165	177
Air Quality Compliance Rate in Jiangsu (%)	60.3	64.2	66.8	70.2	68.0
Number of Cities in Jiangsu that Reached Level II Air Quality Standards	0	0	0	0	0

Therefore, although the official statistics of Jiangsu Province's air quality compliance rate have improved in recent years [20], it is worth studying how to objectively evaluate the air quality and policy effectiveness in Jiangsu Province, given that the emissions of major air pollutants are still increasing [22].

In recent years, the academic community has also carried out various explorations on the measurement and evaluation of Jiangsu Province's air quality. Wang et al. (2016) used the Logarithmic Mean Divisia Index (LMDI) to analyze the driving factors of SO_2 emissions in Jiangsu Province, and found that energy intensity is the main reason for the increase. They believe that the government needs to determine specific emission reduction targets and policy initiatives according to the actual energy structure of different cities [23]. Ge et al. (2017) used the Projection Pursuit Cluster (PPC) Model to analyze the Social Vulnerability for Air Pollution of the Yangtze River Delta (YRD) region represented by Jiangsu Province. By calculating the Social Vulnerability Index (SVI), they concluded that Jiangsu's SVI was higher than that of Shanghai [24]. He et al. (2018) studied the impact of various factors including the industrial structure, energy consumption structure, and energy efficiency on air quality in Jiangsu Province from 2006 to 2015, and further explored the impact of relevant policies on energy consumption and air quality. Their study showed that every 1% optimization of the industrial structure in Jiangsu Province would result in an improvement of 0.0054% in the Air Quality Index [25]. Zhang et al. (2017) analyzed the spatial distribution of acid rain using the recent data of acid rain and urban pollutant emissions in the eastern coastal areas. They concluded that since 2009, the increase of $NH_4{}^+$ and Ca^{2+} has led to an increase in the number of haze days in Jiangsu, and that the long-distance spread of pollutants and alkaline pollutants are key drivers of acid rain and haze problems [26]. Xu et al. (2017) used the Structural Decomposition Analysis (SDA) method to decompose the factors behind the increase of CO_2 emissions in Jiangsu Province. They pointed out that the economic growth of Jiangsu Province has generally contributed to the increase of CO_2 emissions, and that the transfers-out and investment effects are the main reason for the increase of CO_2 emissions [27]. Chen et al. (2017) studied the relations between short-term ozone exposure and daily total mortality using a generalized additive model and univariate random-effects meta-analysis. By studying seven cities in Jiangsu Province from 2013 to 2014, they concluded that there was a significant correlation between premature total mortality and short-term ozone exposure [28]. Wang et al. (2017) divided the factors reducing air pollution into three stages: source prevention, process control, and end-of-pipe treatment based on index decomposition analysis and a whole process treatment perspective. After studying the treatment of energy-related SO_2 emissions in 13 cities of Jiangsu Province, they divided these cities into

4 types: the leading type, process-dependent type, end-dependent type, and lagging type. They also found that the development pattern of "Pollute First, Govern Later" still has not fundamentally changed for Jiangsu Province [29].

The above research on the air quality of cities in Jiangsu were still based on the existing air quality measurement standards of China, and most of them only considered 1–2 major pollutants such as $PM_{2.5}$ and PM_{10}. However, using only China's current national air quality standards—Ambient Air Quality Standards (GB3095-2012) and the Technical Regulation on Ambient Air Quality Index (HJ 633-2012)—it is difficult to make a comprehensive and objective assessment on the air quality of Jiangsu Province in recent years. This is because, first of all, China's current air quality assessment standards essentially only involve one type of pollutant (the pollutant with the highest concentration on the day), so it is difficult to fully reflect the overall air quality [30,31]. Secondly, China's current standards are only based on the average concentration value of each pollutant within a certain period of time, so it is difficult to reflect the extreme concentration values of the pollutant and its fluctuations in that period. Finally, the above standards were established in 2012, which set the upper limit of the "24-h average $PM_{2.5}$" as 500 [30,31]. However, this limit can no longer adapt to the reality because currently, the actual PM2.5 concentration values of many cities in China were far exceeding this upper limit (i.e., off the charts) [32]. Under such circumstances, the current assessment of China's air quality is often simplified into vertical and horizontal comparisons of the $PM_{2.5}$ statistics of different regions.

Therefore, based on above research, and drawing on the research of Cannistraro et al. (2016) [33], this paper has constructed a comprehensive air quality evaluation system that encompasses the six major pollutants (SO_2, NO_2, CO, PM_{10}, $PM_{2.5}$ and O_3) [30,31] covered in China's national routine monitoring and air quality assessment. Based on the statistics of the six air pollutants in cities of Jiangsu Province from April 2015 to March 2018, this paper has also utilized the Fuzzy Comprehensive Evaluation Model in order to measure and evaluate the air quality of various cities in Jiangsu Province, and to enrich the existing academic literature on Jiangsu Province's air quality. Scholars have adopted new approaches to the evaluation of air quality in Jiangsu in recent years. Cao et al. (2018) studied the air pollution caused by the exhaust gas from inland ships in the Jiangsu section of the Beijing-Hangzhou Grand Canal. Their results indicated that dry cargo ships are the most important air pollution sources in the Jiangsu section of the Grand Canal, while ships with a gross tonnage between 200 and 600 tons had the largest exhaust gas emissions [34]. Yang et al. (2018) derived the Variogram Model using the daily average concentration data of PM2.5 in the southern part of Jiangsu Province in 2014, and generated the distribution information maps of pollutants in the southern area of Jiangsu with help of the spatiotemporal ordinary kriging (STOK) technology. Their results showed that in 2014, about 29.3% of the area in southern Jiangsu was polluted by PM2.5, which also showed a spatial trend of the PM2.5 pollutants declining from the west to the east of southern Jiangsu [35]. This paper decided to calculate and evaluate the air quality of Jiangsu by Fuzzy Comprehensive Evaluation Model. Compared with other methods, the Fuzzy Comprehensive Evaluation Model describes the air quality level of evaluation with the membership function, and can evaluate the parameters in the model, which makes the results as close as possible to objective fact [36,37]. Moreover, this paper improves the original model in order to better evaluate the air quality of Jiangsu on the basis of six pollutants (please refer to Part 2).

The structure of this paper is as follows: Part 2 introduces the research methods used; Part 3 lists the calculation results; Part 4 analyzes the air quality of various cities of Jiangsu Province since April 2015; Part 5 summarizes the findings in this paper and provides corresponding policy recommendations.

2. Methodology and Data

2.1. Improved Fuzzy Comprehensive Evaluation Model

The air quality evaluation system constructed in this paper includes six major air pollutants: SO_2, NO_2, CO, PM_{10}, $PM_{2.5}$, and O_3. In this complex system, the evaluation of the air quality of each city in Jiangsu Province is determined by these six pollutants. Therefore, in the calculation of air quality evaluation scores, we first need to calculate the evaluation score of each pollutant, and then obtain the overall air quality evaluation result of that city based on the evaluation scores of each pollutant. In the actual evaluation process, in order to deal with the uncertainties of a complex systems containing six major pollutants of SO_2, NO_2, CO, PM_{10}, $PM_{2.5}$, and O_3, this paper has made a few improvements to the Fuzzy Comprehensive Evaluation Model commonly used in academic circles [38–42] in order to reduce the uncertainty of this evaluation system. The specific steps are as follows:

A city's air quality evaluation object P. $U = \{u_1, u_2, \cdots, u_n\}$ represents a set of pollutant indicators related to this evaluation object P. For each pollutant, there is a Rating Set $V = \{v_1, v_2, \cdots, v_m\}$. After making a fuzzy evaluation on the Rating Set of each pollutant in U, this paper has obtained (1) the fuzzy evaluation matrix about n factors:

$$
R = \begin{bmatrix} R_1 \\ R_2 \\ \vdots \\ R_4 \end{bmatrix} = \begin{bmatrix} r_{11} & r_{12} & \cdots & r_{1n} \\ r_{21} & r_{22} & \cdots & r_{2n} \\ \vdots & \vdots & \ddots & \vdots \\ r_{m1} & r_{m2} & \cdots & r_{mn} \end{bmatrix} \tag{1}
$$

where r_{ij} is determined by the membership function. The calculation steps are as follows. Equations (2)–(4):

(1) Rating Level 1:

$$
r_{i1}(r_i) = \begin{cases} 0 & r_i \geq v_{i2} \\ -\frac{r_i - v_{i2}}{v_{i2} - v_{i1}} & v_{i1} < r_i < v_{i2} \\ 1 & r_i \leq v_{i1} \end{cases} \tag{2}
$$

(2) Rating Level j:

$$
r_{ij}(r_i) = \begin{cases} 0 & r_i \leq v_{ij-1}, r_i \geq v_{ij+1} \\ \frac{r_i - v_{ij-1}}{v_{ij} - v_{ij-1}} & v_{ij-1} < r_i < v_{ij} \\ -\frac{r_i - v_{ij+1}}{v_{ij+1} - v_{ij}} & v_{ij-1} < r_i < v_{ij+1} \end{cases} \tag{3}
$$

(3) Rating Level n:

$$
r_{in}(r_i) = \begin{cases} 0 & r_i \leq v_{in} \\ \frac{r_i - v_{in-1}}{r_{in} - v_{in-1}} & v_{in-1} < r_i < v_{in} \\ 1 & r_i \geq v_{in} \end{cases} \tag{4}
$$

The element r_{ij} in the above matrix represents the fuzzy membership degree of the factor u_i with regard to the rating v_i, that is, a fuzzy relationship from U to V; thus, the determined (U, V, R) constitutes a Fuzzy Comprehensive Evaluation Model.

In order to calculate the comprehensive evaluation value of different cities' air quality, it is also necessary to determine the weight of each factor. Since the degree of harm of the six pollutants has not yet been uniformly quantified, we weigh all pollutants equally in the calculation. Let the weights be $W = \{w_1, w_2, \cdots, w_n\}$, which denotes the weight of each indicator, and satisfies the condition

$\sum_{i=1}^{n} w_i = 1$. By using the matrix and vector algorithm, this paper can obtain Fuzzy Evaluation Set B with the following Equation (5):

$$B = W \times R^T = \begin{bmatrix} r_{11} & r_{12} & \cdots & r_{1n} \\ r_{21} & r_{22} & \cdots & r_{2n} \\ \vdots & \vdots & \ddots & \vdots \\ r_{m1} & r_{m2} & \cdots & r_{mn} \end{bmatrix} \times [w_1, w_2, \cdots, w_n]^T = [b_1, b_2, \cdots, b_m] \quad (5)$$

Finally, based on the principle of maximum membership degree, this paper can obtain a comprehensive evaluation value of different cities' air quality by analyzing the Fuzzy Evaluation Vector B: in this Fuzzy Evaluation Set $B = (b_1, b_2, \cdots, b_m)$, let b_i be the membership degree of Level v_i to this Fuzzy Evaluation Set B. Let $M = max(b_1, b_2, \cdots, b_m)$, M's value represent the Fuzzy Comprehensive Evaluation Score of the evaluation object, i.e., the comprehensive evaluation value of the air quality of the city. Ranging from 0 to 1, the larger the M's value, the better air quality of the city; the smaller the M's value, the worse the air quality.

2.2. Data Sources

In February 2015, the Jiangsu Provincial People's Congress officially published the "Regulations on the Prevention and Control of Air Pollution in Jiangsu Province", which was officially implemented from March 2015 [43]. As the first local regulation reviewed and approved by the Jiangsu Provincial People's Congress since 2001, this regulation has gone through four rounds of review, which was a record in the local legislation of Jiangsu Province [22]. This regulation strictly stipulates the acquisition and disclosure of air pollution monitoring data, the development and publication of the heavy-pollution industrial projects list, as well as the enforcement actions including rectification, production restriction, production suspension, and closure, which provided effective local regulatory guidance on air pollution control to different cities of Jiangsu Province [43]. In March 2018, the second meeting of the Standing Committee of the 13th People's Congress of Jiangsu Province adopted a resolution to formally amend the regulations [44].

This paper has selected April 2015 to March 2018 as the study period, which was after the regulations were first implemented but before the amendment took place in order to analyze the influence and effectiveness of this regulation based on the six major air pollutants. This paper's calculation is based on the monthly air quality and pollutant monitoring data published by the Data Center of China's Ministry of Environmental Protection and China's National Environmental Monitoring Center, covering the monthly average concentration data of six major pollutants, i.e., $PM_{2.5}$, PM_{10}, CO, NO_2, O_3, and SO_2 [45,46].

3. Results

By adopting the Fuzzy Optimization Theory introduced in Part 2.1, this paper has calculated the air quality evaluation results for 13 cities in Jiangsu Province from April 2015 to March 2018 based on the aforementioned air pollutant data (as shown in Tables 2–5 and Figures 1–4 below).

Table 2. Air Quality Evaluation Score of Cities in Jiangsu Province (April 2015–December 2015).

	April 2015	May 2015	June 2015	July 2015	August 2015	September 2015	October 2015	November 2015	December 2015
Nanjing	0.5078	0.4561	0.4464	0.5025	0.4522	0.4712	0.4796	0.5141	0.5031
Nantong	0.5023	0.5110	0.4219	0.4226	0.4595	0.4774	0.4991	0.5425	0.5281
Suqian	0.5671	0.5598	0.4169	0.5073	0.5027	0.5209	0.4746	0.4360	0.4499
Changzhou	0.4251	0.4412	0.5918	0.4965	0.5023	0.4711	0.5129	0.5460	0.4783
Xuzhou	0.4354	0.3811	0.4156	0.5256	0.5413	0.5015	0.4517	0.4554	0.4719
Yangzhou	0.4622	0.4152	0.4165	0.4605	0.4659	0.4731	0.5067	0.5417	0.5522
Wuxi	0.4409	0.4820	0.4952	0.4847	0.4541	0.4278	0.5235	0.5389	0.4879
Tai'an	0.5000	0.4869	0.3814	0.4362	0.4872	0.4746	0.4466	0.5039	0.5069
Huai'an	0.5363	0.5268	0.4034	0.5267	0.5125	0.5612	0.5049	0.5295	0.4825
Yancheng	0.5172	0.5781	0.4812	0.5224	0.5149	0.5674	0.5297	0.5416	0.5152
Suzhou	0.4939	0.4987	0.5106	0.4928	0.4620	0.4509	0.5419	0.5614	0.5243
Lianyungang	0.5527	0.5319	0.4431	0.4999	0.5145	0.5087	0.5427	0.5195	0.4630
Zhenjiang	0.4478	0.4901	0.3907	0.4126	0.4321	0.4375	0.5113	0.5202	0.5232

Table 3. Air Quality Evaluation Score of Cities in Jiangsu Province (January 2016–September 2016).

	January 2016	February 2016	March 2016	April 2016	May 2016	June 2016	July 2016	August 2016	September 2016
Nanjing	0.5017	0.5163	0.4603	0.5163	0.4329	0.4763	0.4889	0.3910	0.4528
Nantong	0.5490	0.5354	0.5671	0.4766	0.4709	0.4742	0.4229	0.5417	0.5318
Suqian	0.4550	0.4272	0.4795	0.5459	0.6062	0.5217	0.5450	0.5195	0.5080
Changzhou	0.5114	0.4952	0.4439	0.4836	0.4601	0.4782	0.4688	0.4450	0.4958
Xuzhou	0.4226	0.4517	0.4518	0.4337	0.4692	0.4708	0.5016	0.4610	0.4870
Yangzhou	0.5316	0.5409	0.4953	0.5111	0.4650	0.5004	0.4484	0.5064	0.5171
Wuxi	0.5212	0.5380	0.4838	0.4485	0.4538	0.5090	0.4596	0.4541	0.4975
Tai'an	0.5124	0.4549	0.4397	0.4915	0.4743	0.4753	0.4302	0.5111	0.4970
Huai'an	0.4966	0.4833	0.4704	0.4948	0.5083	0.4862	0.5006	0.5470	0.5037
Yancheng	0.5021	0.5167	0.5509	0.5418	0.5691	0.5510	0.5927	0.6346	0.5506
Suzhou	0.5570	0.5409	0.5577	0.5176	0.4904	0.4983	0.4752	0.4927	0.5089
Lianyungang	0.4655	0.5381	0.5253	0.4592	0.5008	0.4927	0.5721	0.5498	0.4982
Zhenjiang	0.5190	0.4876	0.4867	0.5554	0.5202	0.5281	0.4402	0.4188	0.4846

Table 4. Air Quality Evaluation Score of Cities in Jiangsu Province (October 2016–June 2017).

	October 2016	November 2016	December 2016	January 2017	February 2017	March 2017	April 2017	May 2017	June 2017
Nanjing	0.4996	0.5976	0.5778	0.5359	0.5573	0.5393	0.5328	0.5593	0.4724
Nantong	0.5831	0.5627	0.6058	0.6149	0.5783	0.5556	0.4985	0.5737	0.4972
Suqian	0.5225	0.5059	0.5055	0.4737	0.4633	0.5306	0.5044	0.4556	0.4465
Changzhou	0.5355	0.5471	0.5404	0.5477	0.5160	0.5265	0.5196	0.5768	0.4594
Xuzhou	0.3773	0.4003	0.3694	0.3725	0.4045	0.4091	0.4556	0.3653	0.3874
Yangzhou	0.5447	0.4927	0.5354	0.5100	0.4575	0.4707	0.4835	0.5255	0.4291
Wuxi	0.4925	0.5243	0.5466	0.5630	0.5632	0.5453	0.5096	0.5667	0.4786
Tai'an	0.5403	0.4616	0.5403	0.5231	0.4827	0.5140	0.4887	0.5428	0.4628
Huai'an	0.5623	0.5181	0.5180	0.5148	0.4980	0.5200	0.4773	0.5208	0.4634
Yancheng	0.5602	0.5527	0.5951	0.5432	0.5130	0.5170	0.5319	0.6203	0.5707
Suzhou	0.5074	0.5637	0.5360	0.5926	0.5615	0.5358	0.4805	0.5951	0.5641
Lianyungang	0.5097	0.4891	0.5521	0.5265	0.5374	0.5189	0.5390	0.5645	0.5893
Zhenjiang	0.5719	0.5598	0.5724	0.5352	0.4760	0.4841	0.5111	0.5376	0.4572

Table 5. Air Quality Evaluation Score of Cities in Jiangsu Province (July 2017–March 2018).

	July 2017	August 2017	September 2017	October 2017	November 2017	December 2017	January 2018	February 2018	March 2018
Nanjing	0.4905	0.5343	0.5277	0.5377	0.5249	0.4723	0.4189	0.4983	0.4879
Nantong	0.3523	0.4103	0.4970	0.6436	0.5820	0.5737	0.6621	0.6210	0.6042
Suqian	0.5265	0.4668	0.5104	0.4812	0.4841	0.4722	0.3686	0.4575	0.4533
Changzhou	0.4253	0.4587	0.4998	0.5359	0.5206	0.4471	0.4539	0.4987	0.4882
Xuzhou	0.4723	0.4301	0.4060	0.3263	0.3928	0.3759	0.3133	0.3469	0.3644
Yangzhou	0.3598	0.4609	0.5262	0.5650	0.4818	0.5013	0.5240	0.5274	0.5033
Wuxi	0.4320	0.4637	0.4931	0.5254	0.5136	0.4806	0.4867	0.5063	0.5002
Tai'an	0.4414	0.5191	0.5353	0.5949	0.5144	0.5005	0.5201	0.5450	0.5189
Huai'an	0.5172	0.5342	0.5391	0.4978	0.4988	0.4858	0.4860	0.4933	0.4925
Yancheng	0.5404	0.5642	0.5101	0.6055	0.5517	0.5234	0.5637	0.5700	0.5521
Suzhou	0.4595	0.4800	0.5303	0.5579	0.5500	0.5039	0.5270	0.5393	0.5340
Lianyungang	0.6237	0.5682	0.5865	0.5468	0.5778	0.5896	0.5184	0.5559	0.5639
Zhenjiang	0.4094	0.4624	0.4674	0.5283	0.4669	0.4650	0.4612	0.4900	0.4700

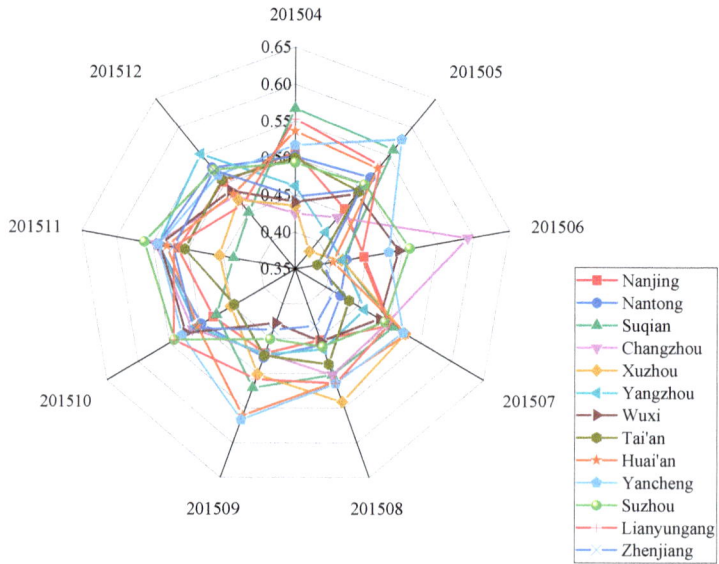

Figure 1. Air Quality Evaluation Score of 13 Cities in Jiangsu Province (April 2015–December 2015).

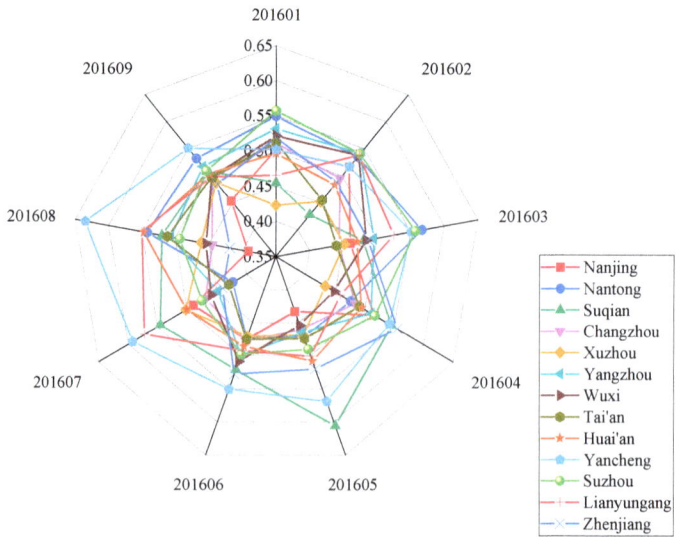

Figure 2. Air Quality Evaluation Score of 13 Cities in Jiangsu Province (January 2016–September 2016).

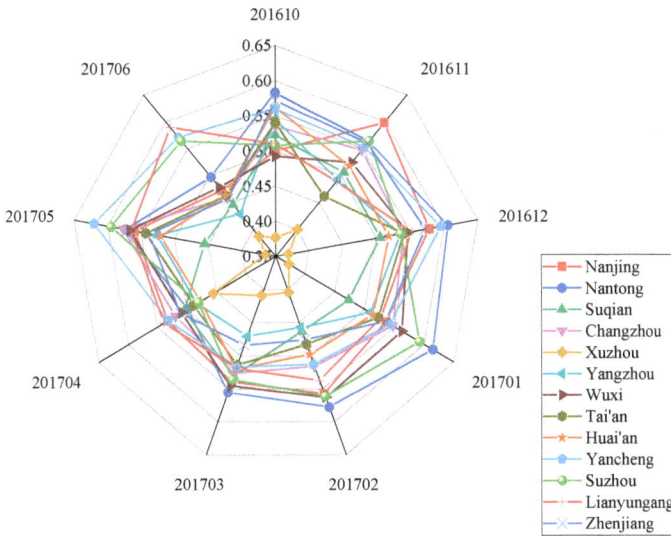

Figure 3. Air Quality Evaluation Score of 13 Cities in Jiangsu Province (October 2016–June 2017).

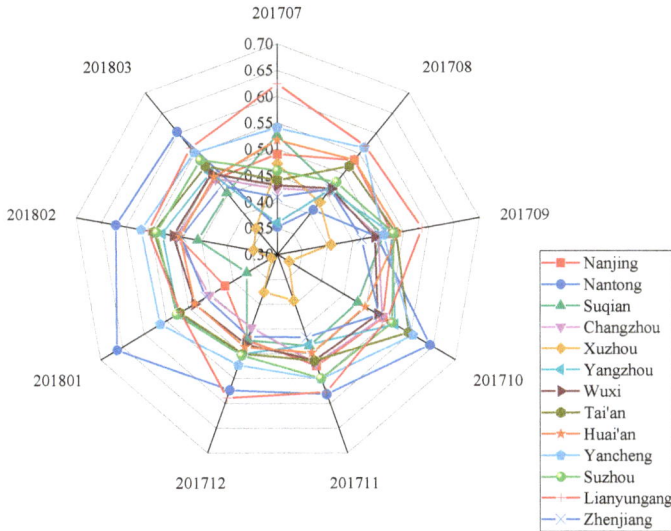

Figure 4. Air Quality Evaluation Score of 13 Cities in Jiangsu Province (July 2017–March 2018).

4. Discussions

Based on the air quality evaluation scores of various cities in Jiangsu Province from April 2015 to March 2018, this paper concludes that:

(1) Although none of the cities in Jiangsu Province has reached the Level II Air Quality Standards stated in China's "Ambient Air Quality Standards (GB3095-2012)" on air pollutants' annual average concentration limits within the study period [20,30], the lowest air quality evaluation score in the study period was that of Xuzhou in January, 2018 (0.3133), which was much higher than 0. Moreover, most cities' air quality scores ranged between 0.4–0.6 in the study period, which indicates that the overall air quality of Jiangsu cities was quite good. This is also why the Chinese central government, especially

the Ministry of Environmental Protection, didn't enforce intensified air pollution control policies (such as those on the Beijing-Tianjin-Hebei region) on Jiangsu Province.

(2) However, there are still some concerns regarding the air quality of cities in Jiangsu Province, such as:

- Although the air quality of the three coastal cities, Jiangsu-Nantong, Yancheng, and Lianyungang, was quite good during the study period—especially Nantong, whose air quality score has ranked top one among all Jiangsu cities for 5 months during the six months from October 2017 to March 2018—there have been large seasonal fluctuations in these cities' air quality. Taking Nantong as an example, according to the statistics of Jiangsu Provincial Academy of Environmental Science, in terms of the pollution sources of the six major air pollutants, local pollution sources accounted for 51%–73% (average 62%). In terms of the types of the pollution sources, coal burning accounts for the largest proportion, 26%; mobile pollution sources account for 24%; industrial pollution sources account for 23%, dust pollution accounts for 18%; other "scattered pollution" sources account for 9% [47]. Therefore, although Nantong has adopted a series of control measures in order to improve air quality, including the 39 so-called "strictest in history" relocation (or closure) projects targeting heavy pollution companies that were completed by the end of 2017 [48], its air quality during summers is still quite poor due to the impact of mobile pollution sources and dust pollution, resulting in significant fluctuations in air quality. In addition, although Yancheng's air quality ranked top among Jiangsu cities for four consecutive years according to official statistics, taking all the six major air pollutants into consideration, Yancheng's air quality has also experienced large fluctuations during the study period. Taking the factors of geographical location and wind direction into account, the air pollution in these three cities is greatly affected by the wind in offshore waters. Because of stronger winds in southern Jiangsu compared with the northern regions, the air quality of Nantong, which is located in the south, is generally better than that of Yancheng and Lianyungang in the north. Due to the clear seasonal pattern in wind direction in coastal areas of Jiangsu Province (east to southeast during spring and summer, and northerly winds in autumn and winter) and stronger winds during winter compared with that of summer [49], the air pollutants would linger for a long time above these three cities in summertime due to weaker east and southeast wind than in winter, resulting in worse air quality during summer than in winter.
- There is a clear difference in the air quality of different cities in Jiangsu Province; except for coastal cities, the air quality of southern cities in Jiangsu is generally better than that of the northern cities. During the study period, the air quality of southern cities (such as Suzhou, Wuxi and Changzhou) is generally better than the northern cities (represented by Xuzhou, Suqian and Huai'an). The reasons for this are, on the one hand, Suzhou, Wuxi, and Changzhou enjoy better economic condition and are less dependent on heavy-pollution energy sources such as coal. On the other hand, these cities have adopted low-carbon and energy-saving policies. Taking Suzhou as an example, it successfully decreased the energy consumption per unit of industrial production to 0.917 tons of standard coal per 10,000 yuan in 2010, and its total energy consumption per unit of GDP has also dropped from 1.043 tons of standard coal per 10,000 yuan of GDP in 2005 to 0.824 in 2010, with an average annual decrease of 4.87% [50], which has laid a good foundation for further implementation of air pollution control policies. In the above comparison, the GDP values are inflation-adjusted. While looking at the northern cities of Jiangsu Province, Xuzhou has long relied on coal resources, and there were once more than 250 coal mines in the city. As of 2017, 70% of the mountains in Xuzhou have suffered severe damage, and there are, in total, 381,900 mu of coal mining subsidence land in the city [51], which has not only caused serious air pollution, but has also caused severe damage to land resources. From 2010 to 2015, in the energy consumption structure of industrial companies of Suqian City, coal has taken a proportion of over 70%. In 2016, industrial smoke and dust emissions mainly from coal burning accounted for 46% of the city's total exhaust gas emissions [52]. Hua'an's average annual standard coal consumption

has increased by nearly 10% since 2008, and coal has taken the largest proportion in its energy structure, almost reaching 65% in 2015 [53].

(3) During the study period, the cities adopted a series of air pollution control policies based on the "Regulations on the Prevention and Control of Air Pollution in Jiangsu Province". These policies have shown different effects in different cities with regard to the changes in air quality evaluation scores. More specifically:

- Despite differences in effectiveness, the air pollution control policies have achieved improvements in the nine cities, i.e., Nantong, Changzhou, Wuxi, Yangzhou, Suzhou, Yancheng, Zhenjiang, Tai'an, and Lianyungang. Among them, Nantong's air quality has seen an improvement of 20.28% when comparing that of the ending period (March 2018) to that of April 2015, when the air pollution control policies were first implemented. Changzhou and Wuxi have improved their air quality by more than 10%, while Yangzhou, Suzhou, and Yancheng have improved theirs by more than 5%. These cities have all initiated their own pollution control policies with local characteristics based on the "Regulations on the Prevention and Control of Air Pollution in Jiangsu Province". For example, Nantong has actively promoted the relocation of heavily polluting companies out of its main urban zones, carried out pilot projects for the ultra-low emission transformation of coal-fired power plants, and upgraded the standards for smoke and dust emissions from cement industries and coal-fired boilers [48]. By the end of 2017, the relocation and transformation of all heavily polluting companies in Nantong's central urban area (Chongchuan District) has been completed [54]. Wuxi City has shut down three coal-fired power plants in its urban area, completed the rectification of more than 1200 small coal-fired boilers in its main urban area and subordinate counties, and implemented ultra-low emission transformation for eight large coal-fired power units in the city [55]. Suzhou and Yancheng have also formulated and implemented their annual work plan for air pollution prevention and control based on the "Regulations on the Prevention and Control of Air Pollution in Jiangsu Province", as well as the characteristics of their own pollutants and industrial structure [56,57], which has achieved remarkable results. Meanwhile, although Yangzhou's air quality has improved by 8.89% at the ending period compared with that of the beginning period, there has been clear decline since May 2017. According to the official environmental quality report issued by Yangzhou Municipal Government, the proportion of days with good air quality in Yangzhou City from January to September 2017 was 59.7%, down 12.2 percentage points year-on-year; meanwhile, the indicators on $PM_{2.5}$, PM_{10}, O_3, and NO_2 have all exceeded the standards by varying degrees [58]. The reason behind is that although Yangzhou has formulated the annual work plan for air pollution prevention and control, the implementation of the work plan has not been detailed enough since the end of 2016, resulting in a decline in air quality in 2017. After realizing this problem, Yangzhou local government revised and approved the "Yangzhou City Heavy Air Pollution Early Warning and Emergency Plan" in October 2017, included 31 high-emissions companies in the municipal-level key emission monitoring list, and incorporated the heavy air pollution warning and emergency response into the environmental performance evaluation of the CPC and local government leaders under a system of responsibility and accountability [59]. In December 2017, the provincial-level inspection team for air pollution prevention and control set up by the Jiangsu Provincial Environmental Protection Department officially started their one-month on-site inspection of Yangzhou [60]. As such, Yangzhou's air quality has shown clear improvements since January 2018.
- The air quality of Nanjing, Huai'an, Xuzhou and Suqian has shown different degrees of decline when comparing their ending score with the beginning score. Among them, Suqian's air quality has declined by 20.07%, and Xuzhou's by almost 20% (16.32%). The reason for this is that, on the one hand, for historical and geographical reasons, these cities rely more on coal burning in terms of energy structure and have more heavy-pollution companies. Taking Nanjing as an example, in the study on pollution sources of key monitoring cities for air pollution prevention

and control completed by China's Ministry of Environmental Protection in 2015, the primary pollution source of Beijing, Hangzhou, Guangzhou, and Shenzhen is motor vehicles, while that of Nanjing and Shijiazhuang is coal burning. The biggest consumers of coal in Nanjing are the four high-energy-consumption industries of electricity, steel, petrochemicals, and cement. From 2015 to 2016, Nanjing's total coal consumption exceeded 35 million tons, its sulfur dioxide emissions per unit of GDP ranked top one among sub-provincial cities, and its average chemical oxygen demand (COD) emissions ranked second [61]. As discussed above, the energy structure of Xuzhou, Suqian, and Huai'an also relies heavily on coal combustion. A major action to control air pollution by Xuzhou is to gradually transform the mining area into ecological parks, such as the ecological restoration project of coal mining subsidence areas around Pan'an Lake in the Jiawang District which was completed in 2017. This was the comprehensive project with the largest individual investment in Jiangsu Province since the establishment of People's Republic of China, covering 17,400 mu [62]. In addition, the fluctuations and decline in these cities' air quality is also partly due to the dust and smog spread from northern China since October 2017. With the cold air in northern China moving southward, the dust and smog in northern China has intensified the air pollution in these four non-coastal cities in northern Jiangsu, especially in 2018 [63–66].

5. Conclusions

This paper has studied the air quality of 13 cities in Jiangsu Province from April 2015 to March 2018 based on an expanded AQI system of six major air pollutants. After expanding the existing air quality evaluation standards of China, this paper has calculated the air quality evaluation scores of various cities in Jiangsu based on the six major air pollutants by using the improved Fuzzy Comprehensive Evaluation Model. This paper has further analyzed the effectiveness of air pollution control policies in Jiangsu Province and its different cities during the study period. The conclusions are: there are distinct differences in air quality of different cities in Jiangsu Province; except for the coastal cities such as Nantong, Yancheng and Lianyungang, the southern cities of Jiangsu generally have better air quality than the northern cities. Apart from natural factors such as geographical location, economic conditions and energy structures are also important causes of this situation. In addition, air pollution control policies have achieved significant results in the cities of Nantong, Changzhou, Wuxi, Yangzhou, Suzhou, Yancheng, Zhenjiang, Tai'an, and Lianyungang. Among them, Nantong has achieved the biggest improvement, i.e., 20.28%; Changzhou and Wuxi have improved their air quality by more than 10%, while Yangzhou, Suzhou and Yancheng have improved theirs by more than 5%. However, the air quality of Nanjing, Huai'an, Xuzhou, and Suqian has declined by different degrees when comparing that of the last period with the beginning period: Suqian's air quality has declined by 20.07%, while Xuzhou's has declined by 16.32%.

Based on above findings, this paper provides the following policy recommendations in order to further improve air pollution control of Jiangsu Province.

(1) Fundamentally change the energy structure of Jiangsu cities that are overly-reliant on coal combustion (especially the northern cities) with the latest revision and implementation of the "Regulations on the Prevention and Control of Air Pollution in Jiangsu Province" in April 2018 [44]. Establish long-term treatment measures against air pollution through industry upgrade and technological advancement in order to achieve a long-term and stabilized pollution control performance as well as to minimize the cyclical fluctuations of air quality, and ensure the sustainability in air pollution control.

(2) Take advantage of the trend of regional economic integration of the YRD region and integrate the formulation and implementation of air pollution control policies in Jiangsu Province. Since 2015, there haven't been many integrated measures for air pollution control by Jiangsu Province except the "Regulations on the Prevention and Control of Air Pollution in Jiangsu Province". Therefore, under the overall trend of regional economic integration in the YRD region, it is suggested that Jiangsu Province further improve the information sharing and decision-making

mechanism for air pollution control and treatment among its different cities by using various air pollution control programs such as the "Yangtze River Delta Regional Air Quality Improvement and Treatment Program (2017–2020)" [67], "Key Emphasis in the Cooperation of Yangtze River Delta Regional Air Pollution Prevention and Control (2018)" [68], etc. and by learning from the successful experience of Shanghai, Anhui, and other provinces, in order to fully realize the integration of air pollution control policies in the province.

(3) It is necessary to fully consider the local differences in air pollution of various cities of Jiangsu Province, and make targeted pollution control policies based on coordinated work and different characteristics of each city's air pollution sources and industrial structure. For northern cities such as Xuzhou, Suqian and Huai'an, it is necessary to change the energy structure which is overly-reliant on coal, to strictly restrict the number of new coal mining and coal-fired plants construction projects, and to prohibit various types of loose coal combustion while accelerating the development of clean energy. For cities such as Suzhou, Wuxi, and Changzhou, it is necessary to further improve and optimize the public transportation system, and strictly control the number of motor vehicles in order to curb the growth of mobile pollution sources.

Author Contributions: X.L. organized the whole research and designed the methodology. As the correspondence author, X.G. analyzed the data and results. He also drew the final figures. X.L. and X.G. approved this final version of manuscript together.

Funding: This research received no external funding.

Acknowledgments: X.L. and X.G. acknowledge the Special Issue "Air Quality Assessment Standards and Sustainable Development in Developing Countries" of *Sustainability*.

Conflicts of Interest: The authors declare no conflict of interest.

References

1. Guo, D.; Yu, J.; Ban, M. Security-Constrained Unit Commitment Considering Differentiated Regional Air Pollutant Intensity. *Sustainability* **2018**, *10*, 1433. [CrossRef]
2. Han, L.; Zhou, W.; Li, W. Growing Urbanization and the Impact on Fine Particulate Matter (PM2.5) Dynamics. *Sustainability* **2018**, *10*, 1696. [CrossRef]
3. Sun, Z.; An, C.; Sun, H. Regional Differences in Energy and Environmental Performance: An Empirical Study of 283 Cities in China. *Sustainability* **2018**, *10*, 2303. [CrossRef]
4. Yang, W.; Li, L. Energy Efficiency, Ownership Structure, and Sustainable Development: Evidence from China. *Sustainability* **2017**, *9*, 912. [CrossRef]
5. Yang, W.X.; Li, L.G. Analysis of Total Factor Efficiency of Water Resource and Energy in China: A Study Based on DEA-SBM Model. *Sustainability* **2017**, *9*, 1316. [CrossRef]
6. Yang, W.; Li, L. Efficiency Evaluation and Policy Analysis of Industrial Wastewater Control in China. *Energies* **2017**, *10*, 1201. [CrossRef]
7. Yang, W.X.; Li, L.G. Efficiency evaluation of industrial waste gas control in China: A study based on data envelopment analysis (DEA) model. *J. Clean. Prod.* **2018**, *179*, 1–11. [CrossRef]
8. Wu, D.; Ma, X.; Zhang, S. Integrating synergistic effects of air pollution control technologies: More cost-effective approach in the coal-fired sector in China. *J. Clean. Prod.* **2018**, *199*, 1035–1042. [CrossRef]
9. Wang, B.; Liu, Y.; Li, Z.; Li, Z. Association of indoor air pollution from coal combustion with influenza-like illness in housewives. *Environ. Pollut.* **2016**, *216*, 646–652. [CrossRef] [PubMed]
10. Li, L.G.; Yang, W.X. Total Factor Efficiency Study on China's Industrial Coal Input and Wastewater Control with Dual Target Variables. *Sustainability* **2018**, *10*, 2121. [CrossRef]
11. Guo, X.; Guo, X.; Yuan, J. Impact Analysis of Air Pollutant Emission Policies on Thermal Coal Supply Chain Enterprises in China. *Sustainability* **2015**, *7*, 75–95. [CrossRef]
12. Zhao, H.; Ma, W.; Dong, H.; Jiang, P. Analysis of Co-Effects on Air Pollutants and CO_2 Emissions Generated by End-of-Pipe Measures of Pollution Control in China's Coal-Fired Power Plants. *Sustainability* **2017**, *9*, 499. [CrossRef]
13. Bai, R.; Lam, J.C.K.; Li, V.O.K. A review on health cost accounting of air pollution in China. *Environ. Int.* **2018**, *120*, 279–294. [CrossRef] [PubMed]

14. Tian, Y.; Liu, H.; Liang, T.; Xiang, X.; Li, M.; Juan, J.; Song, J.; Cao, Y.; Wang, X.; Chen, L.; et al. Ambient air pollution and daily hospital admissions: A nationwide study in 218 Chinese cities. *Environ. Pollut.* **2018**, *242*, 1042–1049. [CrossRef] [PubMed]

15. Liu, M.; Huang, Y.; Ma, Z.; Jin, Z.; Liu, X.; Wang, H.; Liu, Y.; Wang, J.; Jantunen, M.; Bi, J.; et al. Spatial and temporal trends in the mortality burden of air pollution in China: 2004–2012. *Environ. Int.* **2017**, *98*, 75–81. [CrossRef] [PubMed]

16. Chen, X.; Shao, S.; Tian, Z.; Xie, Z.; Yin, P. Impacts of air pollution and its spatial spillover effect on public health based on China's big data sample. *J. Clean. Prod.* **2017**, *142*, 915–925. [CrossRef]

17. Yang, T.; Liu, W. Does air pollution affect public health and health inequality? Empirical evidence from China. *J. Clean. Prod.* **2018**. [CrossRef]

18. Yang, B.; Qian, Z.; Li, S.; Fan, S.; Chen, G.; Syberg, K.M.; Xian, H.; Wang, S.; Ma, H.; Chen, D.; et al. Long-term exposure to ambient air pollution (including PM1) and metabolic syndrome: The 33 Communities Chinese Health Study (33CCHS). *Environ. Res.* **2018**, *164*, 204–211. [CrossRef] [PubMed]

19. Zhu, Y.; Tao, S.; Sun, J.; Wang, X.; Li, X.; Tsang, D.C.W.; Zhu, L.; Shen, G.; Huang, H.; Cai, C.; et al. Multimedia modeling of the PAH concentration and distribution in the Yangtze River Delta and human health risk assessment. *Sci. Total Environ.* **2019**, *647*, 962–972. [CrossRef] [PubMed]

20. Environmental Protection Department of Jiangsu Province. *Jiangsu Province Environmental Status Bulletin, 2013–2017*; Environmental Protection Department of Jiangsu Province: Nanjing, China, 2018.

21. Environmental Protection Department of Jiangsu Province. *Jiangsu Province Environmental Status Bulletin, 2017*; Environmental Protection Department of Jiangsu Province: Nanjing, China, 2018.

22. Dai, J. *Research on the Effects of Financial and Tax Subsidy Policy on the Atmospheric Haze Governance in Jiangsu Province*; Nanjing University of Science and Technology: Nanjing, China, 2017.

23. Wang, Y.; Wang, Q.; Hang, Y. Driving Factors of SO2 Emissions in 13 Cities, Jiangsu, China. *Energy Procedia* **2016**, *88*, 182–186. [CrossRef]

24. Ge, Y.; Zhang, H.; Dou, W.; Chen, W.; Liu, N.; Wang, Y.; Shi, Y.; Rao, W. Mapping Social Vulnerability to Air Pollution: A Case Study of the Yangtze River Delta Region, China. *Sustainability* **2017**, *9*, 109. [CrossRef]

25. He, L.; Zhong, Z.; Yin, F.; Wang, D. Impact of Energy Consumption on Air Quality in Jiangsu Province of China. *Sustainability* **2018**, *10*, 94. [CrossRef]

26. Zhang, G.; Liu, D.; He, X.; Yu, D.; Pu, M. Acid rain in Jiangsu province, eastern China: Tempo-spatial variations features and analysis. *Atmos. Pollut. Res.* **2017**, *8*, 1031–1043. [CrossRef]

27. Xu, S.; Zhang, L.; Liu, Y.; Zhang, W.; He, Z.; Long, R.; Chen, H. Determination of the factors that influence increments in CO2 emissions in Jiangsu, China using the SDA method. *J. Clean. Prod.* **2017**, *142*, 3061–3074. [CrossRef]

28. Chen, K.; Zhou, L.; Chen, X.; Bi, J.; Kinney, P.L. Acute effect of ozone exposure on daily mortality in seven cities of Jiangsu Province, China: No clear evidence for threshold. *Environ. Res.* **2017**, *155*, 235–241. [CrossRef] [PubMed]

29. Wang, Q.; Wang, Y.; Zhou, P.; Wei, H. Whole process decomposition of energy-related SO$_2$ in Jiangsu Province, China. *Appl. Energy* **2017**, *194*, 679–687. [CrossRef]

30. Ministry of Environmental Protection of the People's Republic of China. *Ambient Air Quality Standards: GB3095-2012*; China Environmental Science Press: Beijing, China, 2012.

31. Ministry of Environmental Protection of the People's Republic of China. *Technical Regulation on Ambient Air Quality Index (on Trial): HJ 633-2012*; China Environmental Science Press: Beijing, China, 2012.

32. Xinhuanet. Hebei's Air Quality "Burst" and PM2.5 Index in Shijiazhuang Is Over 1000. Available online: http://www.xinhuanet.com/local/2016-12/19/c_129411383.htm (accessed on 21 April 2018).

33. Cannistraro, G.; Cannistraro, M.; Cannistraro, A.; Galvagno, A.; Engineer, F. Analysis of air pollution in the urban center of four cities Sicilian. *Int. J. Heat Technol.* **2016**, *34*, S219–S225. [CrossRef]

34. Cao, Y.; Wang, X.; Yin, C.; Xu, W.; Shi, W.; Qian, G.; Xun, Z. Inland Vessels Emission Inventory and the emission characteristics of the Beijing-Hangzhou Grand Canal in Jiangsu province. *Process Saf. Environ. Prot.* **2018**, *113*, 498–506. [CrossRef]

35. Yang, Y.; Christakos, G.; Yang, X.; He, J. Spatiotemporal characterization and mapping of PM2.5 concentrations in southern Jiangsu Province, China. *Environ. Pollut.* **2018**, *234*, 794–803. [CrossRef] [PubMed]

36. Loh, H.S.; Zhou, Q.; Thai, V.V.; Wong, Y.D.; Yuen, K.F. Fuzzy comprehensive evaluation of port-centric supply chain disruption threats. *Ocean Coast. Manag.* **2017**, *148*, 53–62. [CrossRef]

37. Wang, M.; Niu, D. Research on project post-evaluation of wind power based on improved ANP and fuzzy comprehensive evaluation model of trapezoid subordinate function improved by interval number. *Renew. Energy* **2019**, *132*, 255–265. [CrossRef]

38. Chen, C.; Qi, M.; Kong, X.; Huang, G.; Li, Y. Air pollutant and CO2 emissions mitigation in urban energy systems through a fuzzy possibilistic programming method under uncertainty. *J. Clean. Prod.* **2018**, *192*, 115–137. [CrossRef]

39. Debnath, J.; Majumder, D.; Biswas, A. Air quality assessment using weighted interval type-2 fuzzy inference system. *Ecol. Inform.* **2018**, *46*, 133–146. [CrossRef]

40. D'Urso, P.; Massari, R.; Cappelli, C.; Giovanni, L. De Autoregressive metric-based trimmed fuzzy clustering with an application to PM10 time series. *Chemom. Intell. Lab. Syst.* **2017**, *161*, 15–26. [CrossRef]

41. Wang, J.; Li, H.; Lu, H. Application of a novel early warning system based on fuzzy time series in urban air quality forecasting in China. *Appl. Soft Comput.* **2018**, *71*, 783–799. [CrossRef]

42. Xu, Y.; Du, P.; Wang, J. Research and application of a hybrid model based on dynamic fuzzy synthetic evaluation for establishing air quality forecasting and early warning system: A case study in China. *Environ. Pollut.* **2017**, *223*, 435–448. [CrossRef] [PubMed]

43. The Twelfth People's Congress of Jiangsu Province. *Air Pollution Prevention Ordinance of Jiangsu Province*; The Twelfth People's Congress of Jiangsu Province: Nanjing, China, 2015.

44. The Thirteenth People's Congress of Jiangsu Province. *Decision on Modifying 16 Local Regulations, including the Air Pollution Prevention Ordinance of Jiangsu Province*; The Twelfth People's Congress of Jiangsu Province: Nanjing, China, 2018.

45. China's National Environmental Monitoring Center. The City Air Quality Publishing Platform. Available online: http://106.37.208.233:20035/ (accessed on 17 July 2018).

46. Data Center of China's Ministry of Environmental Protection. Concentration of Main Pollutants in the PRD Region, 2015–2018. Available online: http://datacenter.sepa.gov.cn/ (accessed on 17 July 2018).

47. Jiangsu Provincial Academy of Environmental Science. *Air Pollution Source List and Air Pollutant Source Analysis Project for Nantong*; Jangsu Provincial Academy of Environmental Science: Nanjing, China, 2015.

48. Nantong Municipal Government. Notice on Printing and Distributing the 2017 Work Plan for Air Pollution Prevention and Control in Nantong. Available online: http://www.nantong.gov.cn//ntsrmzf/szfbwj/content/5ca1e69c-7fa1-4f8c-b74c-6df435090b3d.html (accessed on 26 August 2018).

49. Gao, X.; Zheng, Y.; Wang, J.; Cao, B. Temporal and Spatial Distribution Characteristics of Wind Elements in the Offshore Area of Jiangsu Province. *China Water Transp.* **2015**, *15*, 193–195.

50. Suzhou Municipal Government. Notice on Printing and Distributing the Low Carbon Development Plan of Suzhou. Available online: http://www.zfxxgk.suzhou.gov.cn/sxqzf/szsrmzf/201403/t20140314_366113.html (accessed on 26 August 2018).

51. Department of Land and Resources of Jiangsu Province. Reply to Recommendation No. 3062 of the First Session of the 13th National People's Congress of Jiangsu Province. Available online: http://www.jsmlr.gov.cn/gtxxgk/nrgllndex.action?type=2&messageID=2c90825464622937016464aefa0a0158 (accessed on 26 August 2018).

52. Wu, J.; Sun, D. Thoughts on the Prevention and Control of Environmental Air Pollution in Suqian City. *J. Green Sci. Technol.* **2016**, *4*, 61–62.

53. Shang, Z.; Gong, S. Characteristics of Industrial Energy Consumption and Carbon Emissions in Huai'an. *Territ. Nat. Resour. Study* **2016**, *6*, 75–79.

54. Nantong Municipal Government. Notice on Printing and Distributing the 2018 Work Plan for Air Pollution Prevention and Control in Nantong. Available online: http://www.nantong.gov.cn/ntsrmzf/szfbwj/content/074b86c9-f6a6-4fbd-9613-6ace39c4c45e.html (accessed on 26 August 2018).

55. Wuxi Municipal Government. *The Annual Air Pollution Prevention and Control Plan of Wuxi*; Wuxi Municipal Government: Wuxi, China, 2017.

56. Suzhou Municipal Government. Notice on Printing and Distributing the 2016 Annual Work Task Plan for the Prevention and Control of Air Pollution in Suzhou. Available online: http://www.zfxxgk.suzhou.gov.cn/sxqzf/szsrmzf/201702/t20170204_840900.html (accessed on 26 August 2018).

57. Yancheng Municipal Government. The 2016 Environmental Status Bulletin of Yancheng. Available online: http://www.jsychb.gov.cn/2017/ggtz_0605/5821.html (accessed on 26 August 2018).

58. Yangzhou Municipal Government. *The Environmental Quality Report of Yangzhou City in the Third Quarter of 2017*; Yangzhou Municipal Government: Yangzhou, China, 2017.

59. Yangzhou Municipal Government. *Heavy Pollution Warning and Contingency Plan for Air Pollution of Yangzhou City (Revised in September 2017)*; Yangzhou Municipal Government: Yangzhou, China, 2017.

60. Yangzhou Environmental Protection Bureau. The provincial air pollution inspection group entered Yangzhou. Available online: http://hbj.yangzhou.gov.cn/yzhbjceshi/gzdt/2017-12/12/content_8ea24c0f73404e5680453543c7071ed3.shtml (accessed on 21 April 2018).

61. CPC Jiangsu Provincial Committee. The Ministry of Environmental Protection Announced the Source of Air Pollution in the Nine Cities: The Primary Source of Pollution in Nanjing Is Coal. Available online: http://www.zgjssw.gov.cn/shixianchuanzhen/nanjing/201504/t2086578.shtml (accessed on 26 August 2018).

62. Yao, X. Pan'an Lake in Jiawang District of Xuzhou is transformed into a large garden by coal mining subsidence area. *People's Dly, 15 December 2017*. p. 14. Available online: http://js.people.com.cn/n2/2017/1215/c360304-31034826.html (accessed on 21 April 2018).

63. China Jiangsu Website. Results of Air Pollution Prevention Special Action on 2018 New Year's Day in Nanjing. Available online: http://jsnews.jschina.com.cn/jsyw/201801/t20180101_1311491.shtml (accessed on 21 April 2018).

64. Cannistraro, G.; Cannistraro, M.; Cannistraro, A.; Galvagno, A.; Trovato, G.; Engineer, F. Reducing the Demand of Energy Cooling in the CED, 'Centers of Processing Data', with Use of Free-Cooling Systems. *Int. J. Heat Technol.* **2016**, *34*, 498–502. [CrossRef]

65. Piccolo, A.; Siclari, R.; Rando, F.; Cannistraro, M. Comparative performance of thermoacoustic heat exchangers with different pore geometries in oscillatory flow. implementation of experimental techniques. *Appl. Sci.* **2017**, *7*, 784. [CrossRef]

66. Cannistraro, M.; Castelluccio, M.E.; Germanò, D. New sol-gel deposition technique in the Smart-Windows–Computation of possible applications of Smart-Windows in buildings. *J. Build. Eng.* **2018**, *19*, 295–301. [CrossRef]

67. The Air Pollution Prevention and Control Coordination Group of the YRD Region. *Further Plan for Deepening the Air Quality Improvement Work of the YRD Region (2017–2020)*; The Air Pollution Prevention and Control Coordination Group of the YRD Region: Suzhou, China, 2017.

68. The Air Pollution Prevention and Control Coordination Group of the YRD Region. *Priorities in Air Pollution Prevention and Control Coordination Work of the YRD Region (2018)*; The Air Pollution Prevention and Control Coordination Group of the YRD Region: Suzhou, China, 2018.

sustainability

MDPI

Article

Methane, Nitrous Oxide and Ammonia Emissions from Livestock Farming in the Red River Delta, Vietnam: An Inventory and Projection for 2000–2030

An Ha Truong [1,2], Minh Thuy Kim [1], Thi Thu Nguyen [1], Ngoc Tung Nguyen [1] and Quang Trung Nguyen [1,*]

[1]   Center for Research and Technology Transfer, Vietnam Academy of Science and Technology, 18-Hoang Quoc Viet, Cau Giay, 100000 Hanoi, Vietnam; truonganha87@gmail.com (A.H.T.); thuythuy1107@gmail.com (M.T.K.); thu.nguyen27393@gmail.com (T.T.N.); tungnguyen.vast@gmail.com (N.T.N.)
[2]   University of Science and Technology of Hanoi, Vietnam Academy of Science and Technology, 18-Hoang Quoc Viet, Cau Giay, 100000 Hanoi, Vietnam
*   Correspondence: trungnq@ctctt.vast.vn; Tel.: +84-243-756-8422

Received: 29 September 2018; Accepted: 18 October 2018; Published: 22 October 2018

Abstract: Livestock farming is a major source of greenhouse gas and ammonia emissions. In this study, we estimate methane, nitrous oxide and ammonia emission from livestock sector in the Red River Delta region from 2000 to 2015 and provide a projection to 2030 using IPCC 2006 methodologies with the integration of local emission factors and provincial statistic livestock database. Methane, nitrous oxide and ammonia emissions from livestock farming in the Red River Delta in 2030 are estimated at 132 kt, 8.3 kt and 34.2 kt, respectively. Total global warming potential is estimated at 5.9 MtCO$_{2eq}$ in 2030 and accounts for 33% of projected greenhouse gas emissions from livestock in Vietnam. Pig farming is responsible for half of both greenhouse gases and ammonia emissions in the Red River Delta region. Cattle is another major livestock responsible for greenhouse gas emissions and poultry is one that is responsible for ammonia emissions. Hanoi contributes for the largest emissions in the region in 2015 but will be surpassed by other provinces in Vietnam by 2030.

Keywords: emission inventory; livestock; greenhouse gases; air pollutant

1. Introduction

Economic growth in Vietnam has shifted food consumption patterns to incorporate more livestock products (meat, dairy products, and eggs) [1]. With the growing demand for livestock products, livestock farming is expanding in Vietnam and is among the fastest growing agricultural production subsectors in Vietnam [1]. In 2015, livestock accounted for 28% of value added from the agriculture sector. The development and intensification of this subsector has led to an increase in the total animal population during the past decade. In 2016, Vietnam had 29 million pigs, 5.5 million cattle, 2.5 million buffalos, and 361 million poultry [2]. The largest population increases compared to 2005 have been in poultry (increased by 65% with 142 million head added), followed by pigs (increased by 8% with 2 million head added); while numbers of cattle and buffalo have fluctuated slightly.

The development and intensification of livestock farming helps to ensure national food security and boosts economic growth. However, this sector is also a significant contributor to environmental pollution in general and air pollution in particular. Livestock farming contributes significantly to global greenhouse gas (GHG) emissions [3] by releasing methane (CH$_4$) and nitrous oxide (N$_2$O), as well as air pollutants, mostly ammonia (NH$_3$), into the atmosphere. Livestock farming is the largest emissions source of NH$_3$ [4–6], which plays a major role in eutrophication and acidification [7]. The Food and

Agriculture Organization (FAO) has estimated that 18% of global GHG emissions originate from the livestock sector.

Vietnam is listed among the 20 countries with the highest GHG emissions in the UNFCCC and FAOSTAT databases [8]. Emissions from livestock farming account for approximately 20% of greenhouse gas emissions from agricultural activities in Vietnam according to the National GHG emissions inventory for 2010 [9]. Emissions from enteric digestion are responsible for half of all livestock emissions, with the other half originating from manure management, one of the fastest-growing sources of GHG emissions in Vietnam during 1994–2010 [5]. An inventory of CH_4 emissions from livestock in Asia in 2000 [10] showed that poultry emitted the largest amounts of CH_4 in Vietnam, followed by cattle, buffalo, and pigs. A CH_4 and N_2O emissions inventory for South, Southeast, and East Asia was recently conducted [11] using emissions inventory methodologies from the International Panel on Climate Change (IPCC) 1997 Guidelines for National Emission Inventory, and ranked Vietnam in 6th place for NH_3 emissions and 7th place for CH_4 and N_2O emissions among the 23 countries studied. An estimate of air pollutants and GHGs over Asia aggregated Vietnam within the Southeast Asia region [12]. To the best of our knowledge, no emissions inventory has been conducted for CH_4 and N_2O in Vietnam using IPCC 2006 methodologies. Previous studies estimating livestock farming emissions in Vietnam have been conducted at the provincial scale or for one type of pollutant (such as GHG or air pollutant). Examples of such studies include estimates of CH_4 emissions from cattle in Daklak province [13], CH_4 emissions from cattle in Quang Ngai province, with mitigation scenarios [14], and GHG and pollutants from livestock farming within a ward of Hung Yen province [15].

It is important to develop a historical inventory and projections of future livestock GHG and air pollutants to improve our understanding of the evolution of emissions and their associated impact on air quality. In this study, we focused on the Red River Delta (RRD) region, which is among the largest livestock farming centers in Vietnam. This region contained 8726 livestock farms in 2016, accounting for 42% of all livestock farms in the country [2]. RRD contains the largest number of pigs and poultry, with populations of 7.4 million and 93.7 million head, respectively (account for 26% of country's total). This inventory attempts to quantify emissions of CH_4, N_2O, and NH_3 produced by livestock farming, in RRD from 2000 to 2030 at a 5-years resolution using the IPCC 2006 Guidelines for National Emission Inventory [16] and regional or country-specific emission factors wherever applicable. Its results are designed to provide input to more comprehensive studies about regional air quality, for example using an air dispersion model and the Greenhouse Gases—Air Pollution Interactions and Synergies (GAINS) model.

2. Materials and Methods

2.1. Emission Inventory Methodology

We conducted an emissions inventory for livestock farming for the sources and pollutant species listed in Table 1. We applied emissions inventory methodologies from the IPCC Guidelines for the National Emission Inventory [16]. In general, we applied Tier 1 methods, such that activity data were multiplied by relevant emission factors. Country-specific emission factors were used (Tier 2 method) wherever applicable. The general equation for estimating livestock emission is Equation (1) [17].

$$E_j = \sum_T N_T \times EF_{T,j} \tag{1}$$

where E_j is the emission from animal type T and pollutant j; N_T is the number of animal of type T, $EF_{T,j}$ is the emission factor for animal type T for pollutant j.

Equation (2) from IPCC 2006 Guidelines was used to estimate direct N_2O emissions from manure management.

$$E_{N2O} = \left[\sum_S \left[\sum_T (N_T \times Nex_T \times MS_{T,S}) \right] \times EF_S \right] \times \frac{44}{28} \qquad (2)$$

where N_T is number of animal type T, Nex_T is the annual average Nitrogen excretion per head of animal type T. Nex_T is calculated using Equation (3), where $N_{rate\ T}$ is the default Nitrogen excretion rate; TAM is the typical animal mass for animal type T. Both values are provided in the IPCC 2006 Guidelines. Value of Nex_T for animals in Asia are listed in Table 2. $MS_{T,S}$ is the fraction of total annual nitrogen excretion for each animal of type T in manure management system S. $MS_{T,S}$ is provided in Table 3. EF_S is default emission factor for direct N_2O emission from manure management system S (Table 3). 44/28 is the conversion of N_2O-N emissions to N_2O emissions.

$$Nex_T = N_{rate\ T} \times TAM/1000 \times 365 \qquad (3)$$

Table 1. Activities and pollutant species included in the inventory.

Source/Pollutant	CH$_4$	N$_2$O	NH$_3$
Enteric fermentation	✓		
Manure management	✓	✓	✓

Table 2. Nitrogen (N) excretion per animal type (kgN head^{-1} yr^{-1}).

Animal	$N_{rate\ T}$ (kgN/1000 kg Animal Mass/Day) [16]	TAM (kg/Animal) [16]	Nex_T (kgN/Head/yr) [16]
Dairy cattle	0.47	350	60.043
Other cattle	0.34	319	39.588
Pigs	0.42	28	4.292
Poultry	0.82	1.8	0.539
Goats	1.37	30	15.002
Horses	0.46	238	39.960
Buffalo	0.32	380	44.384

N_{rateT}, default N excretion rate; TAM, typical animal mass for animal of type T; Nex_T, annual average N excretion per head of animal of type T.

Table 3. Fraction of total annual N excretion for each animal type and emission factors by manure management system.

Manure Management System	Fraction of Total Annual N Excretion (kg N Excreted) by Manure Management System							Emission Factor kg N2O-N/kg N Excreted
	Dairy Cattle	Other Cattle	Pig	Horse	Goat	Buffalo	Poultry	
Pasture/range	0.20	0.50	-	1	1	0.50	-	-
Daily spread	0.29	0.02	-	-	-	0.04	0.55	0
Solid storage	-	-	0.15	-	-	-	-	0.005
Dry lot	0.07	0.48	-	-	-	0.46	-	0.02
Liquid/slurry	0.38	-	0.15	-	-	-	-	0.05
Anaerobic lagoon	0.04	-	-	-	-	-	-	0
Pit storage	-	-	-	-	-	-	-	0.002
Anaerobic digester	0.02	-	0.30	-	-	-	-	0
Composting static pile	-	-	0.40	-	-	-	-	0.006
Poultry manure with litter	-	-	-	-	-	-	0.45	0.001

2.2. Data

The RRD region consists of 11 provinces and two cities, including Hanoi, the capital of Vietnam. In our inventory, historical activity data from 2000 to 2015 was acquired at the provincial level and summed to obtain regional data. Projected activities were obtained from approved provincial, regional, and national agricultural development plans.

Historical data on provincial livestock numbers were obtained from the Statistical Yearbook of each province and from the Vietnam Statistical Yearbook [18,19]. More detailed data (e.g., numbers of dairy cattle and laying hens) were obtained from the Department of Livestock Production, Ministry of Agriculture and Rural Development, and are publicly accessible [20]. Livestock is classified into the following groups: Dairy cattle, other cattle, pigs, poultry, horses, and goats. Data on livestock number by province is provided in Table S1, Supplementary Materials.

Projected livestock numbers for 2020 were obtained from the Provincial Agriculture Development Plan for each province. Projections for 2030 were not available for all provinces examined in this study; therefore, we distributed the projected livestock populations for Vietnam in 2030 [21] into these provinces using the proportion of each type of animal of each province over the national total in 2020. Projections for 2025 take the average of 2020 and 2030 values. Historical and projected livestock populations are presented in Table 4.

Table 4. Livestock population data used in this emissions inventory.

Animal (10^3 Head)	2000	2005	2010	2015	2020	2025	2030
Dairy Cattle	13.5	20.0	19.3	48.3	45.3	63.7	77.5
Other cattle	489.4	689.9	604.0	445.4	754.1	757.7	871.4
Poultry	54,742	59,597	76,394	90,829	97,686	109,352	124,153
Horses	1.5	1.3	1.8	0.9	1.0	1.0	0.9
Goat	10.5	10.5	75.6	79.1	96.8	112.1	129.6
Buffalo	278.1	209.1	168.7	130.4	130.2	108.4	108.7
Pig	5688	7796	7301	7061	9326	9906	10,476

Distributions of N excretion for each animal type managed under different manure management systems are provided in Table 3. We estimated the proportion of manure by type of management system for the pig and poultry industries using results from previous studies [22–24]. We used default values from the IPCC 2006 Guidelines for cattle, buffalo, and other animals.

2.3. Emission Factors

A summary of the CH_4 and NH_3 emission factors used in this study is provided in Table 5. We used regional emission factors for CH_4 emission fromenteric fermentation from previous studies for dairy and beef cattle [25,26] and buffalo [15]. We used the IPCC 2006 default values for all other animals. We used the IPCC 2006 Guidelines for N_2O, in which emission factors were specified for manure management systems (Table 3). We used an adjusted European NH_3 emission factors [11], which were also used in a previous study [15] for Vietnam.

Table 5. Methane (CH_4) and ammonia (NH_3) emission factors.

		Dairy Cattle	Other Cattle	Pig	Horse	Goat	Buffalo	Poultry
Enteric fermentation								
CH_4 (kg/head^{-1}yr^{-1})	used in this study	94.5 [a]	41 [a]	1 [c]	18 [c]	5 [c]	82.3 [b]	-
	IPCC 2006 [c]	68	47	1	18	5	55	-
Manure Management								
CH_4 (kg/head^{-1}yr^{-1})		26 [c]	1 [c]	6 [c]	1.64 [c]	0.17 [c]	2 [c]	0.02 [c]
NH_3 (kg/head^{-1}yr^{-1})		5.6 [b]	3 [b]	1.5 [b]	7 [b]	1.1 [b]	3.4 [b]	0.12 [b]

[a] [26]; [b] [15]; [c] [16]

3. Results

3.1. Estimated Total Emissions

Figure 1 shows the estimated CH_4, N_2O and NH_3 emissions from livestock farming in RRD. Total CH_4 emissions in 2015 were 87 kt (Figure 1a), or 2.4 Mt CO_2 equivalent (CO_{2eq}) as determined using the global warming potential (GWP) for 100-year time horizon from IPCC Fifth Assessment Report [27].

Given the current agricultural development plan, the total amount of CH_4 emissions for RRD in 2030 was estimated at 132 kt, or 3.7 Mt CO_{2eq}. Decreases in CH_4 emissions in 2010 and 2015 were due to decreases in numbers of other cattle and buffalo in those years. Although the buffalo population continued to decrease in the subsequent years, increases in the numbers of other animals kept CH_4 emissions on an upward trend from 2020 onward. Enteric fermentation and manure management contributed equally to total CH_4 emissions, which were estimated at 63 and 69 kt, respectively, for 2030. N_2O emissions showed an upward trend, reaching 8.3 kt by 2030 (Figure 1b), or 2.2 Mt CO_{2eq}. Although N_2O emissions were 16-fold lower than those of CH_4, higher GWP limited the global warming impacts of N_2O to 1.6 times lower than those of CH_4.

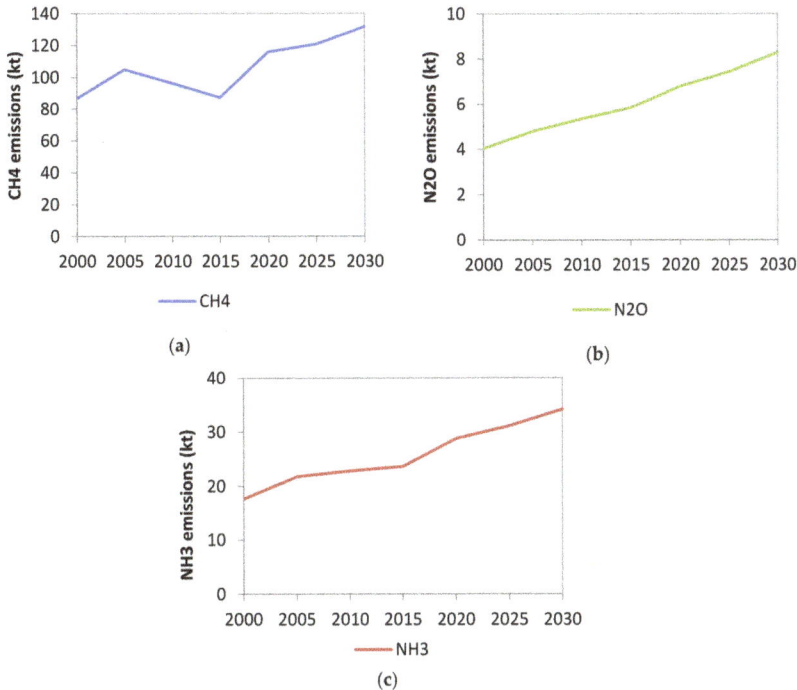

Figure 1. Total CH_4 (a), N_2O (b) and NH_3 (c) emissions from livestock farming in the RRD region.

The total GHG emissions from livestock from our estimation for RRD region in 2015 is 4.0 MtCO_{2eq}. GHG emissions projection for 2020 and 2030 are 5.0 and 5.9MtCO_{2eq}, respectively. Our estimations indicates that RRD region accounts for about one-third of Vietnam's GHG emissions from livestock farming according to the national GHG inventory [9]. This result reflects the fact that RRD is the largest livestock farming center in Vietnam. Compare with a previous estimate [28], as summarized in Reference [1], our estimation results in much higher GHG emissions. Total GHG emissions from livestock in RRD is estimated at 2.1 MtCO_{2eq} in 2012 in the study of Reference [28] while our estimation for 2010 is 4.1 MtCO_{2eq}.

NH_3 emissions increased over time as the animal population expanded during the past decade, and are projected to further increase until 2030 (Figure 1c). By 2030, total NH_3 emissions from livestock in RRD are expected to reach 34 kt.

3.2. Emissions by Animal Type

CH_4 emissions from enteric fermentation and manure management were of the same order of magnitude. However, the contributions differed by animal type in these emissions categories. Cattle contributed the largest proportion of CH$_4$ emissions from enteric digestion (Figure 2a). Modifying diet is one of the mitigation option for methane emissions from enteric fermentation [29]. Several studies have explored the potential for emissions reduction by changing cattle diet at a local scale [13,25,26]. The National Plan for GHG emissions reduction in the agricultural sector by 2020 [30] has proposed two measures to reduce enteric fermentation emissions: (i) changing the feeding proportion in 30% of total amount of animal feed to reduce 0.91 MtCO$_{2eq}$ (3.73% GHG emissions in livestock production projected for 2020) and (ii) supply Molasses Urea Block for 192,000 dairy cattle for a reduction of 0.37 MtCO$_{2eq}$ (1.51% GHG emissions in livestock production projected for 2020). However, the practice of implementing those mitigation measures nation-wide is not yet documented.

CH$_4$ emissions from manure management are produced mainly from pig farming (Figure 2b). Pig husbandry emits 50 kt of methane in 2015, accounts for 57% total methane emission. The dominance of pig farming in CH$_4$ emission suggests that more effort should be made to effectively mitigate emissions in this sector, as RRD has the highest pig farming density in Vietnam. The most common method of emissions mitigation in Vietnam is the production of biogas from pig manure due to subsidization of anaerobic digester construction by the government [31].

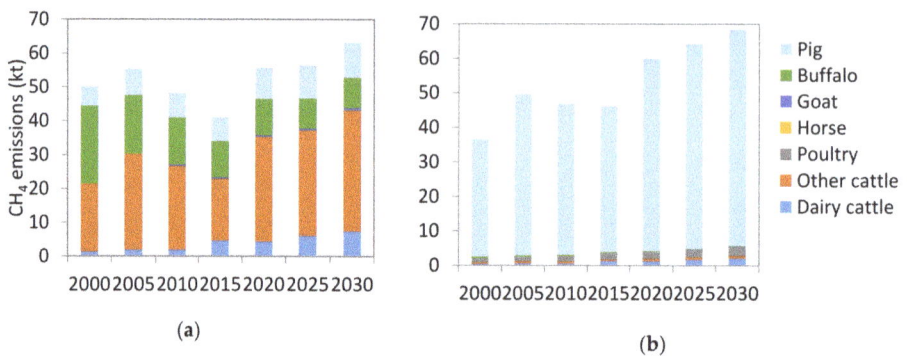

Figure 2. CH$_4$ emissions from enteric fermentation (**a**) and manure management (**b**).

Poultry and pig farming are responsible for about 90% of N$_2$O (Figure 3) and NH$_3$ emissions (Figure 4). Poultry accounted for largest share of N$_2$O emissions (60%) followed by pigs (26%). The farming of these animals contributed equally to NH$_3$ emissions. Although chicken manure is a preferred source of organic fertilizer [1], the remaining uncollected poultry manure apparently has a considerable impact. GHG emissions from poultry husbandry accounted for 27% of total GHG emissions from livestock farming in 2015.

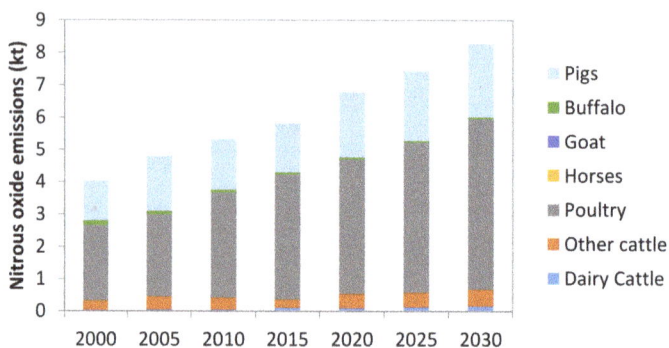

Figure 3. Nitrous oxide (N$_2$O) emissions by animal type.

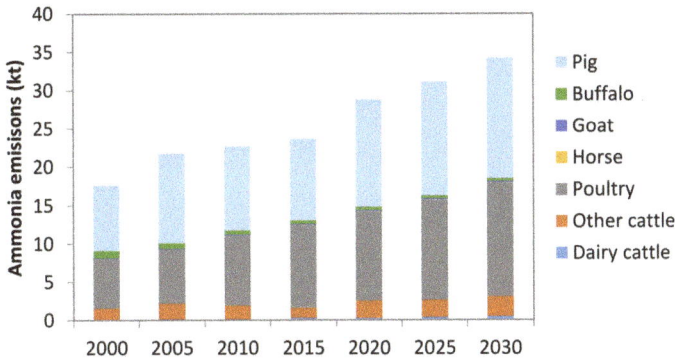

Figure 4. Ammonia (NH₃) emissions by type of animal.

3.3. Emissions by Provinces

Figure 5 presents methane, nitrous oxide and ammonia emissions by provinces in the Red River Delta in 2015 and projection for 2030. In 2015, Hanoi is the dominant city for emissions (Figure 5a) with 21 kt of methane, 1 kt of nitrous oxide and 5.5 kt of ammonia (Tables A1–A3). However, Quang Ninh becomes the highest emission province in the RRD by 2030, followed closely by Hanoi, Vinh Phuc and Thai Binh (Figure 5b, Tables A1–A3). This projection reflects that livestock farming will be developed more in other provinces rather than in the capital city.

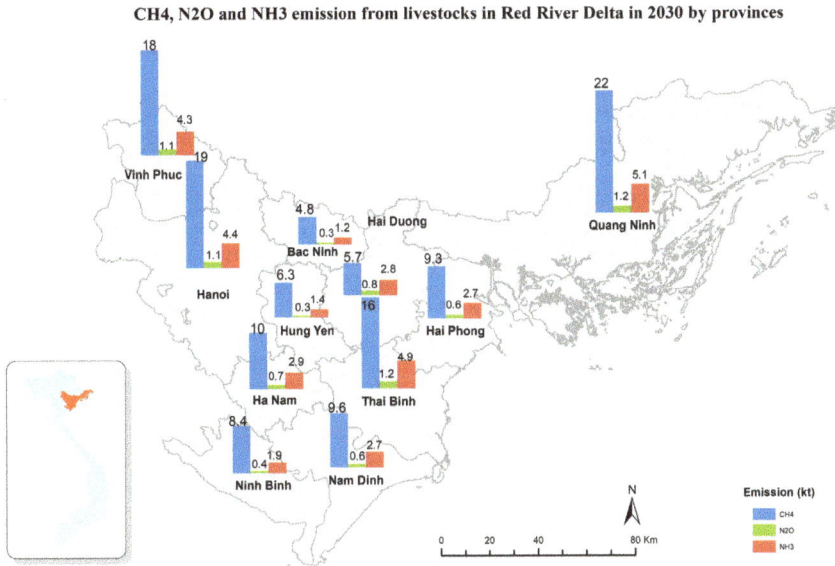

(a)

Figure 5. *Cont.*

CH4, N2O and NH3 emissions from livestocks in Red River Delta in 2015 by provinces

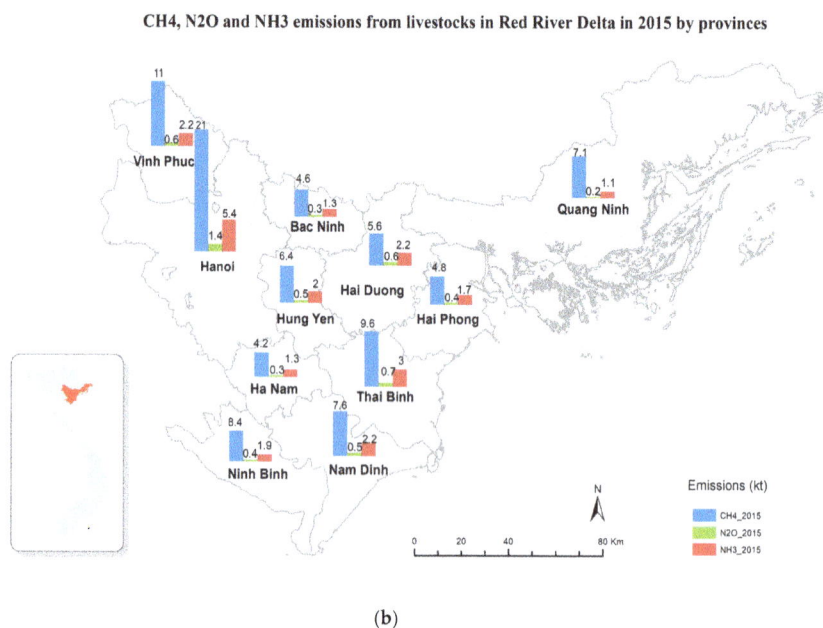

(b)

Figure 5. Emissions from livestock by provinces in 2015 (**a**) and projection for 2030 (**b**).

4. Discussion

Emission factor is a very important element to the accuracy of the estimations in emission inventory. Default methane emission factors for enteric fermentation in IPCC 2006 Guidelines for Asia is 68 kg head^{-1} yr^{-1} for dairy cattle and 47 kg head^{-1} yr^{-1} for other cattle. We used emission factors from studies of References [25,26], which were derived from the RUMINANT model (Tier 3 methodology). These emission factors are higher for dairy cattle and lower for other cattle compared to the default values in IPCC 2006 (see Table A4, Appendix A). These emission factor discrepancies were mainly due to higher milk yields from dairy cattle and lower weight in beef cattle in the studied area. Another study [10] used IPCC 1997 default emission factors, which are lower than IPCC 2006 values for both dairy and non-dairy cattle.

Previous studies [11,15] have used a manure management CH$_4$ emission factor of 16 kg head^{-1} year^{-1} for dairy cattle in a temperate climate region with annual average temperature ranging from 15 to 25 °C. However, the IPCC 2006 guidelines provide CH$_4$ emission factors for temperatures classified at a finer scale. We calculated the annual average temperature for the RRD region to be approximately 25 °C using historical data from three monitoring stations in the region. According to the IPCC 2006 guidelines, the manure management CH$_4$ emission factor for dairy cattle is 26 kg head^{-1} year^{-1}, much higher than the value used in previous inventories.

The N$_2$O emission factors used in this study are presented in Table A4, and expressed in emission per animal head per year to be able to compare with the ones used in previous studies. Some studies have used IPCC 1997 default N$_2$O emission factors for each animal type (e.g., Reference [11]), which were calculated based on proportional regional values of manure production. Our calculation resulted in higher emission factors for all animals except horses and goats; these are "pasture animals", for which N$_2$O emissions are not accounted for as livestock but instead for soil management. Our emission factors for dairy cattle and poultry were an order of magnitude higher than those used previously [11,15]. We used the IPCC 2006 default values for dairy and non-dairy cattle to calculate emission factors, resulting in higher values than those obtained using the IPCC 1997 guidelines due

Sustainability **2018**, *10*, 3826

to the incorporation of different manure management systems and the more detailed classification systems employed in the IPCC 2006 guidelines.

Pig husbandry is the largest GHG and NH_3 emitter in the RRD region, which is responsible for about half of total GHG emissions (in CO_2eq) and about 46% NH3 emissions from livestock in 2015. This is an atypical situation compared to neighboring countries. In the emission inventory for South, South East and East Asia for 2000 [11], cattle was the largest emitter for CH_4, N_2O and NH_3 emissions, with a share of 56%, 30% and 33% respectively. Study of Reference [32] in Indonesia has also shown cattle as the major contributor to GHG emissions in the 2005–2015 period. Our results provide a reminder that regarding agricultural sector emission mitigation, policies in the Red River Delta should not be copied from other countries.

The production of emissions from livestock farming depends on various factors including feeding practices, housing systems, and manure management systems. Detailed historical data on the feed composition for each animal type and the proportions of manure managed by different management systems are needed to obtain more accurate emissions estimates. However, these data are not yet systematically collected or well documented for emissions inventory purposes. Improving the quantity and quality of data and research related to livestock farming will help to improve emissions monitoring in this sector.

Currently, environmental protection regulations for livestock farming in Vietnam mainly focus on water quality, not air quality. There is a national technical standard for wastewater from livestock farming in Vietnam, but no specific regulations with respect to manure management and air quality. In practice, compliance with and enforcement of related environmental regulations in the livestock sector are currently weak [1]. The significant contributions to GHG and air pollutant emissions from this sector deserve more attention.

5. Conclusions

In this study, we estimated CH_4, N_2O, and NH_3 emissions from livestock farming in the RRD, northern Vietnam from 2000 to 2015 and projected future emissions to 2030. This inventory and projection yielded emissions by animal type and by province. The results of our emissions inventory indicate that livestock farming in RRD contributes significantly to GHGs and NH_3 emissions. The emissions inventory and projection showed an upward trend in GHG and NH_3 emissions during 2000–2030. The GWP of CH_4 and N_2O emissions was 5.9 $MtCO_{2eq}$ in 2030, representing 33% of GHG emissions from livestock nationwide. Pig farming contributed the largest proportion of GHG and NH_3 emissions, at 50%. Cattle were responsible for the second largest share of GHG emissions, whereas poultry contributed most of the remaining NH_3 emissions. This study also provides the provincial emissions levels for CH_4, N_2O and NH_3. Understanding the level of emissions emitted in the RRD region and the contribution of different type of livestock as well as the spatial distribution of emissions by province is a first step to developing effective mitigation strategies for reducing GHG and NH_3 emission in the RRD region. Furthermore, this inventory provides an input to implementing regional air dispersion modeling for air pollution impact assessments in the RRD region.

Supplementary Materials: The following are available online at http://www.mdpi.com/2071-1050/10/10/3826/s1, Table S1. Livestock number by provinces in Red River Delta from 2000 to 2030.

Author Contributions: Conceptualization, A.H.T. and Q.T.N.; formal analysis, A.H.T., M.T.K.; data curation, M.T.K., T.T.N.; writing—original draft preparation; A.H.T.; writing—review and editing, N.T.N. and Q.T.N.; visualization, A.H.T.; supervision, N.T.N. and Q.T.N.; funding acquisition, Q.T.N.

Funding: This research was funded by Vietnam Academy of Science and Technology, grant number VAST-QTAT01.01/17-19.

Acknowledgments: The authors would like to thank Minh Ha-Duong for his comments that greatly improved the manuscript.

Conflicts of Interest: The authors declare no conflict of interest.

Appendix A

Table A1. CH$_4$ emissions (kt/y) by provinces.

Province	2000	2005	2010	2015	2020	2025	2030
Hanoi	18.99	24.07	22.88	20.99	21.11	19.89	19.02
Bac Ninh	6.29	6.57	4.88	4.65	4.71	4.77	4.77
Hung Yen	4.71	6.52	6.72	6.38	6.54	6.37	6.32
Vinh Phuc	10.81	13.06	12.34	11.15	16.00	16.83	18.65
Quang Ninh	8.12	9.03	9.01	7.13	15.92	17.60	21.66
Hai Duong	9.01	9.72	6.08	5.64	5.61	5.70	5.70
Hai Phong	5.35	5.85	5.34	4.76	7.60	8.15	9.30
Thai Binh	8.44	11.13	11.28	9.57	15.40	16.20	18.25
Ha Nam	4.12	5.04	4.45	4.17	7.12	8.65	10.00
Nam Dinh	5.67	7.45	7.50	7.60	8.61	9.01	9.56
Ninh Binh	5.11	6.43	5.68	5.16	7.25	7.59	8.37

Table A2. N$_2$O emissions (kt/y) by provinces.

Province	2000	2005	2010	2015	2020	2025	2030
Hanoi	0.83	1.12	1.23	1.38	1.10	1.11	1.05
Bac Ninh	0.256	0.298	0.292	0.31	0.30	0.30	0.30
Hung Yen	0.34	0.44	0.49	0.51	0.32	0.27	0.27
Vinh Phuc	0.40	0.25	0.53	0.57	0.84	0.96	1.11
Quang Ninh	0.20	0.22	0.23	0.23	0.67	0.82	1.21
Hai Duong	0.47	0.57	0.49	0.58	0.70	0.80	0.78
Hai Phong	0.30	0.34	0.39	0.44	0.50	0.56	0.61
Thai Binh	0.47	0.63	0.66	0.70	0.96	1.07	1.22
Ha Nam	0.19	0.25	0.29	0.34	0.49	0.58	0.68
Nam Dinh	0.35	0.42	0.46	0.50	0.53	0.57	0.61
Ninh Binh	0.22	0.25	0.27	0.27	0.35	0.38	0.43

Table A3. NH$_3$ emissions (kt/y) by provinces.

Province	2000	2005	2010	2015	2020	2025	2030
Hanoi	3.68	4.97	5.20	5.45	4.69	4.60	4.35
Bac Ninh	1.19	1.35	1.23	1.28	1.25	1.24	1.23
Hung Yen	1.38	1.82	2.00	2.02	1.53	1.38	1.38
Vinh Phuc	1.73	1.45	2.22	2.23	3.33	3.71	4.26
Quang Ninh	0.96	1.10	1.12	1.08	3.17	3.84	5.05
Hai Duong	1.99	2.45	1.96	2.19	2.54	2.83	2.79
Hai Phong	1.32	1.55	1.63	1.71	2.20	2.41	2.70
Thai Binh	2.05	2.87	2.98	2.95	4.04	4.39	4.93
Ha Nam	0.85	1.12	1.22	1.34	2.11	2.50	2.94
Nam Dinh	1.53	1.94	2.02	2.20	2.34	2.51	2.67
Ninh Binh	0.95	1.14	1.21	1.18	1.59	1.70	1.89

Table A4. Comparision of emission factors used in this study and previous studies.

	Source	Methodology	Type of Animal						
			Dairy Cattle	Other Cattle	Pig	Horse	Goat	Buffalo	Poultry
Enteric fermentation									
	IPCC 2006 [16]	Tier 1	68	47	1	18	5	55	-
	[15]	Tier 2	50.46	64.15	-	-	-	82.3	-
CH$_4$ (kg/head^{-1}yr^{-1})	[25]; [26]	Tier 3, RUMINANT model	94.5	41	-	-	-	-	-
	[10]	Tier 1	47	44.9	1	18	5	53.2	-

Table A4. *Cont.*

	Source	Methodology	Type of Animal						
			Dairy Cattle	Other Cattle	Pig	Horse	Goat	Buffalo	Poultry
Manure Management									
CH$_4$ (kg/head^{-1}yr^{-1})	IPCC 1997 [10,15]	Tier 1	16	1	4	1.6	0.18	2	0.018
	IPCC 2006 temp. 25°C	Tier 1	26	1	6	1.64	0.17	2	0.02
N$_2$O (kg/head^{-1}yr^{-1})	[11,15]	Tier 1	0.29	0.34	0.18	0.87	0.17	0.39	0.0069
	IPCC 1997;	Tier 1	0.29	0.34	0.18	0.77	0.77	0.34	0.0068
	Used in this study		1.92	0.60	0.22	0.00	0.00	0.55	0.0425
NH$_3$ (kg/head^{-1}yr^{-1})	[15]; [11]		5.6	3	1.5	7	1.1	3.4	0.12

References

1. Dinh, X.T.; Cassou, E.; Cao, B.T. *An Overview of Agricultural Pollution in Vietnam: The Livestock Sector*; World Bank Group: Washington, DC, USA, 2017.
2. General Statistic Office of Vietnam. *Statistical Yearbook of Vietnam 2016*; Statistical Publishing House: Hanoi, Vietnam, 2017; ISBN 978-604-75-0553-1.
3. Garnett, T. Livestock-related greenhouse gas emissions: Impacts and options for policy makers. *Environ. Sci. Policy* **2009**, *12*, 491–503. [CrossRef]
4. Bouwman, A.F.; Lee, D.S.; Asman, W.A.H.; Dentener, F.J.; Hoek, K.W.V.D.; Olivier, J.G.J. A global high-resolution emission inventory for ammonia. *Glob. Biogeochem. Cycles* **1997**, *11*, 561–587. [CrossRef]
5. Cassou, E.; Jaffee, S.M.; Ru, J. *The Challenge of Agricultural Pollution: Evidence from China, Vietnam and the Philippines*; Directions in Development; World Bank: Washington, DC, USA, 2017; ISBN 978-1-4648-1202-6.
6. Gu, B.; Sutton, M.A.; Chang, S.X.; Ge, Y.; Chang, J. Agricultural ammonia emissions contribute to China's urban air pollution. *Front. Ecol. Environ.* **2014**, *12*, 265–266. [CrossRef]
7. Fangmeier, A.; Hadwiger-Fangmeier, A.; Van der Eerden, L.; Jäger, H.-J. Effects of atmospheric ammonia on vegetation—A review. *Environ. Pollut.* **1994**, *86*, 43–82. [CrossRef]
8. Wilkes, A.; Reisinger, A.; Wollenberg, E.; Van Dijk, S. *Measurement, Reporting and Verification of Livestock GHG Emissions by Developing Countries in the UNFCCC: Current Practices and Opportunities for Improvement*; CGIAR Research Program on Climate Change, Agriculture and Food Security (CCAFS) and Global Research Alliance for Agricultural Greenhouse Gases (GRA): Wageningen, The Netherlands, 2017.
9. Ministry of Natural Resources and Environment. Greenhouse Gas Inventory of Vietnam 2010. Presented at UN Climate Conference for the United Nations Framework Convention on Climate Change, Lima, Peru, 1–12 December 2014.
10. Yamaji, K.; Ohara, T.; Akimoto, H. A country-specific, high-resolution emission inventory for methane from livestock in Asia in 2000. *Atmos. Environ.* **2003**, *37*, 4393–4406. [CrossRef]
11. Yamaji, K.; Ohara, T.; Akimoto, H. Regional-specific emission inventory for NH$_3$, N$_2$O, and CH$_4$ via animal farming in South, Southeast, and East Asia. *Atmos. Environ.* **2004**, *38*, 7111–7121. [CrossRef]
12. Kurokawa, J.; Ohara, T.; Morikawa, T.; Hanayama, S.; Janssens-Maenhout, G.; Fukui, T.; Kawashima, K.; Akimoto, H. Emissions of air pollutants and greenhouse gases over Asian regions during 2000–2008: Regional Emission inventory in ASia (REAS) version 2. *Atmos. Chem. Phys.* **2013**, *13*, 11019–11058. [CrossRef]
13. Ramírez-Restrepo, C.A.; Van Tien, D.; Le Duc, N.; Herrero, M.; Le Dinh, P.; Van, D.D.; Le Thi Hoa, S.; Chi, C.V.; Solano-Patiño, C.; Lerner, A.M.; et al. Estimation of methane emissions from local and crossbreed beef cattle in Daklak province of Vietnam. *Asian Australas. J. Anim. Sci.* **2017**, *30*, 1054–1060. [CrossRef] [PubMed]
14. Lê, Ð.N.; Ðinh, V.D.; Searchinger, T.D.; Lê, Ð.P. Current situation and scenarios for reducing enteric methane emission from extensive beef cattle production system of smallholders in Quang Ngai province, Vietnam. *Can Tho Univ. J. Sci.* **2016**, *46b*, 1–7. [CrossRef]

15. Hoang, A.L.; Dang, T.X.H.; Dinh, M.C. Emission Inventory for NH$_3$, N$_2$O, and CH$_4$ of Animal Husbandry Activities: A case in Tho Vinh Commune, Kim Dong District, Hung Yen Province. *J. Sci. Earth Environ. Sci.* **2017**, *33*, 117–126.

16. IPCC. *2006 IPCC Guidelines for National Greenhouse Gas Inventories*; Institute for Global Environmental Strategies (IGES): Hayama, Japan, 2006; Volume 4.

17. Shrestha, R.M.; Nguyen Thi, K.O.; Shrestha, R.P.; Rupakheti, M.; Rajbhandari, S.; Permadi, D.A.; Kanabkaew, T.; Iyngararasan, M. *Atmospheric Brown Clouds (ABC) Emission Inventory Manual*; United Nations Environment Programme: Nairobi, Kenya, 2013; ISBN 978-92-807-3325-9.

18. General Statistic Office of Vietnam. *Statistical Yearbook of Vietnam 2015*; Statistical Publishing House: Hanoi, Vietnam, 2016; ISBN 978-604-75-0364-3.

19. General Statistics Office of Vietnam. *Statistical Yearbook of Vietnam 2005*; Statistical Publishing House: Hanoi, Vietnam, 2006.

20. Vietnam Livestock Production. Available online: http://channuoivietnam.com/thong-ke-chan-nuoi/ (accessed on 11 September 2018).

21. Prime Minister. Decision 124/QD-TTg on Approval of National Agriculture Development Plan until 2020 with Vision to 2030. 2012. Available online: https://www.ecolex.org/details/legislation/decision-no-124qd-ttg-approving-the-master-plan-for-agricultural-production-development-through-2020-with-a-vision-toward-2030-lex-faoc112552/ (assessed on 11 September 2018).

22. Bui, H.D. Production and Quality of Industrial Chicken Manure before and after Compositing. *Vietnam. Agric. Sci. J.* **2009**, *7*, 245–252.

23. Vu, T.K.V.; Tran, M.T.; Dang, T.T.S. A survey of manure management on pig farms in Northern Vietnam. *Livest. Sci.* **2007**, *112*, 288–297. [CrossRef]

24. Vu, T.T.H.; Vu, Q.C.; Nguyen, T.H.C.; Le, V.C. Results of study on current status and solutions for environmental management in livestock farming in household and small scale farm in the North of Vietnam. *J. Water Resour. Sci. Technol.* **2013**, *18*, 41–47.

25. Le, D.N.; Dinh, V.D.; Le, D.P.; Le, V.T.; Vu, C.C.; Le, T.H.S.; Ramírez-Restrepo, C.A. Study on enteric methane emission from smallholder semi-intensive beef cattle production system in the Red River Delta: A case study in Dong Anh District, Hanoi. *Sci. Technol. J. Agric. Dev.* **2015**, *7*, 70–79.

26. Lê, Ð.P.; Lê, Ð.N. Scenarios for Reducing Enteric Methane Emission from Small Scale Dairy Production Farms in Ba Vi, Ha Noi. *J. Sci. Dev.* **2015**, *13*, 543–550.

27. IPCC. *Climate Change 2013: The Physical Science Basis. Contribution of Working Group I to the Fifth Assessment Report of the Intergovernmental Panel on Climate Change*; Cambridge University Press: Cambridge, UK; New York, NY, USA, 2013; p. 1535.

28. Vu, T.K.V. *Report on Assessment of Climate Change Impact to Livestock Farming and Disease Prevention, Proposed Adaptation Measures*; National Institute of Animal Sciences: Hanoi, Vietnam, 2013.

29. Monteny, G.-J.; Bannink, A.; Chadwick, D. Greenhouse gas abatement strategies for animal husbandry. *Agric. Ecosyst. Environ.* **2006**, *112*, 163–170. [CrossRef]

30. Prime Minister. Approval of Plan for GHG Emissions Reduction in Agriculture. 2011; Volume 3119/QĐ-BNN-KHCN. Available online: https://www.google.com/url?sa=t&rct=j&q=&esrc=s&source=web&cd=1&ved=2ahUKEwjCr5a6w5veAhXEtYsKHSp2BRgQFjAAegQIABAC&url=https%3A%2F%2Ftheredddesk.org%2Fsites%2Fdefault%2Ffiles%2F3119-qd-bnn-khcn_2.pdf&usg=AOvVaw2jO-t-aZeRBiRle1I12liX (assessed on 11 September 2018).

31. Thien Thu, C.T.; Cuong, P.H.; Hang, L.T.; Chao, N.V.; Anh, L.X.; Trach, N.X.; Sommer, S.G. Manure management practices on biogas and non-biogas pig farms in developing countries—Using livestock farms in Vietnam as an example. *J. Clean. Prod.* **2012**, *27*, 64–71. [CrossRef]

32. Nugrahaeningtyas, E.; Baek, C.-Y.; Jeon, J.-H.; Jo, H.-J.; Park, K.-H. Greenhouse Gas Emission Intensities for the Livestock Sector in Indonesia, Based on the National Specific Data. *Sustainability* **2018**, *10*, 1912. [CrossRef]

sustainability

MDPI

Article

Issuances of Automotive Vehicles and the Impacts on Air Quality in the Largest City in the Brazilian Amazon

Elizabeth Cartaxo [1,*], Ilsa Valois [2], Vladimiro Miranda [3] and Marcia Costa [4]

[1] Technology College, Federal University of Amazonas, Manaus 69080-900, Amazon, Brazil
[2] Nilton Lins University, Manaus 69058-030, Amazon, Brazil; ilsavalois@gmail.com
[3] Institute for Systems and Computer Engineering, Technology and Science (INESC TEC),
    Porto 4200-465, Portugal; vmiranda@inesctec.pt
[4] Metropolitan University of Manaus, Manaus 69050-000, Amazon, Brazil; marciarebelo70@gmail.com
*   Correspondence: elizcartaxo@gmail.com; Tel.: +55-92-98125-0886

Received: 15 October 2018; Accepted: 3 November 2018; Published: 8 November 2018

Abstract: Manaus, a city of more than two million people, suffers problems arising from strong sunlight and aggravated by several factors, such as traffic congestion and greenhouse gas emissions generated by evaporation and burning of fuel. The present study examined Carbon Monoxide (CO) and Nitrogen Dioxide (NO_2) emissions in an urban area of the city using different methodologies. CO and NO_2 were measured using automated and passive analyzers, respectively. Meanwhile, direct monitoring of these pollutants was performed in vehicular sources in the vicinity of sampling locations. Results showed that levels of carbon monoxide vary over time, being higher during peak movement of vehicles. NO_2 values have exceeded the recommendations of the World Health Organization (WHO), and monitoring at source showed high levels of CO and NO_2 emissions to the atmosphere.

Keywords: vehicle; pollution; measurement and environment

1. Introduction

The study of chemical interactions occurring in the atmosphere is very complex. The atmosphere is similar to a natural laboratory. However, unlike in laboratory research, where researchers perform specific experimental reactions with controlled atmospheric variables, the atmosphere involves a series of chemical reactions that are difficult to monitor, mainly owing to low concentrations, altitude variation, and slow reactions. This is why studies on the chemical reactions occurring in the atmosphere and their consequences are growing, despite the difficulties they present [1].

Moreover, in addition to natural components, any portion of atmospheric air contains primary pollutants emitted by pollution sources and as a result of reactions occurring among these, from chemical reactions among them, a relative amount of secondary pollutants. The composition of atmospheric air, the chemical reactions occurring in the atmosphere, the movement of air masses, the energy balance, and meteorological conditions are all factors responsible for observed atmospheric phenomena that have intensified in recent times [2].

Until the Constitution of 1988, environmental concerns in Brazil permeated constitutional standards. Environmental impact studies, for example, were introduced into Brazilian law through Act number 6803/1980, which obliged companies linked to petrochemical, clorochemical, and carbochemical industries and nuclear installations to present special studies of alternatives and impact assessments [3].

After the UN Conference on the Environment held in Stockholm in 1972, there was an increase in ecological awareness, a phenomenon that came to be reflected in legislative processes ensuring protection and preservation of the environment [4].

In Brazil, the environment acquired constitutional status after the 1988 Constitution. The Magna Carta reveals some central axes: The environment as a fundamental right; the conservation of biological diversity and ecological processes; the creation of specially protected territorial areas; the need for prior study of the environmental impact of activities potentially causing significant degradation; and environmental education [5]. It consciously seems to promote the idea that development at any cost causes profound changes in the environment and society. Some noteworthy examples of the most damaging consequences are those resulting from accelerated urbanization processes [6].

Manahan [7], when analyzing the climatic differences between urban and rural areas, confirms the impacts of urbanization on the climate of cities worldwide. According to the author, urbanization induces a strong effect on microclimates. In rural areas, vegetation and water bodies have a moderating effect, absorbing modest amounts of solar energy and releasing it slowly. In contrast, in cities, stone, concrete, and asphalt pavements strongly absorb solar energy and reradiate heat back to the urban microclimate.

There are many studies concerning the consequences of population increases for environmental imbalance, considering that the process of expansion of cities has supported itself through the availability of abundant and cheap energy sources. Reference data from the National Energy Report (BEN) indicates that the domestic supply of energy in Brazil grew by 5.6% in 2007, from 226.1 million tonnes of oil equivalent (TOE) in 2006 to 238.8 million TOE in 2007. This growth was greater than the growth of the economy (5.4%) registered by IBGE (Brazilian Institute of Geography and Statistics) [8].

In 2012, total demand for oil products stayed at 2.274 million barrels of oil equivalent (BOE) per day, 6.6% more than in 2011. The production of oil, at a negative rate of 1.7%-including LNG (liquefied natural gas) and shale oil-reached the amount of 2.16 million bbl/day (barrels per day). In this context, there were net imports of oil and oil products on the order of 211 thousand BOE/day in 2012 [9].

Manaus is the Amazon's largest city and its population has exploded in recent decades as a result of development policies that prioritized the state's capital at the expense of other municipalities. Consequently, as in other cities in Brazil, the process of development occurred unsustainably, when considering the increase in population and the growing number of motor vehicles. Geographically located near the Equator (latitude 03°07′00″ N, longitude 059°57′00″ W, and altitude 67.00 m) and with an area of 11,401 km², Manaus is home to a population of 2,145,444 inhabitants [10].

These data give a population density of 188.18 inhabitants per km². Throughout the Amazonas state, density is only two inhabitants per km², revealing a huge concentration of population in the state's capital. Since 1970, the city's population has grown from 300,000 to over 2 million inhabitants. In 2000, Manaus was the ninth most populous city in Brazil and grew to take eighth place in 2004 [10]. Population growth in Manaus is demonstrated in Table 1.

The increased number of thermoelectric power plants (UTEs), which is also proportional to population growth (since it is the basis of the power generation matrix), contributes directly to indices of carbon monoxide, nitrogen oxides, and others pollutants in the atmosphere caused even by urban traffic of the modern cities [11].

However, it is important to note that the polluting power of UTEs is not greater than the contribution of vehicles, because the transportation sector consumes high levels of energy and is responsible for over 50% of fuel consumption globally. This can be attributed primarily to development processes, which were dependent on petrol and diesel, particularly for the energy industry and transport sector [6].

Table 1. Population growth in Manaus [10,12].

Year	Inhabitants
1970	284,000
1980	635,000
1990	1,100,000
2000	1,405,835
2004	1,592,555
2005	1,644,690
2007	1,646,602
2008	1,709,010
2009	1,738,641
2010	1,802,014

The image of Figure 1 shows the geographic panorama of the location of the UTEs installed in the city of Manaus.

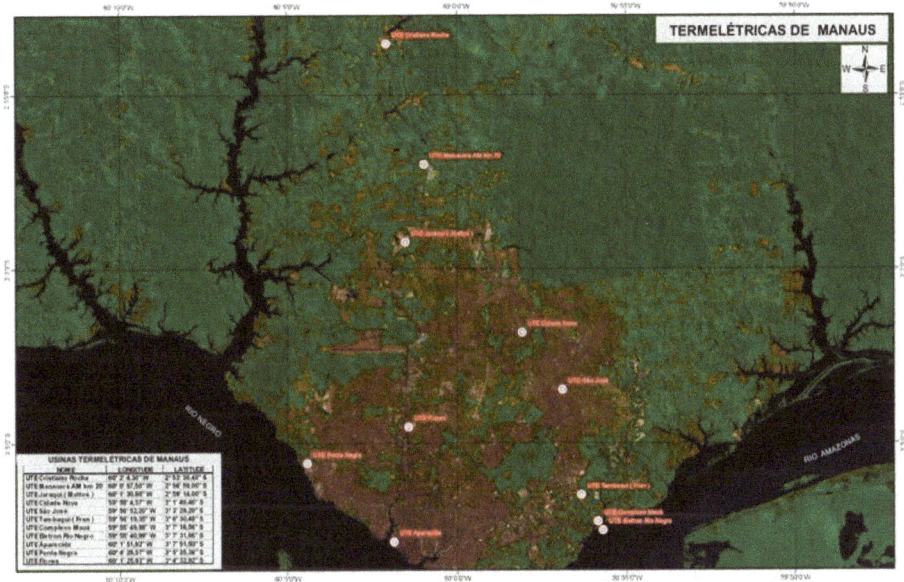

Figure 1. Location of UTE's in the city of Manaus/AM. Source: [13].

Rapid growth has also been observed with regard to the number of vehicles in the city of Manaus. According to data provided by the Ministry of Cities, National Traffic Department (DENATRAN), Manaus has 705,296 motor vehicles, and difficult conditions of transit (slow moving traffic, paving, and other problems) associated with pollutant emissions and the noise can produced [14] may be responsible for the deterioration of air quality in the city (Table 2).

The transport sector's role in increasing emissions has been extensively reported worldwide. In the countries of the European Community, transport contributes 75% of all carbon monoxide (CO), 40% of hydrocarbons (HC), and 48% of nitrogen oxides (NOx) [15]. In the United States, the main source of CO is transport, despite reductions in emissions that have occurred since 1970 owing to increasingly stringent standards of emission control and improvements in energy efficiency [16].

Table 2. Data concerning the vehicle fleet in Manaus. Source: [17].

Type	2010	2012	2013
Automobile	246,265	285,796	309,162
Truck	14,314	15,990	16,406
Truck-tractor	1902	2348	2540
Van	48,537	58,977	83,700
Pickup-truck	18,367	21,607	23,166
Minibus	2280	2777	2894
Motorcycle	80,333	104,819	119,763
Motor-scooter	8269	10,320	11,821
Bus	5626	7307	7714
Trailer	1642	1848	1836
Semi-trailer	9618	10,861	11,236
Tractor wheels	48	50	58
Tricycle	90	342	571
Utility	2269	3473	4078
Other	60	65	67
Total	439,620	526,580	595,012

In Brazil, the situation is broadly similar, justifying concerns relating to air quality. The traffic congestion occurring in virtually all capitals is responsible for 90% of CO, 80 to 90% of emissions of NOx and hydrocarbons, and a considerable portion of particles that constitute a threat to human health [18].

Consequently, the issue of transportation and the environment is a paradoxical one. On one hand, increasing the number of vehicles meets the growing demand for mobility; on the other hand, the effects are significant for society. Hence, transport plays a fundamental role in the lives of city dwellers and is a major environmental concern because in coming decades, transport is expected to remain a significant contributor to air pollution, especially in more populous cities.

Converging factors, such as the deployment of car factories and the construction of highways, led to an increase in oil demand. As a result, the consumption of fuels increased, increasing emissions and impacts on environment and society that are difficult to measure [8].

Undeniably, there has been major technological development of engines and fuels, resulting in significant reductions in emissions of some pollutants. Notwithstanding, factors such as the growth of vehicular fleets, traffic jams, and increases in distances traveled contribute to significant emission increases. Moreover, one should take into account the fact that, although vehicle pollutants can be formed in combustion processes, the production of pollutants per unit of burned fuel is higher in vehicle engines [19].

There are several factors that produce this effect, including no permanent combustion, insufficient fuel atomization, and the engine cooling system that prevents the oxidizing mix from burning equally. Overtaking, stopped traffic, conversions, vehicle speed, and other typical traffic events also have significant impacts on fuel consumption and emissions, caused by changes in operating motor vehicles [19].

Despite the fact that the State of Amazonas represents the largest isolated energy system, its capital, Manaus, does not have an environmental management policy directed at monitoring and supervising air quality. The number of thermoelectric plants and motor vehicles is increasing within its territory, mainly within the state's capital; consequently, the city needs to adapt to the increase in fuel and energy consumption.

Some developed areas such as the United States and Western Europe, have made significant progress in relation to the control of air pollutants. These advances have been globally important; however, poor air quality continues to affect many people in developing countries. This situation is caused by rapid population growth combined with growing energy demand, weak standards for pollution control, dirty fuels, and inefficient technologies. Some governments have begun to address this problem, but very strict measures are needed to reduce the serious impacts of air pollution on public health worldwide [20].

Currently, some consequences are already notable in the atmosphere of Manaus as a result of increased emissions of pollutants. These include smog, which is probably due to the high turnover of cars, abundant sunlight, and frequent temperature inversions [21].

Nevertheless, very little is known about the effects of urban development on the quality of air that local people breathe. This is because the immensity of the Amazon rainforest conveys the idea of infinity of natural resources. It follows that, having abundance of exaggerated features and dimensions, as is the case of the Amazon region, conscious concern with saving nature is minimized and natural resources are used on a large scale, surpassing necessary limits [8].

Among major air pollutants, carbon monoxide is considered to be the vehicular pollutant with greatest influence on the loss of quality of atmospheric air. Its emissions are related to incomplete combustion, both in mobile sources and in stationary sources. The effects caused by the exposure to the pollutant are associated with their affinity to hemoglobin in the blood, which is greater than that of hemoglobin with oxygen, potentially leading to death by asphyxiation.

Nitrogen oxides also play an important role in atmospheric chemistry. These pollutants are not required to be present in fuel composition in order to be formed, with formation occurring naturally during all processes of combustion. These gases form when fuel is burned at high temperatures, mainly from motor vehicle exhaust and stationary sources. Nitrogen dioxide is a strong oxidizing agent that reacts in air to form corrosive nitric acid as well as organic nitrates and other toxic secondary pollutants, which can attack the outer layer of the material of the monuments of the works of art [22]. It also plays a major role in atmospheric reactions by producing ground-level ozone (or smog) [23].

Nitrogen oxides may be involved in a series of reactions commonly associated with the occurrence of peak ozone that produces photochemical smog, reducing visibility. In urbanized areas, one of the obvious effects caused by burning fossil fuels is the occurrence of smog. It is a photochemical phenomenon characterized by the formation of a kind of fog composed of pollution, water vapor, and other chemical compounds [7].

In Manaus, the consequences of these factors added to forest fires in the dry season, result in an environment with a high level of pollution and serious health problems caused by changes in air quality. Currently, some consequences are already evident in the atmosphere of Manaus due to the increase in pollutant emissions. This is the case with *smog*, probably due to high turnover of automobiles, abundant sunlight and frequent thermal inversions.

According to Kuhn et al. [21], the main contribution in the urban pollutant plume in Manaus was attributed to the city's thermal power plant complex. Strong evidence has shown that there are significant amounts of ozone from this source, as well as the relationship between the concentrations of nitrogen dioxide found in the site studied and the activities related to the intense vehicular traffic, which may directly influence the formation of photochemical *smog* in this region.

Figure 2 shows a photochemical *smog* in the city of Manaus, the visible differences of this environment in the comparison of the images: (a) photographic record of an avenue in humid season of the year; and, (b) photographic record of the same avenue in dry period of the year. The dry period occurs between the months of July to December; the humid period occurs in the months January to June.

This phenomenon is unlike events recorded in other countries, such as in many American cities where haze occurs during wet conditions. The American environmental agency, the EPA, has announced a major effort to improve air quality in national parks and wilderness areas [23].

(a)   (b)

Figure 2. Photographic record of Efigênio Sales Avenue, Coroado/Manaus (2014). (**a**) In the humid period; (**b**) In the dry period.

On the contrary, in the Amazon, the atmosphere undergoes great changes during the dry season due to emissions of trace gases and aerosol particles from pasture and forest. The intense deforestation activity and the consequent emission of gases and particles from fires during the dry season have important implications at local, regional and global levels [24]. In Manaus the rains wash the particles, providing a better air quality.

The yellow color in the atmosphere of a city enveloped by smog is due to the presence of nitrogen dioxide, because this gas absorbs some visible light near the limit of violet and, therefore, solar light transmitted through the fog appears yellow.

The consequences of NO_2 atmospheric emissions are diverse, because they cause several harmful effects, direct or indirect, on the health and well-being of humans, fauna and flora, materials, soils, and water bodies. The degree and extent of these effects depends on the scale of these emissions. They can occur at local and regional levels owing to the short residence time of NO_2. Local impacts are limited to the vicinity of sources. Regional impacts comprise a much larger radius of hundreds of kilometers.

According to Kun et al. [21], the main contributors to urban air pollution in the plume of Manaus were the complexes of power plants in the city. Strong evidence showed that there are significant amounts of ozone derived from this source, potentially directly influencing the formation of photochemical smog in the region. The study recorded a rate of ozone in the order of 15 ppb/h within the plume of pollutants.

In the Amazon, there is no air quality monitoring network, the vehicle fleet is increasing exponentially, and the electric sector is made up of 80% thermal generation. These facts motivated this work, which presents the partial results of a survey on air quality carried out by the research group of the Interdisciplinary Center for Energy and Environment (NIEMA) at the Federal University of Amazonas. The study analyzed CO and NO_2 emissions in an urban area of the city of Manaus, using different methodologies.

2. Materials and Methods

As the impact of human activities is increasing on a global scale, the need to recognize and deal with the health risks associated with air pollution has become increasingly urgent [25]. Accordingly, emission monitoring has been adopted as a tool for the management of air pollution, aiming to eliminate or reduce pollutants to acceptable levels.

However, despite the efforts that have been made to clean up the atmosphere, pollution remains a major problem, posing a continuous health risk. Population growth, coupled with an increase in the number of circulating vehicles, traffic conditions, and characteristics of vehicular fuel pollutants, are factors that create a worrying scenario in relation to air quality. This research was conducted by researchers from NIEMA in the context of two research projects supported by National Council for Scientific and Technological Development (CNPq).

2.1. Emissions Monitoring

There are two main types of monitoring: Monitoring emissions directly at the source, and air quality monitoring. In the first type, the concentration of pollutants released into the atmosphere by ducts and chimneys (CME, or Maximum Concentration of Issue) is measured. Conversely, monitoring of air quality deals with the measurement of emissions scattered into ambient air (CMI, or Maximum Concentration Immersion). Both are important because they are chemically related, depicting the pathways that pollutants follow in atmospheric air, dragged by winds, washed by rainfall, or transformed by chemical reactions and solar energy [6].

The weather conditions affect the quality of air, phenomena such as the dispersion and removal of pollutants are closely related to climatic factors, weather conditions, topography, and the use and occupation. For this reason the study of atmospheric chemistry involves knowledge of these conditions, for environmental vitality of any region depends on the ability to exchange energy and matter without accelerating the entropy processes [26].

In this study the variations of meteorological parameters were recorded and considered in the analysis of the measurements. Figure 3 shows the data considered during the study period.

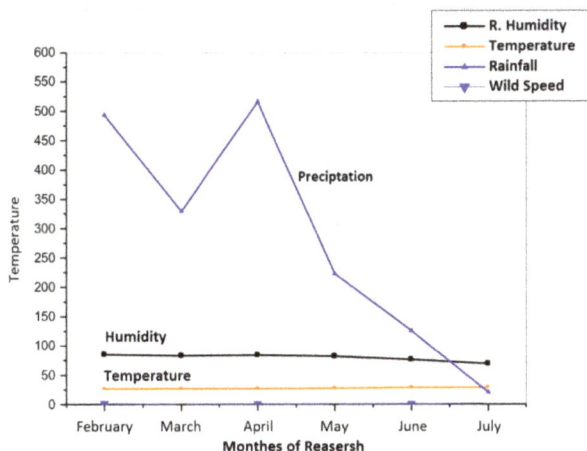

Figure 3. Simultaneous variation of meteorological parameters.

Air quality can be measured using passive or active monitors. Passive samplers are well known because the molecules of the gas of interest are absorbed in the atmosphere by diffusion and/or molecular permeation. These methods were initially used for monitoring indoor environments but are today occasionally used to monitor gases and vapors at low outdoor concentrations [27].

Two sampling sites (passive and active) were installed in a space belonging to the Ministry of Agriculture, Livestock and Supply (MAPA) at a location with coordinates 03°02'36" S and 60°04'28" W (Figure 4). This particular location was selected owing to its situation in a neighborhood of busy avenues, where it is consequently influenced by both mobile and urban sources. The two points were located in parallel. Direct monitoring of vehicular sources was conducted at a distance of 3 m from the samplers [27].

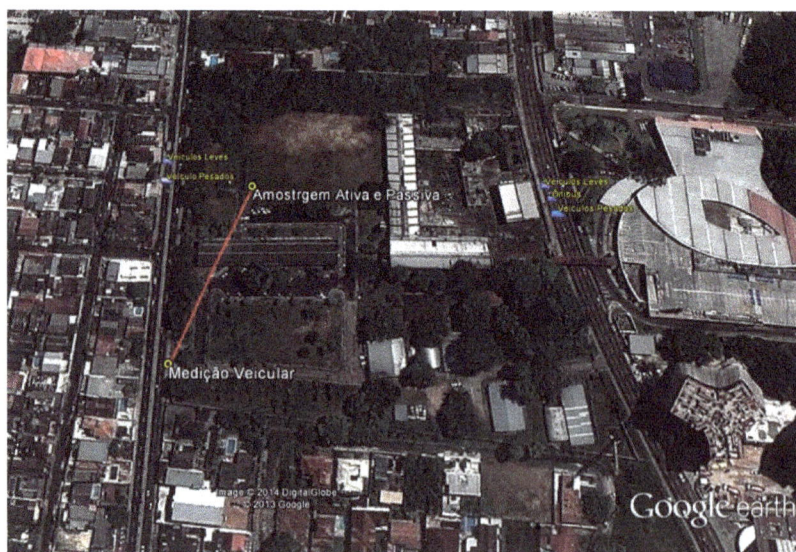

Figure 4. Image of passive and active sample mesh.

2.2. Data Stratification

The sample was made up of vehicles manufactured between 1982 and 2013, as shown Table 3.

Data collection took place through stratified sampling. Strata have been classified as follows: Light vehicles, heavy vehicles (trucks and buses), and motorcycles, with the total size of 262 vehicles. The strata are detailed in Table 3. In this sample we selected one hundred and seventy light vehicles, sixty seven motorcycles, eight trucks, eight buses and eight minibus.

According to the type of vehicles, the sample was stratified as follows: Two hundred and eleven vehicles powered by gasoline, twenty one vehicles running on ethanol and 30 Diesel.

Table 3. Year of Manufacture.

Year	N°. of Vehicles	Year	N°. of Vehicles	Year	N°. of Vehicles
1985	2	2000	5	2007	15
1987	2	2001	4	2008	26
1990	2	2002	2	2009	22
1995	2	2003	2	2010	44
1996	2	2004	11	2011	44
1998	2	2005	14	2012	34
1999	2	2006	18	2013	3

2.3. Passive Monitoring

Concentrations of NO_2 were provided by passive analyzers. The methodology adopted was described by Ugucione et al. [28], and consisted of the use of passive samplers installed three meters above the ground on a stand with eight samplers; three of these samplers were used as blank and remained sealed during sampling.

Passive monitoring took place over three months; subsequently, chemical analysis was performed in the laboratory of environmental analytical chemistry of the National Institute of Amazon Research (INPA), employing a molecular UV-Vis spectrophotometer with absorbance at a wavelength of 540 nm. The calibration curve was prepared for each analysis using the standard solution of sodium nitrite.

The concentration of NO_2 was calculated using the first integration of Fick's Law [29]. The Figure 5 shows the image of the measurement sites and the Figure 6 is the photographic record of one of the installed passive analysers.

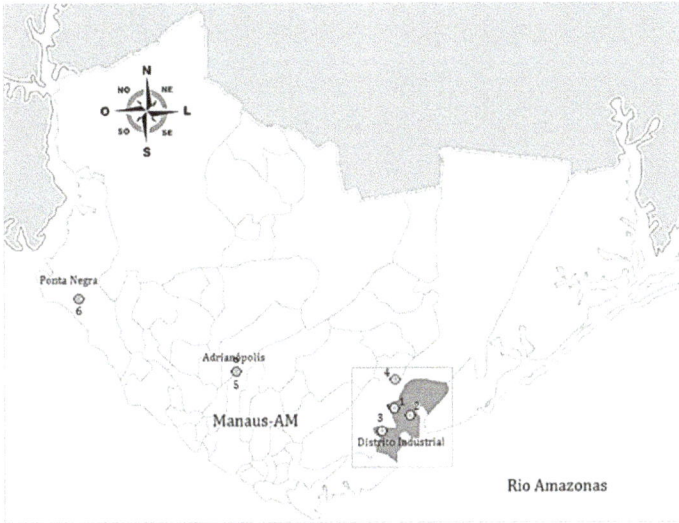

Figure 5. Map of the passive sampling mesh.

Figure 6. Support with samplers installed at the study site.

2.4. Automatic Monitoring

In addition, an air quality monitoring station, installed a few meters away from the point of passive samplers (Figure 7), provided data on the carbon monoxide concentrations. The Model 48I CO Analyzer from Thermo Scientific was used, with this being widespread and tested throughout the world by environmental agencies.

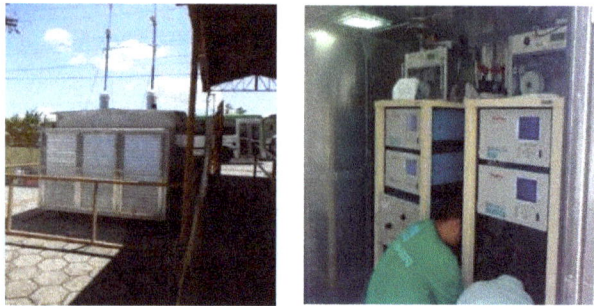

Figure 7. Air Quality Monitoring Station installed.

2.5. Vehicular Source Monitoring

The instrument of measurement used for vehicular sources was the exhaust gas analyzer for engines, BRIDGE MODEL 900403. This analyzer uses a proven methodology for the measurement of gases, NDIR (Non Dispersive Infra-Red) for the gases CO, HC, and CO_2, and electrochemical sensors for measuring O_2 and NOx. The collection of data for vehicular sources was carried out for three consecutive days during November 2012. This study made use only of data for CO and NO_2. The Figure 8 shows photographic records of the measurements made.

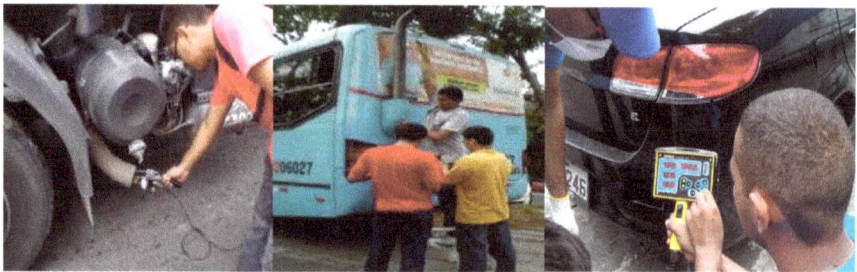

Figure 8. Measurement in vehicular sources.

3. Results and Discussion

Direct NO_2 emissions from vehicular sources were grouped according to the corresponding year of manufacture of each vehicle. The results indicate that average NO_2 concentration was 89.01 ppm (Figure 9).

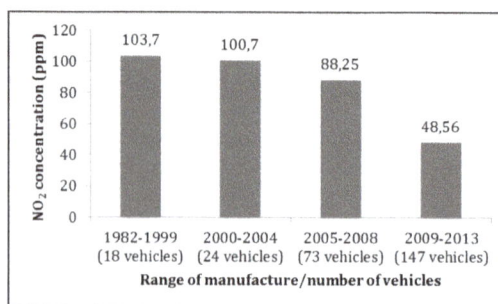

Figure 9. Emission of NO_2 according to the year of manufacture of the vehicle.

It was also evident that older vehicles are the largest emitters, because the largest quantity of emissions corresponds to a smaller number of old vehicles, while a large number of new vehicles have lower emissions.

According to CONAMA Resolution 418, the maximum exhaust emission of CO was fixed at 0.3% (3000 ppm). Dependent on vehicle type, it was observed that measured values were above legal limits. The findings also showed that motorbikes have average CO emissions much greater than those of other types of vehicles. The lack of space for installation of efficient equipment such as filters, for example, seems to be one of the difficulties faced in making this means of transport more ecologically friendly (Table 4).

Table 4. CO average emissions according to vehicle type obtained by the research sample.

Vehicle Type	CO (%)
Light Vehicle	0.61 (6100 ppm)
Motorcycles	12.64 (126,400 ppm)
Trucks	0.71 (7100 ppm)
Bus/Minibus	0.75 (7500 ppm)

In 2009, PROMOT (Program for Control of Air Pollution by Motorcycles and Similar Vehicles) imposed fairly strict restrictions on the emission of pollutants, but these were restricted only to new motorcycles.

It was found that gasoline is the largest emitter of CO (Table 5). Owing to the large number of gasoline-powered vehicles, high concentrations of carbon monoxide are generally found in cities, mainly in areas of large moving vehicles.

Table 5. Average emissions according to fuel type obtained by the research sample.

Fuel Type	CO (%)	NOx (ppm)
Gasoline	1.95 (19,500 ppm)	82.2
Ethanol	1.521 (15,210 ppm)	113.93
Diesel	0.94 (9400 ppm)	34.03

It also became evident that ethanol-fueled vehicles emit more nitrogen oxides than diesel and gasoline-powered ones. Measures have been taken for stationary vehicles and those at average acceleration that facilitate an increase in evaporation losses. With the vehicle stationary and the engine running, the conditions under which measurements were conducted, CO emissions are higher. This is because, at the beginning of the combustion process, the fuel quantity in the mix (fuel $\times$ air) is greater, resulting in inefficiency of combustion and in the formation of CO in large quantities. This fact confirms that traffic congestion conditions are decisive for pollutant emissions.

It also became evident that ethanol-fueled vehicles emit more nitrogen oxides than diesel and gasoline-powered ones. Measures have been taken for stationary vehicles and those at average acceleration that facilitate an increase in evaporation losses. With the vehicle stationary and the engine running, the conditions under which measurements were conducted, CO emissions are higher. This is because, at the beginning of the combustion process, the fuel quantity in the mix (fuel $\times$ air) is greater, resulting in inefficiency of combustion and in the formation of CO in large quantities. This fact confirms that traffic congestion conditions are decisive for pollutant emissions.

In addition to measuring the exhaust gases of vehicles, the automatic monitoring station provided data on the atmospheric concentration of CO. The verification of these results showed that the levels of carbon monoxide vary over time, being lower during morning hours and higher during periods of intense traffic movement. Hourly average emissions of carbon monoxide provided by the monitoring station during the first three days of sampling are shown in Figure 10.

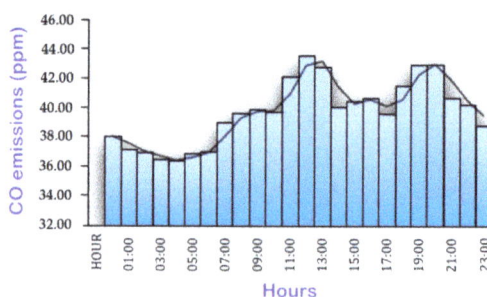

Figure 10. Information from the emission of CO by hours of the day. Source: Research NIEMA. Note: Measurements of NO_2 in atmospheric air were performed with passive analyzers, installed a short distance from the site of vehicular source monitoring. Results obtained at this sampling point during the month in which the survey was conducted were as follows: Sample 1: 0.117 ppm; Sample 2: 0.153 ppm; Sample 3: 0.118 ppm; Sample 4: 0.135 ppm; Sample 5: 0.174 ppm.

The Brazilian legal standard (CONAMA 03) defines the maximum NO_2 limit as being 0.170 ppm, which was exceeded only in one sample. However, the World Health Organization reduces this limit to 0.100 ppm. This standard was exceeded at all sampling points [30].

4. Conclusions

The aim of this study was to quantify emissions of nitrogen oxides and carbon monoxide in vehicle exhaust gases in an urban area. In recent decades, as a consequence of the development process and of limited investment in public transport, the choice of individual means of transport has been responsible for most of the impact on air quality. Saturated urban roads and the idling of vehicle engines result in incomplete combustion of fuel, resulting in substances that are not fully oxidized, such as carbon monoxide and unburned combustible material (HC).

Even if we discard the utopian theories of sustainable development, the increasing number of automotive vehicles seems to make sustainability almost a chimera. Considering population growth and heavy reliance on cars, there are no doubts about the urgency of finding energy alternatives to ensure the possibility of a reasonably sustained system.

This problem is tending toward becoming chronic. The analysis developed here highlights the fact that sustainable development goals are unattainable in the face of breakneck-speed economic growth and social and regional differences, creating dangerous imbalances for future generations. The results from these mismatches are evident, demanding that we reflect on the growing use of energy, increased greenhouse emissions, and investments in public transport, not to mention the need to look at each region with a different perspective, seeking in diversity the solutions for equality.

In Manaus, there more than two million people and almost 500,000 vehicles circling the city. That means a motorization of 4 people per vehicle. By analyzing these facts and the data presented in this study about pollution, we conclude that the city of Manaus is not yet one of the most polluted Brazilian cities, but that there is a problem in the making that cannot be ignored.

The results from the study suggest that gasoline-powered lightweight vehicles are the main emitters of CO and that ethanol-fueled vehicles emit more nitrogen oxides than gasoline-driven vehicles. It was also noted that the emerging segment of motorcycles, which has grown substantially, needs greater control.

In fact, pollution abatement should demand monitoring of emissions mainly from mobile sources in order to reduce impacts on the environment and on people's health. Monitoring is an important environmental management tool in the assessment of air quality, allowing the establishment of measures of prevention and control, which may suggest energy redevelopment interventions consisting of new plant and building technologies, in addition to offering subsidies for traffic.

Sustainability **2018**, *10*, 4091

Above all, so that public policies can be created to reduce emissions, it is important to find ways to measure actual vehicular emissions. Moreover, it is necessary that pollutant emissions are estimated with precision to ensure the appropriate implementation of environmental management policies.

The Program for Air Pollution Control by Automotive Vehicles (PROCONVE) of the Ministry of Environment, based on values measured in laboratory tests, quantified emissions in g/km. The resolution CONAMA number 415 (2009) established the PROCONVE phase L6 to control such emissions from 2013 onwards. To reduce air pollution in urban centers and save fuel, the legal norm in question lays down emission ceilings for the following pollutants from exhaust of motor vehicles.

- Carbon monoxide (1.3 g/km)
- Total hydrocarbons (THC) only for natural gas vehicles (0.3 g/km)
- Non-methane hydrocarbons (NMHC) (0.05 g/km)
- Oxides of nitrogen (NOx) (0.08 g/km)
- Aldehyde (CHO) for Otto cycle (0.02 g/km)
- Particulate matter (PM) to Diesel cycle (0.025 g/km)
- Carbon monoxide at idling to Otto cycle (0.2% in volume)

These tests do not refer to emissions of used vehicles. Moreover, even if they are conducted under strictly controlled conditions, they do not reflect the real use conditions of vehicles.

The processes of dispersion and diffusion of pollutants in air are affected by various environmental characteristics that complicate the process of measurement that is not conducted at source. Measurements taken directly at the source, i.e., at the output from vehicle exhaust, seem to better reflect actual conditions of use of the vehicle.

Many challenges must be faced by developing countries, because these countries must achieve rapid economic growth without compromising air quality. This seems to be an unattainable goal considering global limits; however, an excellent model to follow is that of the developed world, in which progress in air quality has been pursued more effectively.

Author Contributions: Conceptualization, Methodology, Investigation, E.C., I.V. and M.C.; Writing–Original Draft Preparation, E.C. and I.V.; Writing-Review & Editing, E.C. and V.M.; Supervision: E.C. and V.M.; Project Administration, E.C.

Funding: This research received funding by National Council for Scientific and Technological Development (CNPq), Agency of the Ministry of Science, Technology, Innovation and Communications (MCTIC) of the Brazilian government, grant Processes Numbers 5756602008-2 and 5529912005-8.

Acknowledgments: The authors would like to thank CNPq; Superintendence in Manaus of the Ministry of Agriculture, Livestock and Supply (MAPA); and National Institute of Amazon Research (INPA).

Conflicts of Interest: The authors declare no conflict of interest.

References

1. Schirmer, W.N.; de Melo Lisboa, H.J.T.-L. Química da atmosfera: Constituintes naturais, poluentes e suas reações. *Tecno-Lógica* **2008**, *12*, 37–46.
2. De Nevers, N. *Air Pollution Control Engineering*; Waveland Press: Long Grove, IL, USA, 2010.
3. Araújo, S.M. Licenciamento Ambiental e Legislação, Consultoria Legislativa da Câmara dos Deputados. Câmara dos Deputados, Brasília, Brazil. 2002. Available online: http://www2.camara.leg.br/documentos-e-pesquisa/publicacoes/estnottec/tema14/-pdf/208195.pdf (accessed on 10 June 2014).
4. Silva, R.M.P.d. O meio ambiente na Constituição Federal de 1988. *Revista Jus Navigandi*, October 2013.
5. Varella, M.D.; Leuzinger, M.D. O meio ambiente na Constituição de 1988: Sobrevoo por alguns temas vinte anos depois. *Revista de Informação Legislativa* **2008**, *45*, 397–402.
6. Coelho, I.M.H.d.V. Estudo das concentrações de óxidos de nitrogênio em área urbana do município de Manaus por analisadores automáticos. Ph.D. Thesis, Universidade Federal do Amazonas, Manaus, Brazil, 2012.
7. Manahan, S.E. *Environmental Chemistry*, 8rd ed.; CRC Press LLC: Boca Raton, FL, USA, 2005.

8. Tolmasquim, M. *Balanço Energético Nacional 2008*, 2008th ed.; Empresa de Pesquisa Energética (EPE): Rio de Janeiro, Brazil, 2008; p. 44.
9. Resenha Energética Brasileira—Exercício 2012. Available online: http://www.mme.gov.br (accessed on 23 October 2015).
10. Database, States and Cities. Estimative of the Resident Population According to the Municipalities. Available online: https://cidades.ibge.gov.br/brasil/am/manaus/panorama (accessed on 8 July 2018).
11. Cannistraro, G.; Cannistraro, M.; Cannistraro, A.; Galvagno, A.; Engineer, F. Analysis of air pollution in the urban center of four cities Sicilian. *Int. J. Heat Technol.* **2016**, *34*, S219–S225. [CrossRef]
12. Profile of Brazilian Municipalities 2012. Available online: https://ww2.ibge.gov.br/english/estatistica/economia/perfilmunic/2012/default.shtm (accessed on 1 July 2015).
13. Gomes, E.P. Survey of the Main Sources of Atmospheric Emissions in the City of Manaus. Ph.D. Thesis, Federal University of Amazonas, Manaus, Brazil, 2009.
14. Cannistraro, G.; Cannistraro, A.; Cannistraro, M.; Engineer, F.J.H. Evaluation of the sound emissions and climate acoustic in proximity of one railway station. *Hospitals* **2016**, *34*, 589–596. [CrossRef]
15. De Vasconcellos, E.A. *Transporte urbano nos países em desenvolvimento: Reflexões e propostas*, 3rd ed.; Annablume: São Paulo, Brazil, 2000.
16. Spiro, T.G.; Stigliani, W.M. *Chemistry of the Environment*, 2nd ed.; Prentice Hall: Upper Saddle River, NJ, USA, 2002; p. 489.
17. Department, D.N.T. *Fleet of Vehicles, by Type and with License Plate, According to the Municipalities of the Federation*; Denatran: Brasília, Brazil, 2014.
18. Teixeira, E.C.; Feltes, S.; Santana, E.R.R.d. Estudo das emissões de fontes móveis na região metropolitana de Porto Alegre, Rio Grande do Sul. *Química Nova* **2008**, *31*, 244–248. [CrossRef]
19. Lima, E.P.; Giimenes, M.L.; Lima, O.C. Estimação das emissões originadas de veículos leves na cidade de Maringá para o ano de 2005. *Acta Sci. Technol.* **2009**, *31*. [CrossRef]
20. Braga, A.; Pereira, L.A.A.; Böhm, G.M.; Saldiva, P.J.R.U. Poluição atmosférica e saúde humana. *Revista USP* **2001**, *30*, 58–71. [CrossRef]
21. Kuhn, U.; Ganzeveld, L.; Thielmann, A.; Dindorf, T.; Schebeske, G.; Welling, M.; Sciare, J.; Roberts, G.; Meixner, F.; Kesselmeier, J.J.A.C. Physics, Impact of Manaus City on the Amazon Green Ocean atmosphere: Ozone production, precursor sensitivity and aerosol load. *Atmos. Chem. Phys.* **2010**, *10*, 9251–9282. [CrossRef]
22. Cannistraro, G.; Cannistraro, M.; Piccolo, A.; Restivo, R. *Potentials and Limits of Oxidative Photocatalysisand Possible Applications in the Field of Cultural Heritage*; Advanced Materials Research; Trans Tech Publ: Pfaffikon, Switzerland, 2013; pp. 111–117.
23. EPA. Nitrogen Dioxide (NO2) Pollution. Available online: https://www.epa.gov/no2-pollution (accessed on 14 July 2014).
24. Lara, L.L.; Pauliquevis, T.M.; Procópio, A.S.; Rizzo, L.V. *Química atmosférica na Amazônia: A floresta e as emissões de queimadas controlando a composição da atmosfera amazônica*; SciELO: São Paulo, Brasil, 2005.
25. Briggs, D. Environmental pollution and the global burden of disease. *Br. Med. Bull.* **2003**, *68*, 1–24. [CrossRef] [PubMed]
26. Banerjee, T.; Barman, S.; Srivastava, R.K. Application of air pollution dispersion modeling for source-contribution assessment and model performance evaluation at integrated industrial estate-Pantnagar. *Environ. Pollut.* **2011**, *159*, 865–875. [CrossRef] [PubMed]
27. Costa, M.P.D. Avaliação da concentração de dióxido de nitrogênio no entorno de uma Usina Termelétrica em Manaus-AM. Master's Thesis, Universidade Federal do Amazonas, Manaus, Brazil, 2013.
28. Ugucione, C.; Gomes Neto, J.D.A.; Cardoso, A.A. Método colorimétrico para determinação de dióxido de nitrogênio Atmosférico com preconcentração em coluna de C-18. *Química Nova* **2002**, *1*, 352–357. [CrossRef]
29. Palmes, E.; Lindenboom, R.J.A.C. Ohm's law, Fick's law, and diffusion samplers for gases. *Anal. Chem.* **1979**, *51*, 2400–2401. [CrossRef]
30. The National Environment Council. *CONAMA Resolution No. 003/1990*, Special Edition; Volume Resolution No. 003/1990; Institute for the Environment and Renewable Natural Resource: Brasília, Brazil, 2012.

sustainability

MDPI

Article

Spatial and Temporal Variabilities of PM$_{2.5}$ Concentrations in China Using Functional Data Analysis

Deqing Wang [1], Zhangqi Zhong [2], Kaixu Bai [3] and Lingyun He [1,*]

[1] School of Management, China University of Mining and Technology, Daxue Road 1, Xuzhou 221116, Jiangsu, China; dekinywang@cumt.edu.cn
[2] School of Economics, Zhejiang University of Finance & Economics, Hangzhou 310018, Zhejiang, China; zzhongz@zufe.edu.cn
[3] Key Laboratory of Geographic Information Science (Ministry of Education), East China Normal University, Shanghai 200241, China; kxbai@geo.ecnu.edu.cn
[*] Correspondence: Lingyun_he@cumt.edu.cn

Received: 1 February 2019; Accepted: 14 March 2019; Published: 18 March 2019

Abstract: As air pollution characterized by fine particulate matter has become one of the most serious environmental issues in China, a critical understanding of the behavior of major pollutant is increasingly becoming very important for air pollution prevention and control. The main concern of this study is, within the framework of functional data analysis, to compare the fluctuation patterns of PM$_{2.5}$ concentration between provinces from 1998 to 2016 in China, both spatially and temporally. By converting these discrete PM$_{2.5}$ concentration values into a smoothing curve with a roughness penalty, the continuous process of PM$_{2.5}$ concentration for each province was presented. The variance decomposition via functional principal component analysis indicates that the highest mean and largest variability of PM$_{2.5}$ concentration occurred during the period from 2003 to 2012, during which national environmental protection policies were intensively issued. However, the beginning and end stages indicate equal variability, which was far less than that of the middle stage. Since the PM$_{2.5}$ concentration curves showed different fluctuation patterns in each province, the adaptive clustering analysis combined with functional analysis of variance were adopted to explore the categories of PM$_{2.5}$ concentration curves. The classification result shows that: (1) there existed eight patterns of PM$_{2.5}$ concentration among 34 provinces, and the difference among different patterns was significant whether from a static perspective or multiple dynamic perspectives; (2) air pollution in China presents a characteristic of high-emission "club" agglomeration. Comparative analysis of PM$_{2.5}$ profiles showed that the heavy pollution areas could rapidly adjust their emission levels according to the environmental protection policies, whereas low pollution areas characterized by the tourism industry would rationally support the opportunity of developing the economy at the expense of environment and resources. This study not only introduces an advanced technique to extract additional information implied in the functions of PM$_{2.5}$ concentration, but also provides empirical suggestions for government policies directed to reduce or eliminate the haze pollution fundamentally.

Keywords: PM$_{2.5}$ concentrations; functional principal component analysis; adaptive clustering analysis; functional ANOVA; spatial and temporal difference

---

1. Introduction

With the rapid development of industrialization and urbanization in China, haze pollution characterized by particulate matter smaller than 2.5 μm occurs more frequently and widely, which has seriously endangered the physical and mental health of residents, and threatened the sustainable development of China's economy. According to statistics, the severe haze events that occurred in the first quarter of 2013 affected about 13.5% of the land area and 800 million people in China [1]. It is estimated that without a pollution control policy, the particulate matter pollution in China will lead to a 2% GDP loss and 25.2 billion USD in health expenditure in 2030 [2]. Thus, the prevention and control of haze pollution is not only a major livelihood project, but also an important way to assist the transformation of China's economic development model and the optimization and adjustment of China's economic structure. Since China's State Council released the "Air Pollution Prevention and Control Action Plan" in September 2013, which was a milestone for reducing $PM_{2.5}$ concentrations, local governments have promulgated their own air pollution control action plans. However, due to the multiple effect of various complex factors, such as an extensive development mode, unbalanced industry structure, and inefficient energy utilization, the fluctuations of $PM_{2.5}$ concentrations in different provinces exhibits obvious regional disparities and temporal characteristics [3–8]. Therefore, understanding the dynamic behavior of $PM_{2.5}$ concentrations is beneficial to further formulate and implement targeted environmental protection policies.

As a developing country with a dual structure, China is characterized by an unbalance of regional economic development and deteriorating environmental problems which resulted from its extensive mode of economic development and over-consumption of energy. Many researchers have pointed that haze pollution has become an obstacle for China to attract foreign investment, talent and tourists, and even threatens sustainable development in China [9,10]. Since $PM_{2.5}$ concentrations always change with time and fluctuate diversely across regions, intensive studies have been carried out on interpreting the spatial and temporal variability of $PM_{2.5}$ concentrations in China, both from city-level and national-scale perspectives. For example, taking Weifang city as a research object and based on the data of controlled monitoring stations, Li et al. concluded that the annual $PM_{2.5}$ concentrations reached a peak in 2013, while the seasonal and monthly $PM_{2.5}$ concentrations formed a U-shaped trend [11]. Considering Beijing and six surrounding cities as main research areas and based on correlation analysis of geo-statistics techniques, Zhai et al. studied the relevant relationship of $PM_{2.5}$ concentrations in Beijing [12] and found that the pollutant concentrations exhibit obvious cyclical fluctuation patterns with significant spatial correlation. Studies on spatial-temporal characteristics of $PM_{2.5}$ concentration on the national-scale includes references [13–16], their common conclusions are that China's haze pollution presented an obvious spatial spillover effect, and that $PM_{2.5}$ emissions had strong positive spatial autocorrelation with a certain spatial heterogeneity.

In light of the fact that $PM_{2.5}$ concentrations are the combined result of various factors, numerous literatures focus on exploring its primary cause via advanced methods. For example, Guan et al. presented an interdisciplinary study to measure the magnitudes of socio-economic factors in driving primary $PM_{2.5}$ emission changes in China between 1997–2010 [17]. According to the latest air quality standards of China, Wang et al. characterized the spatial and temporal variations of the concentrations of PM10, $PM_{2.5}$ and PM1 in China, their conclusion showed that the ratios of $PM_{2.5}$ to PM10 showed a clear increasing trend from northern to southern China, and that both emissions and meteorological variations dominate the long-term PM concentration trend, while meteorological factors played a leading role in the short term [18]. In order to monitor $PM_{2.5}$ by remote sensing in the Yangtze delta, Xu and Jiang constructed a $PM_{2.5}$ concentration model based on MODIS AOT, $PM_{2.5}$ concentration data of the 36 ground air quality observation sites and meteorological data, and empirical results proved their model estimation was higher than classical methodology [19]. Through the CAMx model, Cheng et al. examined spatial-temporal variations of $PM_{2.5}$ concentrations during two alerts based on multiple data sources, their results suggested that the implementation of emission reduction measures 1–2 days before red alerts could lower the peak of $PM_{2.5}$ concentrations significantly [20]. Using $PM_{2.5}$ concentrations data at China's provincial level over 1998–2012, Shao et al. adopted a dynamic spatial panel model and SGMM to empirically identify the key determinants of smog pollution, their results indicated that there was a significant U-shape curve relationship between smog pollution and economic growth, and smog pollution was worsening with economic growth in most eastern provinces [16]. With

PM10 and PM$_{2.5}$ concentration data collected from five air-quality monitoring sites in Lanzhou from October 2014 to October 2015, Guan et al. investigated the primary transport path using Hybrid Single Particle Lagrangian Integrated Trajectory Model (HYSPLIT) and the PM$_{2.5}$-to-PM10 ratio model [21]. Noticeably in these studies, all model constructions and empirical results were based on discrete and equal-sampled observations without any error disturbance. Additionally, the spatial-temporal characteristics of the PM$_{2.5}$ concentrations are also the major issues for air pollution investigations in many other countries, including developing and developed countries or regions. An array of literature focuses on assessing PM$_{2.5}$ spatial-temporal variability. For example, based on data from biophysical remote sensing and GIS, Famoso F, et al. conducted the measurement and modeling of ground-level ozone concentration of Catania in Italy [22]. Using PM$_{2.5}$ concentrations at 71 EPA monitoring stations from 2006 to 2011, Wu et al. applied a hybrid kriging/LUR model to assess the spatial-temporal variability of PM$_{2.5}$ for Taiwan [23]. In order to identify the local and long-range sources of PM$_{2.5}$ and their relationships with other air pollutants and meteorology, Mukherjee et al. investigated the local and distant sources of PM$_{2.5}$ from 2014 to 2017 in Varanasi city located in middle Indo-Gangetic plain (IGP) of India using various statistical modeling methods [24], such as conditional bivariate probability function (CBPF), land use regression (LUR) and trajectory statistical models (TSM) like potential source contribution function (PSCF),concentration weighted trajectory (CWT) and trajectory cluster analysis. Considering LUR models may fail to capture complex interactions and non-linear relationships between pollutant concentrations and land use variables, Brokamp et al. developed a novel land use random forest (LURF) model and compared its accuracy and precision to a LUR model for elemental components of PM in the urban city of Cincinnati, Ohio [25]. The comprehensive comparison showed that these methodological approaches provide efficient means to better assess PM$_{2.5}$ spatial-temporal variations and prediction levels, and usually work well with large scale pollution dispersion.

Although the existing studies on PM$_{2.5}$ concentrations have provided many meaningful suggestions, their shortcomings are also obvious. Firstly, most of the empirical methods were statistical descriptions or econometric modeling using discrete noisy data, which cannot mine the continuous trajectory and dynamical information implied in the changing process of PM$_{2.5}$ concentrations. Secondly, most studies focused on the research scale of mainland China and metropolitan areas which neglected the increase in regional differentiation, or analyzed the individual district separately, with little consideration of the homogeneity of different regions. Thirdly, the existing studies used mostly rough and historical data collected by ground monitoring stations. Unlike the air pollution index, PM$_{2.5}$ concentrations have only been recorded since 2012 in China, thus having too short or too old time scales that result in a low temporal resolution.

It should be noted that data in many scientific experiments are recorded repeatedly through time or space and have been seen to arise as a continuous process. Examples of such kinds of observations are hourly records of PM$_{2.5}$ concentrations and daily records of air quality. The classical discrete data modeling approaches are found to be inadequate in understanding the underlying process of the pollutant and hence prevent the implicit information from being revealed [26,27]. The coming era of big data makes it possible to analyze these discrete noisy data by converting them into continuous and smoothing functions, then we can explore the dynamic information implied in the original data from multiple derivative functions [28]. The new modern statistical methodology which considers discrete time point values as observations of continuous functions over a continuum is termed as *Functional Data Analysis* (FDA) [29]. The functional concept may bring additional insight by looking at the pattern and temporal variation of pollutant variables in the form of smoothing curves or functions. A previous study by Shaadan et al. highlighted the advantages of an FDA approach in assessing and comparing the PM10 behavior [27], while several studies that focus on using FDA to analyze the pollution behavior have proved the merits of FDA in environmental pollution research [27,30–32]. To the best of our knowledge, there is little research studying the spatial-temporal variability of PM$_{2.5}$ concentrations in China within the framework of continuous functions. Thus, using PM$_{2.5}$ concentrations data at provincial level from 1998 to 2016, this study will employ FDA to classify the fluctuation patterns of PM$_{2.5}$ pollution for 34 provinces, and dynamically compare their evolving trajectories. The empirical results is helpful for enhancing the recognition of the spatial distributions and dynamic changes of PM$_{2.5}$ concentrations in China, and can provide quantitative support for governments to formulate and implement air pollution prevention and control measures.

2. Methodology

In this subsection, we introduce the framework of FDA, which mainly includes smoothing PM$_{2.5}$ pollution functions with roughness penalty, classifying categories of fluctuations via adaptive weighting clustering analysis, and testing the significance of difference among different regions using functional ANOVA. Data processing and analysis are conducted using the free R software (R Development Core Team, 2018), together with package "*fda.usc*" (Febrero-bande et al., 2016) [33] and package "*fda*" (Ramsay et al., 2013) [34].

2.1. Smoothing with or without Roughness Penalty

PM$_{2.5}$ concentrations data is often recorded at discrete time intervals, and is usually analyzed within the framework of traditional time series or multivariate statistical approaches. But in the context of functional data analysis, the PM$_{2.5}$ concentration data is essentially assumed to be continuous with time, even though the concentration data is collected at a daily, monthly or annual frequency. The primary goal of FDA is to convert discrete data, such as $y_{i1}, \cdots, y_{iT_i}$, to a smooth function $f_i(t_j)$, which is computable for any values of t_j with $j = 1, \cdots, T_i$. There are two ways to convert the discrete data into continuous functions, their core difference lies in the presence or absence of disturbance factors. If the data is assumed to be errorless, that is $y_{ij} = f_i(t_j)$, the interpolation method may be employed. However, if there are observational errors that need removing, the smoothing process will be used. In reality, the PM$_{2.5}$ concentrations data is always contaminated by random noise ε_{ij}, that is $y_{ij} = f_i(t_j) + \varepsilon_{ij}$. Considering the universality of practical problems and our intention of converting the discrete noisy data into quadratic differentiable functions, we mainly discuss the smoothing functional method with roughness penalties to error disturbances. Assuming $\Phi(t) = \{\phi_1(t), \cdots, \phi_L(t)\}$ to be the optimal basis function in Hilbert space, the *sum of squared fitting residuals for the roughness penalty* (*PENSSE$_\kappa$*) [29,34,35] is given as follows:

$$PENSSE_\kappa = \sum_{i=1}^{n} \{\sum_{j=1}^{T_i} [y_{ij} - f_i(t_j)]^2 + \kappa \int_T [f_i''(t)]^2 dt\} \quad (1)$$

The intrinsic continuous function $f_i(t)$ in Equation (1) is a linear approximation of the basis function to meet the criterion of minimizing the $PENSSE_\kappa$, i.e., $f_i(t) = \sum_{l=1}^{L} \beta_{il}\phi_l(t)$, where β_{il} denotes the coefficients of the basis function expansion. The smoothing parameter κ specifies the proportion between the goodness of model fitting and the smoothing amount of the function curve. Large values of κ will increase the amount of smoothing. The best value for the smoothing parameter κ is determined by the minimum generalized cross-validation $GCV(\kappa)$ [34]. The criterion is given as follows:

$$GCV(\kappa) = \left(\frac{n}{n - df(\kappa)}\right)\left(\frac{PENSSE_\kappa}{n - df(\kappa)}\right) \quad (2)$$

where the degree of freedom $df(\kappa) = trace\{\Phi(\Phi'\Phi + \kappa R)^{-1}\Phi'\}$ and the roughness penalty matrix R is expressed as $R = \int D^2\phi(s) \cdot D^2\phi'(s)ds$. Based on the above symbols, solving Equation (1) for β will give us $\hat{\beta} = (\Phi'\Phi + \kappa R)^{-1}\Phi'y$, then 34 provinces with 18 yearly measurements will be transformed into 34 PM$_{2.5}$ concentrations curves. A complete theoretical review of the penalty smoothing method can be found in Kokoszka et al. (2017) [36], and the steps of the algorithm are detailed in Ramsay et al. (2009) [34]. It should be noted that FDA does not restrict all samples to be sampled at regular intervals or same frequency on the observing interval, that is $T_i \neq T_j$. Thus, the relaxed structure of data collection and hypothesis of distribution enable FDA to depict practical problems more comprehensively and flexibly [37]. Particularly, once the intrinsic functions are reconstructed from the discrete noisy data, we can not only display the continuously changing trajectory of PM$_{2.5}$ concentrations statically from the holistic perspective, but also can analyze their dynamic process interactively from multiple derivative functions.

2.2. Significance Test of Difference via Functional Analysis of Variance

The functional analysis of variance (F-ANOVA) is used to test whether two or more sets of functional data are identical, independent, and come from the same population. The verification was done by comparing their functional means. Let g represent the number of groups or zones, with $f_{ij}(i = 1, \cdots, g; j = 1, \cdots, n_i)$ as the jth-functional data for i groups, and n_i is the number of curves in group i. As a first step in F-ANOVA, the classical F statistic in the form of functional data is considered and is given as:

$$F_n = \frac{\sum_{i=1}^{g} n_i \left\| \overline{f_{i.}} - \overline{f_{..}} \right\|^2 / (g-1)}{\sum_{i,j} \left\| f_{ij} - \overline{f_{i.}} \right\|^2 / (n-g)} \tag{3}$$

where $\|\cdot\|$ denotes the usual L^2 norm as $\|f\| = (\int f^2(t)dt)^{1/2}$. The expressions used in Equation (3) are described by $f_{ij} = (f_{ij}(t_1), \cdots, f_{ij}(t_T))\prime$, $\overline{f_{i.}} = (\overline{f_{i.}}(t_1), \cdots, \overline{f_{i.}}(t_T))\prime$ and $\overline{f_{..}} = (\overline{f_{..}}(t_1), \cdots, \overline{f_{..}}(t_T))\prime$, which can be computed as $\overline{f_{i.}}(t) = \sum_j^{n_i} f_{ij}(t)/n_i$, $n = \sum_{i=1}^{g} n_i$ and $\overline{f_{..}}(t) = \sum_{i=1}^{g} n_i \overline{f_{i.}}(t)/n$. $\overline{f_{..}}$ is the global functional mean and $\overline{f_{i.}}$ is the functional mean in the ith groups, respectively, at time t. With the above symbols, the equivalent statistic of Equation (3) can be rewritten as:

$$V_n = \sum_{i<j}^{g} n_i \left\| \overline{f_{i.}} - \overline{f_{j.}} \right\|^2 \tag{4}$$

Given the null hypothesis of having the same functional means for each i group, that is, $H_0 : \overline{f_{1.}} = \cdots = \overline{f_{g.}}$, calculate the critical value $P_{H_0}\{F > F_{n,\alpha}\} = \alpha$ and $P_{H_0}\{V > V_{n,\alpha}\} = \alpha$ at the specified significance level α respectively. H_0 should be rejected if the variability between groups, which are measured by the difference in the sample means F_n and V_n, is large enough to be expressed as $F_n > F_{n,\alpha}$ and $V_n > V_{n,\alpha}$. In other words, the test is found to be statistically significant if the p-value is less than the α significance level. The detailed steps of algorithm can be found in Cuevas et al. (2004) [38]. This procedure uses a point-wise critical value obtained using a permutation test for reference lines [39].

2.3. Functional Principal Component and Adaptive Clustering Analysis

The intrinsically infinite dimensionality of functional data poses challenges to traditional clustering methods used for classifying discrete data, both for theory and computation [40–42]. In order to reduce the cost of calculation and elevate the accuracy of classification, we employ the adaptive weighting clustering analysis to classify the fluctuation patterns of PM$_{2.5}$ concentrations curves, and use bootstrap sampling methods to test the significance and robustness of difference among groups.

Let $V(s, t) = (N-1)^{-1} \sum_{i=1}^{N} [f_i(s) - \overline{f}(s)][f_i(t) - \overline{f}(t)]$ be a continuous covariance operator on $[0, T]^2$, by Mercer's lemma [43], there exists an series of orthogonal functions $\varphi_k(t)$ with their corresponding non-negative decreasing eigenvalues λ_k satisfying:

$$\int_0^T V(s, t)\varphi_l(s)ds = \lambda_l \varphi_l(t) \quad t \in [0, T], l \in N \tag{5}$$

with respect to

$$\int_0^T \varphi_l(t)\varphi_m(t)dt = \delta_{lm} = \begin{cases} 1, & m = l \\ 0, & m \neq l \end{cases} \tag{6}$$

Further, for the second-order continuous stochastic process $\{f(\cdot), t \in [0,T]\}$ on $L^2(T)$, the realization of the process for the ith subject is $f_i(t)$. Denote $\mu(t)$ and $V(s,t)$ as the mean and covariance of $f_i(t)$, respectively. Then the Karhunen-Loève expansion of $f_i(t)$ [44] is given as:

$$f_i(t) = \mu(t) + \sum_{k=1}^{\infty} \zeta_{ik}(f_i)\varphi_k(t) , \quad t \in [0,T] \tag{7}$$

where $\zeta_{ik}(f) = \int_T (f_i(t) - \mu(t))\varphi_k(t)dt$ are the functional principal components (FPCs), sometimes referred to as *scores*. The $\zeta_{ik}(\cdot)$ are independent across i for a sample of independent trajectories and are uncorrelated across k with $E(\zeta_{ik}) = 0$ and $var(\zeta_{ik}) = \lambda_k$. Furthermore, the covariance of $\zeta_{ik}(\cdot)$ satisfies

$$E[\zeta_k(f)\zeta_l(f)] = \lambda_k \delta_{kl} \quad k,l \in N \tag{8}$$

From the Karhunen-Loève expansions of stochastic process, we can infer that $\zeta_{ik}(f)$ are the projection scores of centered functions $(f_i(t) - \mu(t))$ to the direction of a standard orthogonal basis function $\varphi_k(t)$, which is objectively derived from the information implied in original PM$_{2.5}$ concentrations data. Based on the Karhunen-Loève expansion of Equation (7), the difference among categories of different functional data is entirely reflected by the difference between their projected scores $\zeta_{\cdot k}(f)$. Since λ_k is also the variance of $\zeta_{\cdot k}(f)$, and without loss of generality, assume their sequence order satisfying $\lambda_1 \geq \lambda_2 \geq \cdots \geq 0$. In order to reflect the objective difference of classification information implied in $\zeta_{\cdot k}(f)$, define $\beta_k = \lambda_k/\sum_{l \geq 1}\lambda_l$ as the weight of $\zeta_{\cdot k}(f)$, we reconstruct the adaptive weighting distance between $\zeta_i(f)$ and $\zeta_j(f)$ as:

$$d[f_i(t), f_j(t)|q] = [\sum_{l=1}^{\infty} (\beta_l|\zeta_l(f_i) - \zeta_l(f_j)|)^q]^{\frac{1}{q}} \tag{9}$$

The distance parameter q is analogous to the classical definition of similarity, with $q = 2$ corresponding to the Euclidean distance. In practice of conducting adaptive clustering analysis, it is unnecessary to choose all the FPCs. Without a loss of core information, the criteria for selecting the number of FPCs is the minimum value M that reaches a certain level of the proportion of total variance explained by the M leading components, such as $\sum_{l=1}^{M} \lambda_l / \sum_{l \geq 1} \lambda_l 1_{\{\lambda_l > 0\}} \geq 90\%$. Further information on the theoretical foundation and applications of functional adaptive clustering method could be obtained from our previous works [45–47].

3. Data Sources and Empirical Results

3.1. Data Sources

The reliable data source of PM$_{2.5}$ concentrations is crucial for this study. After China's Ministry of Environmental Protection issued the new environmental air quality standard in February 2012, local governments began to routinely record and release the data of PM$_{2.5}$ concentrations. Due to lacking data of a long-term time span, it is difficult to extract the dominant patterns of evolution for PM$_{2.5}$ concentrations. Besides, because the number of ground monitoring stations is small and its distribution is uneven, the rough reflection using sparse points to denote the whole area cannot exactly measure the real situation of PM$_{2.5}$ concentrations. In order to solve the data deficiency of historical and regional PM$_{2.5}$ concentrations, this paper adopts the data sets regarding the raster data of the annual average PM$_{2.5}$ concentrations at a global level using satellite-based environmental surveillance, which is published by the socio-economic research center at Columbia University. The data sets used here are obtained from the study by van Donkelaar et al. (2016) [48], which had calibrated each AOD source using AERONET observations. Based on the data sets, using geographic information system technology, we could obtain the corresponding raster data of the annual average PM$_{2.5}$ concentrations in China for the period 1998–2016. Notably, however, compared with that directly from actual monitoring data on the ground, although the data sets collected from satellite-based monitoring process could be affected by meteorological factors, which thereby led to a lower accuracy, the data sets from actual monitoring data on the ground could only roughly provide $PM_{2.5}$ concentrations in a region using area object other than point one based on point source data, and thus it's difficult to accurately measure global $PM_{2.5}$ concentrations in the region. Being an important non-point source data, satellite-based monitoring data sets have more advantages than the traditional methods in terms of reflecting the value of the $PM_{2.5}$ concentration and its change trend in a region. Actually, the research based on satellite-based monitoring data has won the recognition of the academics, owing to the works of Nordhaus et al. [49,50], who won the Nobel Prize Economics in 2018. Thus, the satellite-based monitoring data employed by this study is reliable. Additionally, from the technical perspective of empirical analysis, FDA owns the congenital advantage of modeling noisy data when smoothing with roughness penalty, even when the data is sparse or sampled unequally. Thus, having combined the reliable data source with the advanced methodology, it is reasonable to draw reliable conclusions.

3.2. Reconstructing PM$_{2.5}$ Concentrations Functions and Summary Statistics

As a rule of thumb, it is safer to smooth only when necessary if we want to retain the maximal information [51,52]. In order to verify the necessity of roughness penalty in reconstructing $PM_{2.5}$ concentrations functions, we firstly select the optimal smoothing parameter which minimizes the GCV. Figure 1 shows how the GCV criterion varies as a function of $\log_{10}(k)$ for the mean of $PM_{2.5}$ concentrations. The minimizing value of k is found to be 1.25, and at that value $df(k) = 3.81 \approx 4$. Next, we plot the penalized $PM_{2.5}$ concentrations curve with the selected smoothing parameter, and the comparison object, that is the mean of un-penalized $PM_{2.5}$ concentrations curve without roughness penalty, is also plotted in Figure 2. Taking the trajectory of the penalized curve as benchmark, we can clearly see that the mean of $PM_{2.5}$ concentrations experienced a fluctuation, increased rapidly and then declined slowly, and reached its maximal value round 2007. Though there is a slight rebound during the descending process, the $PM_{2.5}$ concentrations kept a downward trend at the end of the research interval, which can be attributed to the synthetic effect of environmental protection policies [53]. In contrast, the trajectory of the un-penalized $PM_{2.5}$ concentrations curve fluctuated frequently with a cycle about every two years, but the dominant changing trend of $PM_{2.5}$ concentrations was obscured by those slight fluctuations with various amplitudes. Thus, we decided to smooth the $PM_{2.5}$ concentrations with a roughness penalty at the value of $k = 1.25$.

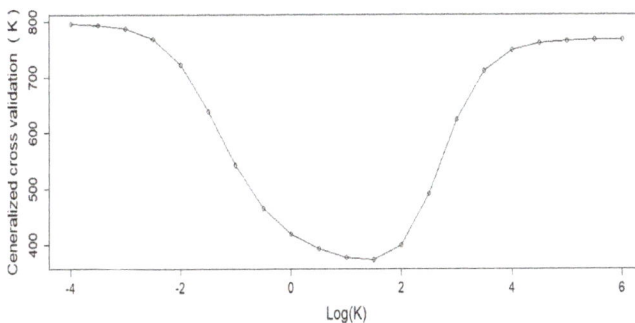

Figure 1. The values of generalized cross validation or GCV criterion for choosing the smoothing parameter for fitting the mean of $PM_{2.5}$ concentrations.

Figure 3 displays the summary statistics for the functional information of $PM_{2.5}$ concentrations in terms of their mean and standard deviation for all regions. It shows that generally, the highest mean $PM_{2.5}$ concentrations were recorded around 2007, the year during which environmental protection policies were formulated and implemented intensively in China, such as a campaign for energy-saving.

The trajectory of the standard deviation function also follows the same pattern as the functional mean of $PM_{2.5}$ concentrations. That is, the $PM_{2.5}$ concentrations variability increased rapidly since 1998, and reached its maximal value around 2007, then kept a high level with a slight rebound. It should be noted that the value of standard deviation is larger when the level of $PM_{2.5}$ concentrations is high. For the increasing deviation, we ascribe it to the differentiated reactions from different regions when facing the dilemma between environmental protection and extensive economic development.

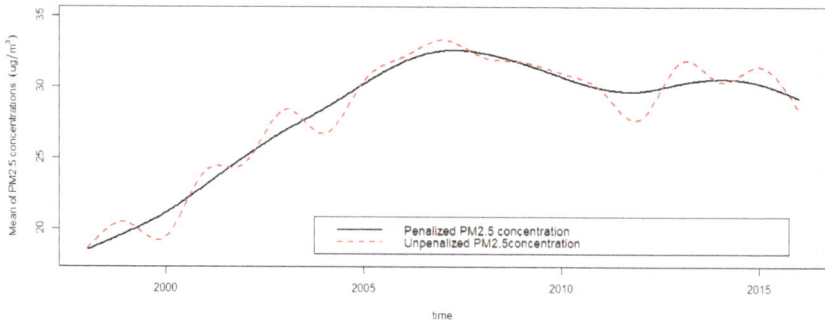

Figure 2. Smoothing curves with roughness penalty and without roughness penalty for the mean of $PM_{2.5}$ concentrations.

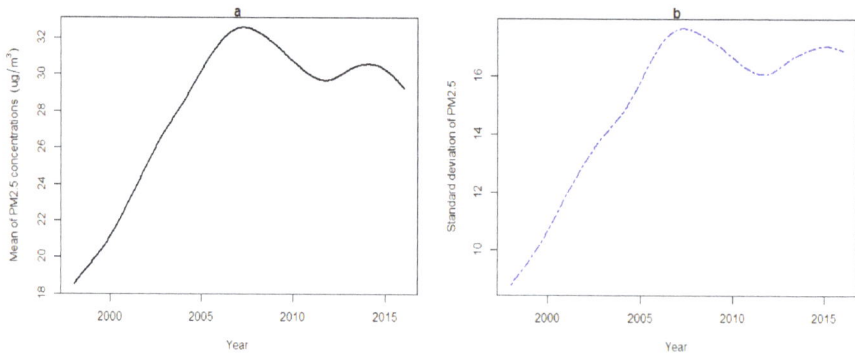

Figure 3. (**a**) The average of mean and (**b**) the standard deviations for yearly $PM_{2.5}$ concentrations curves of all provinces.

Information about the first and second derivatives from the smoothing function can give information on the rate of change and the acceleration in $PM_{2.5}$ concentrations according to time compared to the traditional multivariate statistical approaches which could not possibly capture this kind of information [24,25]. In order to dynamically analyze the evolving process of $PM_{2.5}$ concentrations from 1998 to 2016, we can extract more information by studying how derivatives relate to each other, which is often called a *phase-plane plot* (PPP) [54]. The energy transferring between the first order derivative of $PM_{2.5}$ concentrations which is called average velocity and the second order derivative which is called average acceleration, was shown in Figure 4. The numbers along the curve indicate the year of $PM_{2.5}$ concentrations. The trajectory of PPP exhibits several interesting features. There were two obvious cycles of energy transferring between velocity and acceleration, with the year 2007 as a landmark. During the first cycle, although the sign of growth acceleration for $PM_{2.5}$ concentrations alternated from positive to negative frequently, the growth velocity remained positive all the time, and the largest growth velocity occurred between 2001 and 2002. During the second cycle from 2007 to 2016, both the sign of growth velocity and acceleration alternated between positive and negative, with a larger oscillation. The first cycle corresponded to the period during which the decoupling indicators of China's resources consumption and GDP growth is much lower. The key reason for this phenomenon is that China was in the process of industrialization, particularly in the process of heavy industrialization, which caused the rapid growth of infrastructure construction and consumed vast amounts of basic materials. The second cycle corresponded to the period during which the PM$_{2.5}$ concentrations fluctuated with a high frequency, due to the intensive formulation and implementation of environmental protection policies.

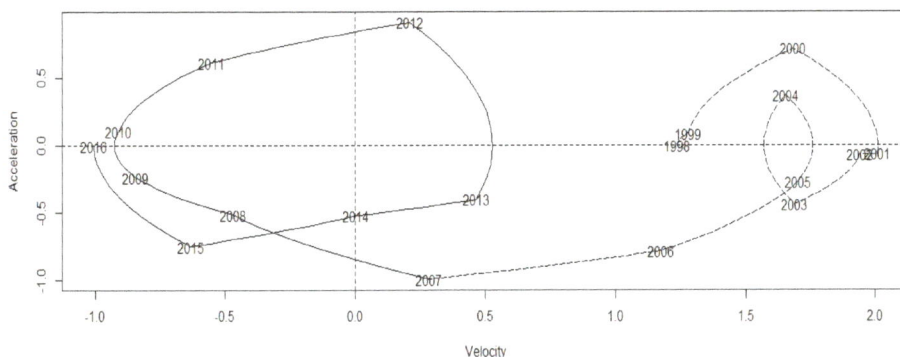

Figure 4. The phase-plane plot for the average PM$_{2.5}$ concentrations curve: the second derivative (acceleration) versus the first derivative (velocity).

3.3. Temporal Variability Decomposition

As one of the most important advantages for FDA, the temporal variance-covariance surface as well as the corresponding contour in functional data gives new ways to gather information, more than a single value or matrix obtained in the traditional univariate and multivariate contexts [55]. The estimated variance surface of PM$_{2.5}$ concentrations from 1998 to 2016 with its corresponding contour plot are presented in Figure 5. We can see the variability becoming larger and larger since 1998, and the highest variability occurs around 2007, the period which also corresponds to the highest mean of PM$_{2.5}$ concentrations. In order to further explore the potential variation from curve to curve, we employ functional principal components analysis (FPCA) to decompose the covariance function. Figure 6 displays the result of covariance decomposition via FPCA for PM$_{2.5}$ concentrations after varimax rotation. For each of the first three principal components, three curves are plotted. The solid curve is the overall smoothed mean which is the same in all provinces just for reference purposes, and the other two curves show the effect of adding and subtracting a suitable multiple of the principal component weight function. The accumulative percentage of variance explained by the first three components is 99.7%, indicating that there was almost no valuable information lost.

It can be seen that, each of the three principal component functions quantifies variability corresponding to a particulate period, thus the trajectories of the varimax rotated FPCs give good interpretations. Specifically, the first principal component function, which accounts for 69.2% of the total variation in the original PM$_{2.5}$ concentrations observations, mainly depicts the variability from 2003 to 2012. Actually, the period from 2003 to 2012 was called the "golden ten years" for the coal industry, which also are the "golden ten years" of China's rapid economic growth. However, restricted to various factors such as industrial structure and resources endowment, each province can only choose the suitable development mode according to its own situation. As a result, the emissions level of particulate matter for each province deviated greatly from the overall mean. Consequently, the covariance function of PM$_{2.5}$ concentrations among 34 provinces oscillated drastically during the period of fossil fuel energy being highly consumed. In contrast, the second and the third principal component function mainly reflect variability located at the end and beginning of the research period, respectively. The proportions of total variation they accounted for is nearly equal, that is 15% and

15.5%, which is even less than the one fourth of the amount explained by the first principal component function. In light of this, the vast disparity in variance contribution rate for each principal component function requires differentiated treatment when conducting functional clustering analysis on the scores of principal components.

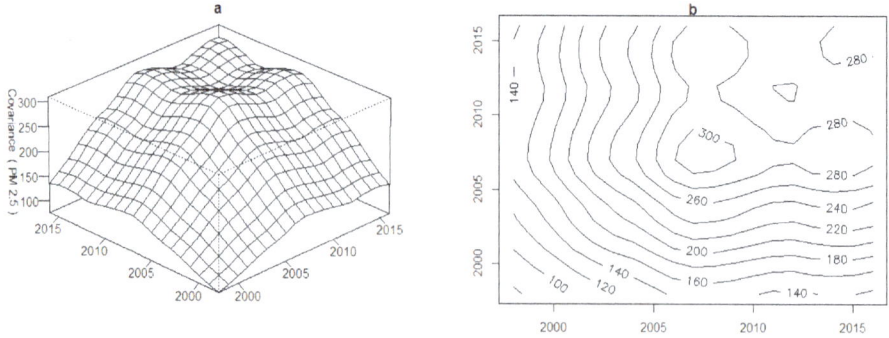

Figure 5. (**a**) Estimated variance surface of PM$_{2.5}$ from 1998 to 2016 and (**b**) the corresponding contour map.

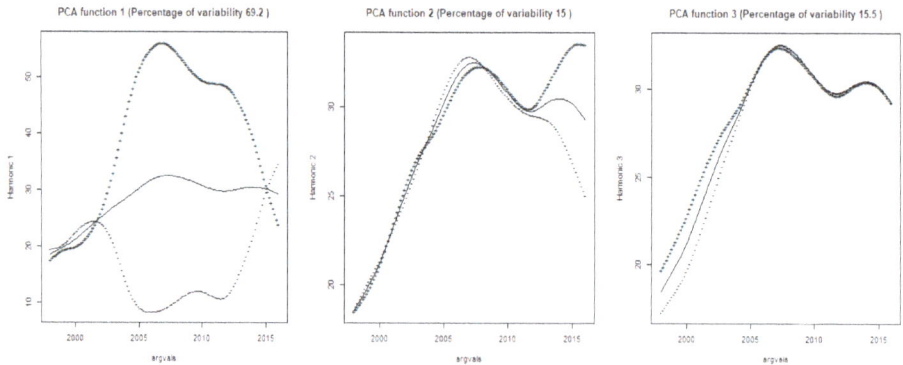

Figure 6. The first three varimax-rotated principal components of PM$_{2.5}$ concentrations.

3.4. Region Classification and Significance Test

In order to visually explore how curves clustering within the three-dimensional subspace spanned by the first three principal component functions, Figure 7 displays the scatter plots of scores on pairs of weight functions for each province. It shows that there is essentially no correlation among these scores, so the three principal components can be considered as uncorrelated variables within 34 provinces. Although the three scatter plots show no very distinctive features, the distribution range for each of the three component differs vastly. It can be seen that the scores on the first principal component ranges from about −100 to 150, with a considerable lager amount of variability. However, the scores on the other two components distribute with a nearly equal range, which is far less than that of the first component. In view of the vast disparity of information amount, different weights for the three principal components should be taken into account when employing clustering analysis to classify the categories of fluctuation.

As a preliminary step of unsupervised classification, it is necessary to determine the number of clusters before conducting adaptive weighting clustering analysis. The optimal number of clusters in unsupervised classification is still an open question [56]. In this study, we adopt the *wssplot*( ) and *NbClust*( ) functions to objectively choose the number of clusters [57]. The selecting criterion presented in Figure 8 indicates that there is a distinct drop in the within-groups sum of squares when moving from one to eight clusters. After eight clusters, this decrease drops off, suggesting that an eight-cluster solution may be a good fit to the $PM_{2.5}$ concentrations data in 34 provinces. Besides, 14 of 24 criteria provided by the *NbClust* package suggest an eight-cluster solution. So we chose eight as the optimal number of clusters, and the initial classification via adaptive weighting clustering was listed in the second column of Table 1, the spatial distribution of $PM_{2.5}$ concentrations for each group was illustrated in Figure 9.

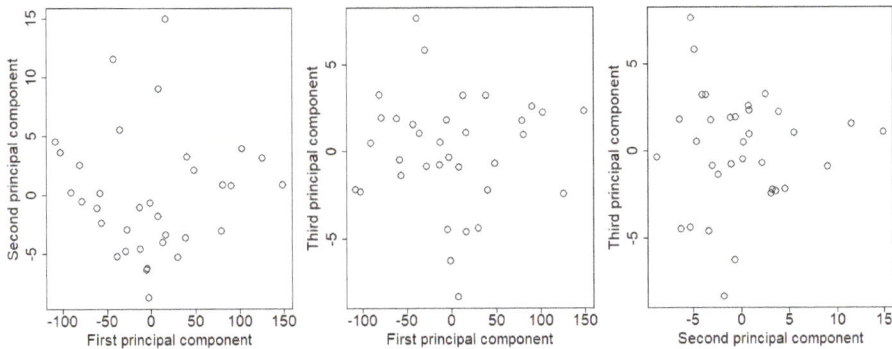

Figure 7. Plot of the first three principal components scores of $PM_{2.5}$ concentrations.

Although we have objectively classified the $PM_{2.5}$ concentrations curves of 34 provinces into eight clusters, it is necessary to quantitatively conduct a further test in the robustness of the initial classification. In other words, we should prove the hypothesis that there indeed was significant difference between the eight groups. To address the above problem, the F-ANOVA based on 1000 bootstrap sampling is performed on original functions as well as their velocity and acceleration, respectively. Figure 10 illustrated the test results of the original $PM_{2.5}$ concentration functions, and the robust test results corresponding to the first order and the second order derivatives were presented in Figures 11 and 12, respectively. Using the test results of F-ANOVA from Figures 10–12, we can safely draw the conclusion that, the fluctuation patterns between the eight groups of $PM_{2.5}$ concentration functions was significantly different at the level of 1%, whether from the static perspective or from multiple dynamic perspectives. Thus, on a credible quantitative analysis basis, we are confident in excavating more reliable and deeper information by further comparing the different trajectories of $PM_{2.5}$ concentration curves in each groups.

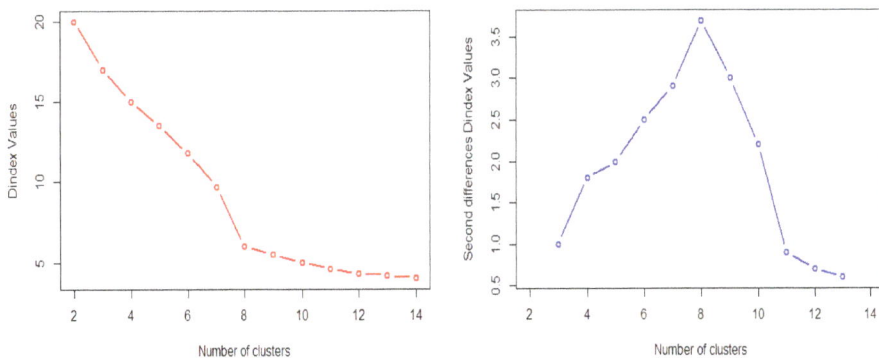

Figure 8. Dindex graphic for determining the best number of clusters.

125

Table 1. The classification of PM$_{2.5}$ concentrations fluctuation.

Group	Provinces	Characteristics	Reasons
1	Liaoning, Jilin, Zhejiang, Guangdong, Guangxi	the dominant fluctuation pattern of PM$_{2.5}$ concentrations in China with slightly more than the national average level and a moderate deviation in the end stage	sparsely-populated provinces with developed heavy industry, or intensively-populated provinces of highly developed tertiary industry Pearl River Delta
2	Heilongjiang, Hainan, Sichuan, Yunnan	the second lowest level with a slightly growing trend and an increasing deviation	provinces with tourism as their pillar industry
3	Shanxi, Jiangxi, Chongqing, Guizhou, Hong Kong, Macao	the dominant fluctuation pattern of PM$_{2.5}$ concentrations in China with slightly less than the national average level	intensively-populated provinces with steady and humid atmospheric
4	Fujian, SHANXI, Ningxia, Taiwan	the third lowest level with a nearly constant deviation	provinces in the southeast coast of China strongly influenced by maritime monsoon, or provinces in western with stable atmospheric circulation throughout the year
5	Neimenggu, Tibet, Gansu, Qinghai, Xinjiang	the lowest level, without obvious growth or deviation.	sparsely-populated provinces in western frontier of China, with traditional agriculture and livestock farming
6	Tianjin, Shandong	the highest level and largest fluctuation amplitude, with obvious turning points corresponding to government environmental policies	energy-intensive industries with enriched, high-frequency use of diesel freight vehicles and non-road machinery
7	Shanghai, Jiangsu, Anhui, Henan	the second highest level, mainly located at Yangtze River Delta with obvious secondary pollution	the most active economic area in China, labor-intensive and enriched industries, resulting in a large quantity of fumes discharged from vehicles
8	Beijing, Hebei, Hunan, Hubei	the third highest level with a growing deviation	highly intensive-populated region, or inland region with secondary pollution from their surrounding neighborhood

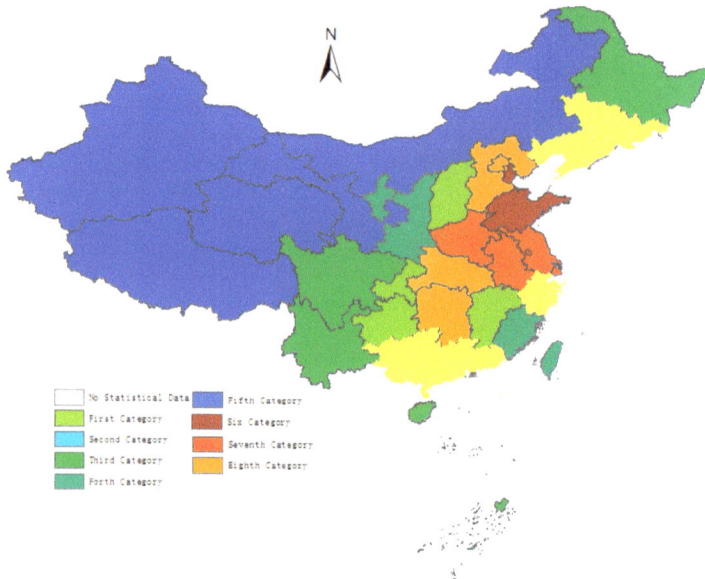

Figure 9. Spatial distribution of PM$_{2.5}$ concentrations for eight groups in China.

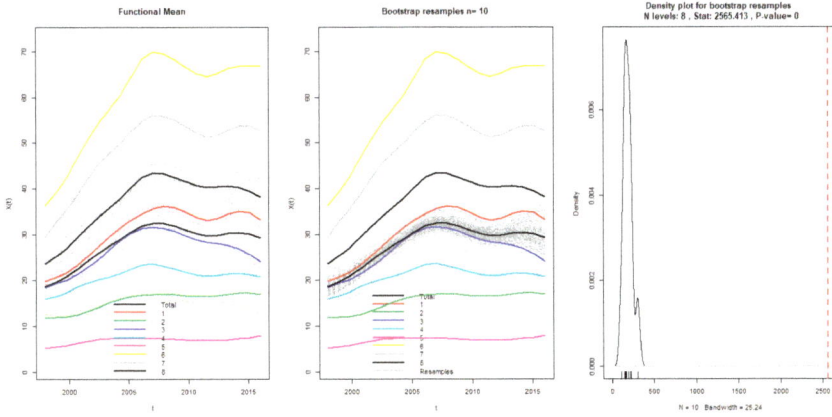

Figure 10. F-ANOVA test for absolute level of PM$_{2.5}$ concentration functions.

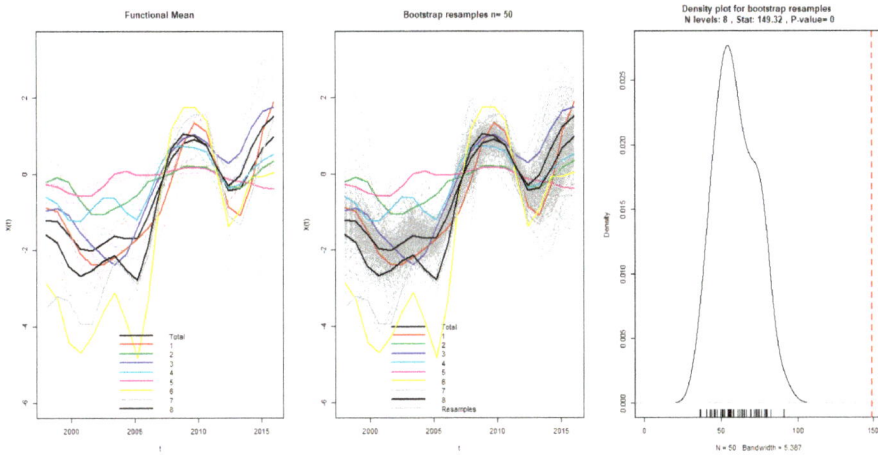

Figure 11. F-ANOVA test for the velocity of PM$_{2.5}$ concentration functions.

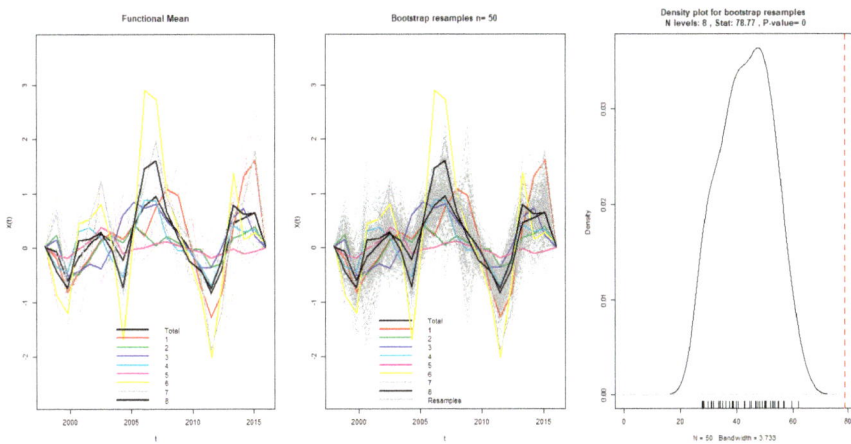

Figure 12. F-ANOVA test for the acceleration of PM$_{2.5}$ concentration functions.

3.5. Comparing the Fluctuation Patterns of PM$_{2.5}$ Concentration in Each Group

Due to multiple differences in industrial structure and topography, together with the different coping strategies toward influence of various environmental policies, the fluctuation process of PM$_{2.5}$ concentrations between provinces has typical category features. In order to interactively display the disparity of fluctuation process, we have taken the overall mean function of China as the benchmark for comparison (blue dashed line), Figure 13 displays how the PM$_{2.5}$ concentration functions vary from province to province, with the mean function of each category in a red solid line. From the perspective of absolute level, we can see the average value of PM$_{2.5}$ concentration for the sixth, the seventh and the eighth category far outweighed the overall mean and their highest value occurred around 2007. However, the average value of the second, the fourth and the fifth group is far less than the overall mean, especially the fifth group which exhibited nearly a horizontal fluctuation trajectory, meaning that there were almost no substantial changes in the PM$_{2.5}$ concentration fluctuations. The mean curves of the first and the third category seemed to be overlapping with the trajectory of the overall mean, indicating that the level of PM$_{2.5}$ concentration for the two categories represented the overall situation of PM$_{2.5}$ concentration in China.

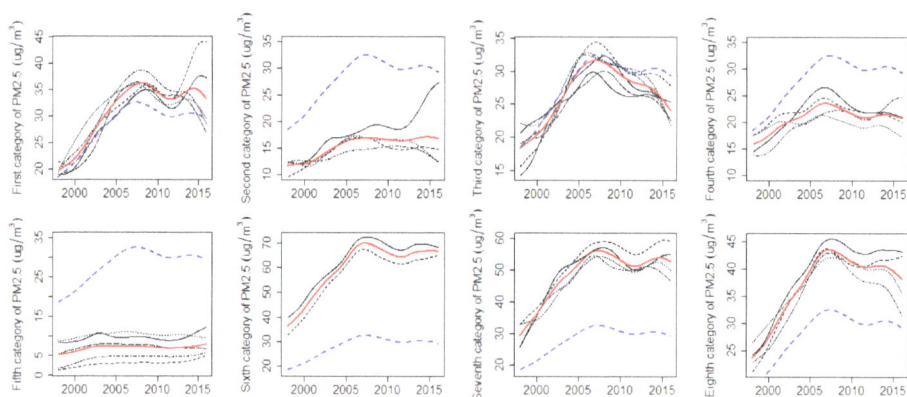

Figure 13. Mean curves of eight groups (**red**) with the benchmark of national average (**blue**).

Since the PM$_{2.5}$ concentrations usually originate from multiple sources, besides motor vehicle usage and static atmosphere flow, we focus on tracking the major cause for regional difference in PM$_{2.5}$ concentration from the perspectives of industrial activities and energy structures. According to the spatial distribution of each group in Figure 9 and data from the "Statistical Yearbook of China (1998–2016)" [58], we found that the provinces with highest level in groups six, seven and eight were mainly located in the Beijing-Tianjin-Hebei region and the Yangtze River Delta region, as well as their surrounding provinces. As is well known, the above regions are the leaders in social and economic development in China, and their prosperity was established on the massive consumption of fossil fuels (coal and oil), especially in colder seasons. The sources of PM$_{2.5}$ in Yangtze River Delta could be attributed to the secondary pollution and active economic activities. Actually, most of traditional manufacturing industries, such as electronics industry and transportation service, located at Yangtze River Delta in China, and a large labor force including ordinary workers and high-tech talent resides in this region. The labor-intensive industries whose layout focused on upstream and intermediate products of industrial chains, produced large quantities of volatile organics, which are the main components of PM$_{2.5}$ in Yangtze River Delta. Besides, the global night-time light data from 1992 to 2012 indicates that the Yangtze River Delta is still the most active economic area [59]. According to environmental statistics from 1998 to 2016 [60], the proportion of fumes, such as SO2 and NOx, discharged from vehicles is closing in on that from factories, and have an exceeding tendency. After chemical reactions in atmosphere, the fumes transmuted into smaller particulate pollutants, such as sulphates and nitrate. Although the pollutants from factories are declining due to the campaign of "Desulphurization and Denitrification" launched in all industrial sectors, the growing number of vehicles is increasing the emission of pollutants in the Yangtze River Delta of China.

In contrast, the provinces with lowest $PM_{2.5}$ concentrations in the second, fourth and fifth group mainly located in two kinds of regions, that are the provinces of tourism and regions in western China. We can see that the fifth group was mainly composed of frontier provinces in western China, which is a major exporter of labor force due to its low economic development or its short industrial chain. It should be noted that the trajectories of $PM_{2.5}$ concentration in the fifth group is almost horizontal with constant deviations. The reason for this is that their highly homogenous economic development was supported by traditional agriculture and livestock farming. Thus, the level of $PM_{2.5}$ concentration in the fifth group is the lowest, seldom effected by adjustments of the industrial structure. Different to provinces in the fifth group, tourism is the pillar industry of provinces in the second group. In order to keep appealing to tourists with their beautiful environments, these provinces have to adopt environmentally-friendly sustainable economic development modes. However, the improving economical development of the second group as well as their comfortable living environment, attracts more and more residents and results in a growing quantity of vehicles. Thus, the $PM_{2.5}$ concentrations of the second group exhibit a slowly growing trend, with an increasing deviation. As for provinces in the fourth group, the $PM_{2.5}$ concentration of Fujian and Taiwan are closely related to human activity and highly developed manufacturing industries. Located at the southeast coast of China and strongly influenced by maritime monsoon, it is hard to form high concentrations of particle pollution in Fujian and Taiwan. As for Shanxi and Ningxia, the main source of $PM_{2.5}$ is dust aerosols resulting from soil erosion and the smoke discharged from energy bases. Due to their open topography, the pollutants of Shangxi and Ningxia can rapidly diffuse due to being influenced by the stable atmospheric circulation in these regions. Except for differences in fluctuation amplitude, the time of turning points corresponding to the fourth group is consistent with that of the national mean, meaning that provinces in the fourth group can adjust their industrial structures quickly according to environmental protection policies.

The $PM_{2.5}$ concentrations of provinces in the first and third group represents the average level and dominant (tendency) of China. These provinces can be classified into two categories, one category located in northeast China is characterized as developed heavy industry, such as Liaoning and Jilin. The other category is located in southeast China with the highest population density, including Chongqing and Hong Kong. The region classification in this paper indicates that the spatial distribution of $PM_{2.5}$ concentration has obvious characteristics of spatial agglomeration. Besides, the classification of $PM_{2.5}$ concentration for 34 provinces in our study is basically consistent with the regional definition, "three districts and ten groups", of 12th Five-Year Plan for Air Pollution Prevention and Control in Key Regions jointly issued by the Ministry of Ecology and Environment, the State Development and Reform Commission and the Ministry of Finance of China [61].

In order to further analyze the differences in the growth of $PM_{2.5}$ concentrations from dynamic perspectives, which also is the advantage of FDA, we plot the trajectories of velocity and acceleration for eight groups in Figure 14. Upon the comparison of fluctuation trajectories between every group, it can be found that the provinces in the sixth group not only possess the largest level of $PM_{2.5}$ concentration, but their fluctuation amplitudes of velocity and acceleration are also the largest ones. Besides, by comparing the turning points in the curves of velocity and acceleration with the issued time of environment protection policies, we found that the provinces in the sixth group could adjust their industrial structure and pollution emissions in time in accordance with the policy requirements. The absolute level and amplitudes of velocity and acceleration for $PM_{2.5}$ concentration of the seventh and eighth group ranked the second and the third, respectively, and their turning points are also highly concurrent with the issued time of environment protection policies. Compared to the regions with the highest $PM_{2.5}$ concentration, the amplitudes of velocity and acceleration for $PM_{2.5}$ concentration of the second, the fourth and the fifth group were remarkably small, but there was few turning points at the issued time of environment protection policies.

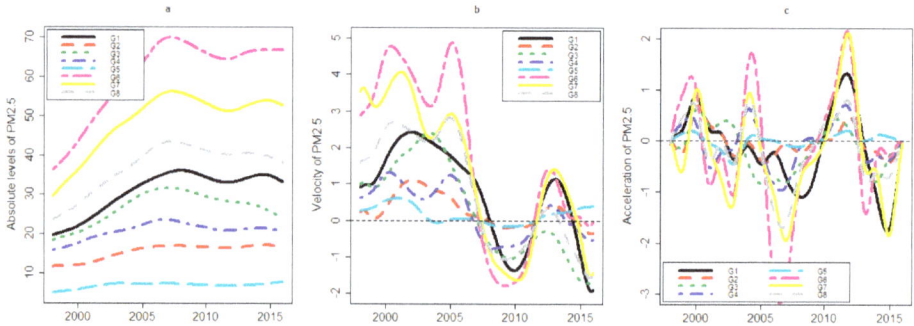

Figure 14. (**a**) PM$_{2.5}$ concentration functions with (**b**) firs-order and (**c**) second-order derivatives of eight groups for the years from 1998 to 2016.

The dynamic analysis of PM$_{2.5}$ concentration indicates that, although the environment protection policies issued by government sectors in China could have dramatic influence on reducing the overall PM$_{2.5}$ concentration, especially in the high pollution regions, the rebound effect would also be obvious after the control periods of regulations. However, the regulating effect of policies was negligible in the low pollution regions because of their environmentally friendly economic development modes. The implication of our empirical results is that the relationship between China's existing economy development mode and environmental protection is still in an irreconcilable stage, and it is hard to eliminate or reduce PM$_{2.5}$ concentrations by just relying on the government's administrative intervention. As low pollution areas have the subjective motivation of protecting the environment to sustain their pillar industry, the government should fundamentally devote its efforts to reducing pollution levels in high pollution areas.

4. Conclusions and Discussion

As a developing country with vast territory and a typical dual economic structure, the rapid development of China occurs at the expense of environment and energy, which has resulted in serious air pollution. Accurately identifying the spatial and temporal patterns of haze pollution is a prerequisite for rational formulation and effective implementation of haze control policies. This study employed FDA techniques to represent PM$_{2.5}$ concentration data in the form of a smoothing curve for each province. Based on the continuous curves reconstructed from discrete noisy PM$_{2.5}$ concentration data with roughness penalty, the FPCA was adopted to decompose the temporal variability of PM$_{2.5}$ concentration curves, and the patterns of PM$_{2.5}$ concentration in 34 provinces was determined using adaptive weighting clustering analysis. The analysis continued with a functional ANOVA to verify the significance of differences between eight groups, and with further exploration in their spatial differences, both from static and multiple dynamic perspectives. The conclusions with policy implications obtained from this study are as follows.

(1)	Imposing roughness penalty on the curves' reconstruction of PM$_{2.5}$ concentration could emphasize the dominant trend of fluctuation, thus enhancing the interpretability of variability implied in PM$_{2.5}$ concentration curves. The standard deviation trajectory of PM$_{2.5}$ concentrations perfectly followed the growing pattern of the overall mean function, which means that facing the opportunity for rapidly developing economy at the expense of environment pollution, the decision-making of different provinces differed vastly, whether for subjective reasons of excessively pursuing GDP or for objective reasons of industrial structure and resource endowment. The above conclusions imply that quite a few provinces could rationally balance extensive economic development with ecological sustainability. Consequently, the feasible approach to eliminate haze pollution should emphasis on optimizing, upgrading and transferring of industrial structure. In particular, the government should encourage low pollution regions, through cutting their taxes or increasing their subsidies, to sustain their environmentally-friendly economic development.

(2) The temporal variability of $PM_{2.5}$ concentration from 1998 to 2016 could be decomposed into three distinctive sub-fluctuation modes by FPCA, which depicts the variations in the beginning, the middle interval and the end of the research period, respectively. Remarkably, the middle interval with largest variation portrayed by the first FPC perfectly matches with the period of the "ten golden years" for coal, and the variance contribution rate of the first FPC far outweighs that of the other two, meaning that the fluctuation of $PM_{2.5}$ concentrations for 34 provinces was mainly located at the period of extensive economic growth. The empirical result again verifies the different coping strategies among the 34 provinces when facing the choice of developing the economy at expense of the environment and energy. The contribution to empirical methodology derived from this study is that the huge disparity in classification information among the three FPCs requires different weights when conducting clustering analysis on 34 $PM_{2.5}$ concentrations curves. Therefore, the same inputs or approaches might not be useful in modeling the pollution processes for different regions.

(3) The fluctuation patterns of $PM_{2.5}$ concentration functions were classified into eight groups via adaptive weighting cluster analysis, and the effect of spatial and geographical locations was analyzed using functional ANOVA. The test results indicate that the differences between the eight groups was significant, whether from the static perspective or dynamic potential. The reason of differences in the $PM_{2.5}$ concentration patterns could possibly be due to the effect of geographic and industrial factors, as well as the different coping strategies of environmental policies. Multiple comparisons of fluctuation patterns show that the heavy pollution areas not only have the highest level of $PM_{2.5}$ concentration, but also have the largest longitudinal amplitude of velocity and acceleration. The tuning points of $PM_{2.5}$ concentration curves for the heavy pollution areas highly matched the issued time of environmental policies, whereas the effect of environmental policies in low pollution areas was not obvious. The findings reveal that the characteristics of $PM_{2.5}$ concentration are very dependent on the industrial structures of the provinces. As such, it is hard to eliminate haze pollution by relying solely on the government's administrative intervention. Thus, the direct way of reducing $PM_{2.5}$ concentration in the short term is to maintain the continuity of environmental policies. In the long run, how to encourage enterprises to transform or upgrade industrial structure via revenue decrease or financial subsidy is an important and unavoidable issue for government to eliminate haze pollution fundamentally.

Compared with the existing literature, the main contribution of this study is focused on how the FDA technique can be used for $PM_{2.5}$ concentrations data analysis. This paper has significance for both empirical methodology and important policy implications. Instead of utilizing discrete noisy $PM_{2.5}$ concentration data, we can create a functional form for the data which could be analyzed over any time interval. So we are able to extract additional information contained in the smoothing curve and its derivatives which may not be normally available from traditional statistical methods. The findings from this study, such as significant differences in $PM_{2.5}$ concentration patterns between regions, not only provide a guideline for analyzing the effectiveness of current air quality control regulations, but also provide information for the environment management for provinces, as well as suggestions on sustainable development for China's government. As a future research direction, significant differences in $PM_{2.5}$ concentration patterns between regions signify that a different approach in modeling the process should be employed, especially linking the change of $PM_{2.5}$ concentration to policy-related implications using functional concurrent models.

Sustainability **2019**, *11*, 1620

Author Contributions: Conceptualization, D.W. and L.H.; Methodology, D.W.; Software, D.W. and K.B.; Validation, D.W., Z.Z. and L.H.; Formal analysis, Z.Z.; Investigation, K.B.; Resources, D.W. and K.B.; Data curation, R.L.; Writing—original draft preparation, D.W.; Writing—review and editing, L.H.; Visualization, D.W.; Supervision, L.H.; Project administration, D.W.; Funding acquisition, D.W. and L.H.

Funding: This study was supported by the Fundamental Research Funds for the Central Universities (Project nos. 2015WA01).

Acknowledgments: We thank the editor and anonymous referees for helpful comments. All errors are our own.

Conflicts of Interest: No potential conflict of interest was reported by the authors.

References

1. Huang, R.J.; Zhang, Y.; Bozzetti, C.; Ho, K.F.; Cao, J.J.; Han, Y.; Daellenbach, K.R.; Slowik, J.G.; Platt, S.M.; Canonaco, F.; et al. High secondary aerosol contribution to particulate pollution during haze events in China. *Nature* **2014**, *514*, 218–222. [CrossRef] [PubMed]

2. Xie, Y.; Dai, H.; Dong, H.; Hanaoka, T.; Masui, T. Economic impacts from $PM_{2.5}$ pollution-related health effects in China: A provincial-level analysis. *Environ. Sci. Technol.* **2016**, *50*, 4836–4843. [CrossRef] [PubMed]

3. Guo, H.; Cheng, T.H.; Gu, X.F.; Wang, Y.; Chen, H.; Bao, F.W.; Shi, S.Y.; Xu, B.R.; Wang, W.N.; Zuo, X.; et al. Assessment of $PM_{2.5}$ concentrations and exposure throughout China using ground observations. *Sci. Total Environ.* **2017**, *601–602*, 1024–1030. [CrossRef] [PubMed]

4. Li, J.M.; Han, X.L.; Li, X.; Yang, J.P.; Li, X.J. Spatiotemporal Patterns of Ground Monitored $PM_{2.5}$ Concentrations in China in Recent Years. *Int. J. Environ. Res. Public Health* **2018**, *15*, 114. [CrossRef] [PubMed]

5. Wang, G.L.; Xue, J.J.; Zhang, J.Y. Analysis of Spatial-temporal distribution Characteristics and Main Cause of Air Pollution in Beijing-Tianjin-Hebei Region in 2014. *Meteorol. Environ. Sci.* **2016**, *39*, 34–42.

6. Ye, W.F.; Ma, Z.Y.; Ha, X.Z. Spatial-temporal patterns of $PM_{2.5}$ concentrations for 338 Chinese Cities. *Sci. Total Environ.* **2018**, *631–632*, 524–533. [CrossRef] [PubMed]

7. Liu, H.M.; Fang, C.L.; Huang, J.J.; Zhu, X.D.; Zhou, Y.; Wang, Z.B.; Zhang, Q. The Spatial Temporal Characteristics and influencing factors of air pollution in Beijing-Tianjin-Hebei urban agglomeration. *J. Geog. Sci.* **2018**, *73*, 177–191.

8. Feng, J.L.; Hu, J.C.; Xu, B.H.; Hu, X.L.; Sun, P.; Han, W.L.; Gu, Z.P.; Yu, X.M.; Wu, M.H. Characteristics and seasonal variation of organic matter in $PM_{2.5}$ at a regional background site of the Yangtze River Delta region, China. *Atmos. Environ.* **2015**, *123*, 288–297. [CrossRef]

9. Fu, J.Y.; Li, L.H. FDI, Environmental Regulation and Pollution Haven Effect—Empirical Analysis of China's Provincial Panel Data. *J. Public Manag.* **2010**, *7*, 65–74.

10. Li, C.; Tong, H.Z.; Yeung, F.U. Urban Residents' Cognition of Haze-fog Weather and Its Impact on Their Urban Tourism Destination Choice. *Tour. Trib.* **2015**, *30*, 37–47.

11. Li, Y.; Dai, Z.; Liu, X. Analysis of Spatial-Temporal Characteristics of the $PM_{2.5}$ Concentrations in Weifang City, China. *Sustainability* **2018**, *10*, 2960. [CrossRef]

12. Zhai, B.; Chen, J.; Yin, W.; Huang, Z. Relevance Analysis on the Variety Characteristics of $PM_{2.5}$ Concentrations in Beijing, China. *Sustainability* **2018**, *10*, 3228. [CrossRef]

13. Fang, C.; Wang, Z.; Xu, G. Spatial-temporal characteristics of $PM_{2.5}$ in China. *J. Geogr. Sci.* **2016**, *26*, 1519–1532. [CrossRef]

14. Peng, J.; Chen, S.; Lü, H.; Liu, Y.; Wu, J. Spatiotemporal patterns of remotely sensed $PM_{2.5}$ concentration in China from 1999 to 2011. *Remote Sens. Environ.* **2016**, *174*, 109–121. [CrossRef]

15. Lin, G.; Fu, J.Y.; Jiang, D.; Hu, W.S.; Dong, D.L.; Huang, Y.H.; Zhao, M.D. Spatio-Temporal Variation of $PM_{2.5}$ Concentrations and, Their relationship with Geographic and Socioeconomic, Factors in China. *Int. J. Environ. Res. Public Health* **2013**, *11*, 173–186. [CrossRef]

16. Shao, S.; Li, X.; Cao, J.H.; Yang, L.L. China's Economic Policy Choices for Governing Smog Pollution Based on Spatial Spillover Effects. *Econ. Res. J.* **2016**, *9*, 73–88.

17. Guan, D.; Su, X.; Zhang, Q.; Peters, G.P.; Liu, Z.; Lei, Y.; He, K. The socioeconomic drivers of China's primary $PM_{2.5}$ emissions. *Environ. Res. Lett.* **2014**, *9*, 024010. [CrossRef]

18. Wang, Y.Q.; Zhang, X.Y.; Sun, J.Y.; Zhang, X.C.; Che, H.Z.; Li, Y. Spatial and temporal variations of the concentrations of PM_{10}, $PM_{2.5}$ and PM_1 in China. *Atmos. Chem. Phys.* **2015**, *15*, 13585–13598. [CrossRef]

19. Xu, J.H.; Jiang, H. Estimation of PM$_{2.5}$ Concentration over the Yangtze Delta Using Remote Sensing: Analysis of Spatial and Temporal Variations. *Environ. Sci.* **2015**, *36*, 3119–3127.
20. Cheng, N.; Zhang, D.; Li, Y.; Xie, X.; Chen, Z.; Meng, F.; Gao, B.; He, B. Spatio-temporal variations of PM$_{2.5}$ concentrations and the evaluation of emission reduction measures during two red air pollution alerts in Beijing. *Sci. Rep.* **2017**, *7*, 8220. [CrossRef]
21. Guan, Q.Y.; Li, F.C.; Yang, L.Q. Spatial-temporal variations and mineral dust fractions in particulate matter mass concentrations in an urban area of northwestern China. *J. Environ. Manag.* **2018**, *222*, 95–103. [CrossRef] [PubMed]
22. Famoso, F.; Wilson, J.; Monforte, P.; Lanzafame, R.; Brusca, S.; Lulla, V. Measurement and modeling of ground-level ozone concentration in Catania, Italy using biophysical remote sensing and GIS. *Int. J. Appl. Eng. Res.* **2017**, *12*, 10551–10562.
23. Wu, C.D.; Zeng, Y.T.; Lung, S.C.C. A hybrid kriging/land-use regression model to assess PM$_{2.5}$ spatial-temporal variability. *Sci. Total Environ.* **2018**, *645*, 1456–1464. [CrossRef] [PubMed]
24. Mukherjee, A.; Agrawal, M. Assessment of local and distant sources of urban PM$_{2.5}$ in middle Indo-Gangetic plain of India using statistical modeling. *Atmos. Res.* **2018**, *213*, 275–287. [CrossRef]
25. Brokamp, C.; Jandarov, R.; Rao, M.B.; LeMasters, G.; Ryan, P. Exposure assessment models for elemental components of particulate matter in an urban environment: A comparison of regression and random forest approaches. *Atmos. Environ.* **2017**, *151*, 1–11. [CrossRef] [PubMed]
26. Gao, H.O. Day of week effects on diurnal ozone/NOx cycles and transportation emissions in Southern California. *Transp. Res. Part D Transp. Environ.* **2007**, *12*, 292–305. [CrossRef]
27. Shaadan, N.; Jemain, A.A.; Latif, M.T. Anomaly detection and assessment of PM$_{10}$ functional data at several locations in the Klang Valley, Malaysia. *Atmos. Pollut. Res.* **2015**, *6*, 365–375. [CrossRef]
28. Tsay, R.S. Some Methods for Analyzing Big Dependent Data. *J. Bus. Econ. Stat.* **2016**, *34*, 673–688. [CrossRef]
29. Ramsay, J.O.; Silverman, B.W. *Functional Data Analysis*, 2nd ed.; Springer: New York, NY, USA, 2005.
30. Torres, J.M.; Nieto, P.J.G.; Alejano, L. Detection of outliers in gas emissions from urban areas using functional data analysis. *J. Hazard. Mater.* **2011**, *186*, 144–149. [CrossRef] [PubMed]
31. Martinez, J.; Saavedra, Á.; García-Nieto, P.J.; Piñeiro, J.I.; Iglesias, C.; Taboada, J.; Sanchoa, J.; Pastor, J. Air quality parameters outliers detection using functional data analysis in the Langreo urban area (Northern Spain). *Appl. Math. Comput.* **2014**, *241*, 1–10. [CrossRef]
32. Liang, Y.; Liu, L. PM$_{2.5}$ pollution characteristic in Beijing-Tianjin-Hebei region based on the perspective of functional data analysis. *Oper. Res. Trans.* **2018**, *22*, 105–114.
33. Febrero-Bande, M.; Fuente, M.O.D.L. Statistical Computing in Functional Data Analysis: The R Package fda.usc. *J. Stat. Softw.* **2012**, *51*, 1–28. [CrossRef]
34. Ramsay, J.O.; Hooker, G.; Graves, S. *Functional Data Analysis with R and MATLAB*; Springer: New York, NY, USA, 2009.
35. Craven, P.; Wahba, G. Smoothing noisy data with spline functions. *Numer. Math.* **1978**, *31*, 377–403. [CrossRef]
36. Kokoszka, P.; Reimherr, M. *Introduction to Functional Data Analysis*; Chapman and Hall/CRC Press: London, UK, 2017.
37. Wang, J.L.; Chiou, J.M.; Mueller, H.G. Review of Functional Data Analysis. *arXiv*, 2015; arXiv:1507.05135.
38. Cuevas, A.; Febrero, M.; Fraiman, R. An anova test for functional data. *Comput. Stat. Data Anal.* **2004**, *47*, 111–122. [CrossRef]
39. Suhaila, J.; Jemain, A.A.; Hamdan, M.F. Comparing rainfall patterns between regions in Peninsular Malaysia via a functional data analysis technique. *J. Hydrol.* **2011**, *411*, 197–206. [CrossRef]
40. Chiou, J.M.; Li, P.L. Functional clustering and identifying substructures of longitudinal data. *J. R. Stat. Soc.* **2007**, *69*, 679–699. [CrossRef]
41. Jacques, J.; Preda, C. Functional data clustering: A survey. *Adv. Data Anal. Classif.* **2014**, *8*, 231–255. [CrossRef]
42. Jacques, J.; Preda, C. Model-based clustering for multivariate functional data. *Comput. Stat. Data Anal.* **2014**, *71*, 92–106. [CrossRef]
43. Bosq, D. Linear processes in function spaces: Theory and applications. *Lect. Notes Stat.* **2000**, *149*, 181–202.

44. Karhunen, K. *Zur Spektraltheorie Stochastischer Prozesse.* Annales Academiae Scientiarum Fennicae. Series A. I, Mathematica. 1946, Volume 7. Available online: https://katalog.ub.uni-heidelberg.de/cgi-bin/titel.cgi?katkey=67295489 (accessed on 14 March 2019).

45. Wang, D.Q.; Liu, X.W.; Zhu, J.P. Research on Clustering Analysis for Functional Data based on Adaptive Iteration. *Stat. Res.* **2015**, *32*, 91–96.

46. Wang, D.Q.; Zhu, J.P.; Wang, J.D. Research of Clustering Analysis for Functional Data based on Adaptive Weighting. *J. Appl. Stat. Manag.* **2015**, *34*, 84–91.

47. Wang, D.Q.; Liu, X.W.; Zhu, J.P. Deeper Extension of Adaptive Weighting Functional Clustering Analysis. *J. Appl. Stat. Manag.* **2016**, *35*, 81–88.

48. Donkelaar, A.V.; Martin, R.V.; Brauer, M. Use of Satellite Observations for Long-Term Exposure Assessment of Global Concentrations of Fine Particulate Matter. *Environ. Health Perspect.* **2014**, *123*, 135–143. [CrossRef]

49. Nordhaus, W.D. Geography and Macroeconomics: New Data and New Finding. *Proc. Natl. Acad. Sci. USA* **2006**, *103*, 3510–3517. [CrossRef]

50. Chen, X.; Nordhaus, W.D. Using luminosity data as a proxy for economic statistics. *Proc. Natl. Acad. Sci. USA* **2011**, *108*, 8589–8594. [CrossRef]

51. Clarkson, D.B.; Fraley, C.; Gu, C.C.; Ramsey, J.O. S+ Functional Data Analysis. *Int. Encycl. Soc. Behav. Sci.* **2006**, *40*, 5822–5828.

52. Ramsay, J.O. Estimating Smooth Monotone Functions. *J. R. Stat. Soc. Ser. B (Stat. Methodol.)* **1996**, *60*, 365–375. [CrossRef]

53. Hou, F.L.; Xin, Y.; Yi, Z.C. A Review of Meteorological Effects on Heavy Haze Pollution in China. *Ecol. Environ. Sci.* **2015**, *24*, 1917–1922.

54. Ramsay, J.O.; Ramsey, J.B. Functional data analysis of the dynamics of the monthly index of nondurable goods production. *J. Econom.* **2002**, *107*, 327–344. [CrossRef]

55. Suhaila, J.; Yusop, Z. Spatial and temporal variabilities of rainfall data using functional data analysis. *Theor. Appl. Climatol.* **2016**, *129*, 1–14. [CrossRef]

56. Baker, F.; Hubert, L. Measuring the Power of Hierarchical Cluster Analysis. *Publ. Am. Stat. Assoc.* **1975**, *70*, 31–38. [CrossRef]

57. Malika, C.; Nadia, G.; Véronique, B.; Azam, N. NbClust: An R Package for Determining the Relevant Number of Clusters in a Data Set. *J. Stat. Softw.* **2014**, *61*, 1–36.

58. NBS (National Bureau of Statistics of China). *China Statistical Yearbook*; China Statistical Press: Beijing, China, 2017.

59. Kangning, X.; Fenglong, C.; Xiuyan, L. The Truth of China Economic Growth: Evidence from Global Night-time Light Data. *Econ. Res. J.* **2015**, *9*, 17–29.

60. NBS (National Bureau of Statistics of China). *China Statistical Yearbook on Environment*; China Statistical Press: Beijing, China, 2017.

61. Ministry of Ecology and Environment, State Development and Reform Commission and Ministry of Finance of the People's Republic of China). The 12th Five-Year Plan for Air Pollution Prevention and Control in Key Regions. 2012. Available online: http://www.gov.cn/gongbao/content/2013/content_2344559.htm (accessed on 29 December 2012).

sustainability

MDPI

Article

Application of Fuzzy Optimization Model Based on Entropy Weight Method in Atmospheric Quality Evaluation: A Case Study of Zhejiang Province, China

Xiaodan Wang [1] and Zhengyu Yang [2],*

1   School of Management, Shandong University of Technology, Zibo 255000, China; xiaodan_sut@163.com
2   School of Public Economics and Administration, Shanghai University of Finance and Economics, Shanghai 200433, China
*   Correspondence: zhengyuyang116@126.com; Tel.: +86-21-6590-4466

Received: 27 January 2019; Accepted: 2 April 2019; Published: 10 April 2019

Abstract: A fuzzy optimization model based on the entropy weight method for investigating air pollution problems in various cities of Zhejiang Province, China has been proposed in this paper. Meanwhile, the air quality comprehensive evaluation system has been constructed based on the six major pollutants (SO_2, NO_2, CO, PM_{10}, $PM_{2.5}$ and O_3) involved in China's current air quality national standards. After analyzing the monthly data of six pollutants in 11 cities of Zhejiang Province from January 2015 to April 2018 by the above method, the authors found that, although the air quality of cities in Zhejiang Province did not reach the long-term serious pollution of Beijing, Tianjin, and Hebei, the air quality changes in the northern cities of Zhejiang were worse than those in southern Zhejiang. For example, the air quality of Shaoxing in northern Zhejiang has dropped by 14.85% in the last study period when compared with that of the beginning period, and Hangzhou, the provincial capital of Zhejiang, has also seen a decrease of 6.69% in air quality. The air quality of Lishui, Zhoushan and Wenzhou in southern Zhejiang has improved by 8.04%, 4.67% and 4.22% respectively. Apart from the geographical influence, the industrial structure of these cities is also an important cause for worse air quality. From the local areas in southern Zhejiang, cities have developed targeted air pollution control measures according to their own characteristics, including adjusting the industrial structure, changing the current energy consumption structure that heavily relies on coal, and improving laws and regulations on air pollution control, etc. In the four cities in central Zhejiang, the air quality at the end of the period (April 2018) has decreased from the beginning of the period (January 2015), given that there were no fundamental changes in their industrial structure and energy pattern.

Keywords: entropy weight method; fuzzy optimization model; air quality

---

1. Introduction

Environmental pollution refers to the matters released from human activities (such as during the process of production and living) into the environment that are harmful to living organisms [1–3]. The ecosystem has self-purifying functions on the pollutants, but if the pollutants released have exceeded the limits allowed by the ecosystem, they would cause great harm to the human society. If the environment is polluted by harmful matters, the growth and reproduction of organisms will be affected, disturbing the normal life of human beings and endangering human health and the sustainable development of human beings [4–8]. Since the 1970s, China has achieved rapid economic development, but its growth model is characterized by extensive production and high pollution. The labor-intensive industry structure and relatively backward production technologies have made the pollution problem of China, especially its air pollution issues keep deteriorating, which has caused serious threats to China's sustainable development [9–13].

In 2016, the total emissions of SO_2, nitrogen oxides and smoke/dust of China reached 11.03 million tons, 13.94 million tons and 10.11 million tons, respectively, which caused great pressure on the environment and sustainable development [14]. According to international standards, 38% of the Chinese people are breathing unhealthy air every day; and about 1.6 million people die of heart disease, lung disease and stroke every year due to air pollution (especially the particulate pollutants in haze) [15]. According to the statistics of the World Bank, China is the most negatively affected country in the world by air pollution, with various types of air pollutants and shocking amounts of emissions. The annual economic losses caused by air pollution are as high as 10% of China's GDP (Gross Domestic Product), which mainly includes premature death, loss of working time and increase of related welfare expenses [16]. These figures and research results have driven us to reflect on the interrelations between China's air pollution and sustainable development.

This paper has selected China's Zhejiang Province as the research object of air pollution evaluation standards and air pollution governance policies (please refer to Figures 1 and 2). Zhejiang Province is on the southeast coast of China, with the Yangtze River Delta in the north, Shanghai in the southeast, Anhui and Jiangxi in the west, and Fujian in the south. It is one of the most developed provinces in China, and has 11 prefecture-level cities. Zhejiang Province is not rich in natural resources and mainly relies on industrial production to support rapid economic growth [17,18]. In 2000, Zhejiang's GDP was 641.10 billion yuan, and its total inflation-adjusted GDP in 2017 after rough calculation was 5176.80 billion yuan, which has achieved an increase of 842.99% [19,20].

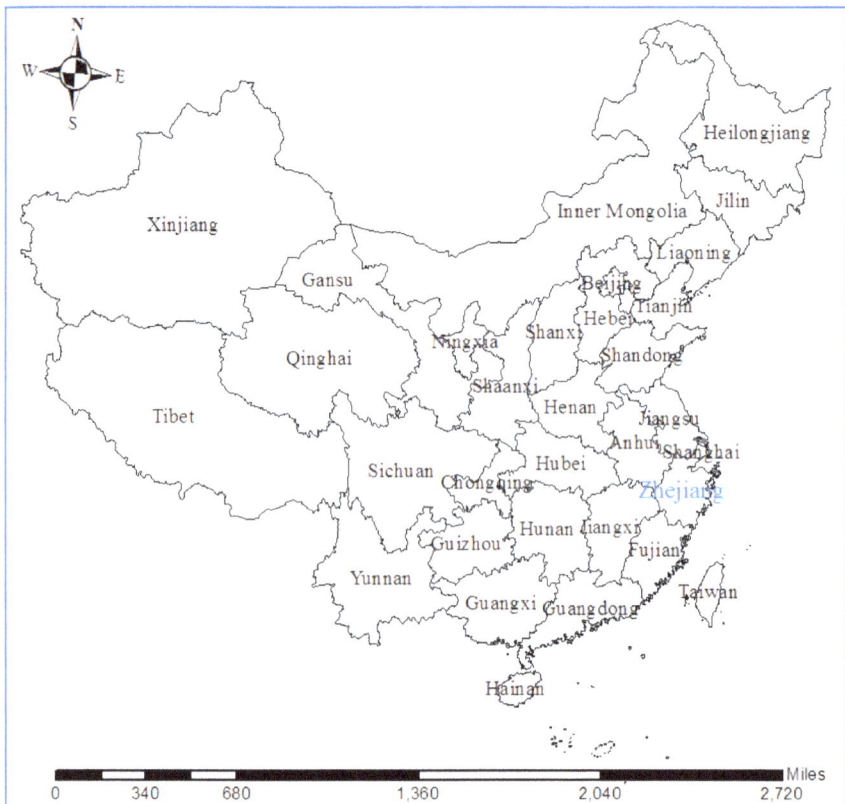

Figure 1. Map of Zhejiang Province in China.

Figure 2. Map of cities in Zhejiang Province, China.

However, just like China's overall economic development model, Zhejiang's rapid economic growth has always been driven by the consumption of limited natural resources, and its economic development is still at the stage of extensive growth characterized by quantity expansion and low-cost competition [21]. In addition, the enterprises with high energy consumption, high emissions and backward production technologies are still in operation, thus driving the growth rate of Zhejiang's industrial waste gas emissions to continuously rise and causing serious environmental pollution. In 2013, the industrial waste gas emissions of Zhejiang Province totaled 2456.5 billion cubic meters, and this number rose to 2695.8 billion cubic meters in 2014 with an increase of almost 9%. At the same time, the proportion of major air pollutants in the waste gas emissions has always been high. In 2016, the total emissions of SO_2, nitrogen oxides and smoke/dust in Zhejiang Province reached 268,400 tons, 380,400 tons and 182,300 tons, respectively.

Therefore, the choice of Zhejiang Province as a research object has the following important meanings: (1) Zhejiang is the most active area of China's private economy, covering an area of 105,500 square kilometers which exceeds South Korea [22,23], with many small and medium-sized private enterprises. However, while Zhejiang's economy has grown rapidly since the reform and opening up, the protection of the environment, especially of the air quality, is seriously inadequate; (2) Unlike in Jiangsu and Shanghai, which are the other two provinces located in the Yangtze River Delta region, air pollution control in Zhejiang is faced with the task of adjusting the economic structure and many small and medium-sized private enterprises. The change in air quality reflects not only the effectiveness of the air pollution control policy, but also the effectiveness of Zhejiang's environmental adjustment of the industrial structure and the private economy; (3) Under the background of China's implementation of the "Integration Development of Yangtze River Delta" strategy in recent years, the governance of air pollution in Zhejiang will represent the implementation of environmental protection and sustainable development strategies in the most developed areas of China's private economy.

The academic circle has given great attention to the air pollution issue of Zhejiang Province. In the field of outdoor air pollution research, Ni et al. used the Weather Research and Forecasting with Chemistry Model to study the air pollution characteristics and its root causes in Hangzhou (the provincial capital of Zhejiang Province) during the second World Internet Conference in the winter of 2015. Their results showed that the control measures implemented one week before the meeting did help reduce the $PM_{2.5}$ pollution to some extent, with the total $PM_{2.5}$ concentration in Hangzhou decreased by 15% [24]. Feng et al. used the WRF/CMAQ (Weather Research and Forecasting/Community Multi-scale Air Quality) Model to analyze the air pollution level in Hangzhou based on observation data from five local environmental monitoring stations in downtown Hangzhou. According to their findings, in 2017, the local pollution sources in Hangzhou accounted for 15.8%, 68.6%, 48.3% and 59.2% of the total concentrations of SO_2, NO_2, $PM_{2.5}$ and PM_{10}, respectively [25]. $PM_{2.5}$ refers to particulate matter with an aerodynamic equivalent diameter of 2.5 µm or less in ambient air, and PM_{10} refers to particulate matter with an aerodynamic equivalent diameter of 10 µm or less in ambient air [26]. Based on the daily PM_{10} and $PM_{2.5}$ concentration data from 50 monitoring stations in Zhejiang Province from 1 February 2015 to 28 February 2017, Wu et al. conducted a quantitative study on the relationship between PM_{10} and $PM_{2.5}$ concentration and green spaces and landscape structure through Principal Composition Cluster Analysis (PCA) and Hierarchical Cluster Analysis (HCA). Their results showed that the increase of urban green space can reduce PM pollution, and the correlation between green space and $PM_{2.5}$ concentration is stronger than that between green space and PM_{10} on the scale of 5 km or less [27]. Xu et al. selected four representative locations (two cities, one suburb site and one rural site) in Hangzhou and Ningbo from December 2014 to November 2015 in order to study the seasonal and spatial variation in terms of fine particle pollution in Zhejiang Province. With help of the Principal Component Analysis (PCA) method, they found that industrial emissions, biomass burning, and formation of secondary inorganic aerosols are the major sources of fine particles in Zhejiang Province [28]. Fu et al. studied the potential correlation between conjunctivitis and air pollution based on the air pollutant data from the Environmental Protection Department of Zhejiang Province from 1 July 2014 to 30 June 2016 and data of 9737 outpatient visits for conjunctivitis at the Eye Center of the Second Affiliated Hospital of Zhejiang University School of Medicine. Their results indicated significant correlations between the number of conjunctival outpatient visits and air pollution in Zhejiang [29]. Xu et al. conducted a sample survey on air pollutants in Ningbo, Zhejiang Province from 3 December 2012 to 27 June 2013 in order to study the chemical characteristics of highly polluting aerosols in the air of Zhejiang Province. By analyzing the meteorological conditions, air mass backward trajectories, distribution of fire spots in surrounding areas and various categories of aerosol pollutants, they concluded that stagnant weather conditions and long-range transport of air masses from heavy industries and biomass burning from northern China to Ningbo are the main contributors to the high aerosol pollution during their study period [30].

In the field of indoor air pollution study, Sun et al. studied the indoor air pollution in Hangzhou. Through high-frequency detection of indoor and outdoor homogenous pollutants, they found that air-conditioning filters play a significant role in the indoor propagation of outdoor pollutants, especially the home and office dust [31]. Mestl et al. studied the relationship between indoor air pollution and deaths in Zhejiang, Shaanxi and Hubei provinces. By analyzing the PM concentrations in kitchens and living rooms, they believed that the premature mortality in these three provinces should reach 60,600 instead of the current estimate of 46,000 [32].

Most of the above studies on air pollution in Zhejiang Province only considered 1–2 major pollutants (such as $PM_{2.5}$, PM_{10}, SO_2, etc.), and their study period was relatively short. In order to conduct a comprehensive and objective assessment on the air quality of the 11 cities in Zhejiang Province in recent years, this paper has included the six major air pollutants (SO_2, NO_2, CO, PM_{10}, $PM_{2.5}$ and O_3) covered by China's regular monitoring and routine air quality evaluation based on China's current National Ambient Air Quality Standards (GB3095-2012) [33], and Technical Regulation on Ambient Air Quality Index (HJ 633-2012) [26]. This paper would also like to illustrate the overall change and movements of these air pollutants over a longer period. Therefore, based on the above-mentioned domestic and overseas research, this paper has further incorporated the six major pollutants (SO_2, NO_2, CO, PM_{10}, $PM_{2.5}$ and O_3) into the air quality assessment indicator system, and studied the air quality of the 11 cities in Zhejiang from January 2015 to April 2018 based on related data in order to present a comprehensive picture of air quality of various cities.

Therefore, the research topic of this paper has the following important significance for sustainability, especially for China's future sustainable development:

(1) The important symbol of sustainable development is the sustainable use of resources and a good ecological environment [34–36]. In recent years, the serious air pollution in China has caused tremendous damage to the atmospheric environment in which human beings depend for survival and development. Therefore, this paper takes Zhejiang Province as the research object and quantitatively evaluates the air pollution status of the city under its jurisdiction, which provides a scientific basis for Chinese governments to formulate air pollution prevention policies and achieve sustainable development.

(2) Sustainability requires economic and social construction under conditions that protect the environment and sustainably use resources [37,38]. China's future sustainable development should maximize the quality of life of the people without exceeding the capacity to maintain the capacity of the ecosystem, and must not destroy the environment and deplete resources for the cost of economic growth [39,40]. Hence, the evaluation index system for the construction of major pollutants provides scientific tools for measuring the carrying capacity of China's atmospheric environment and achieving sustainable development in the future.

In the following parts of this paper, Part 2 introduces the research method; Part 3 provides the calculation results and analysis on the air quality of cities in Zhejiang Province since January 2015; Part 4 offers conclusions of this paper and related policy recommendations.

2. Materials and Methods

In order to take the six major air pollutants into consideration, this paper has introduced the Entropy Weight Method based on the traditional Fuzzy Optimization Model to construct a Fuzzy Optimization Measurement and Evaluation Model for the air quality in Zhejiang Province. In its application, the key of successful modeling is how to reasonably determine the weight of different indicators [41–44] and objective function [45,46]. To this end, this paper has adopted the Entropy Weight Method to determine the weight of indicators in the Fuzzy Optimization Model. Entropy is a concept in thermodynamics that represents a measure of the degree of disorder in a system. When the Entropy Weight Method is applied to the Fuzzy Optimization Model, the smaller the information entropy of an indicator is, the larger amount of information is contained in that indicator; the greater role it plays in this model, and therefore the higher weight it should have in the model. Otherwise, the larger the information entropy of an indicator is, the smaller role it plays in this model, and therefore the lower weight it should have [47–50].

The specific steps of applying the Entropy Weight Method to the Fuzzy Optimization Model are as follows:

(1) Construct a Fuzzy Comprehensive Evaluation Matrix

Given the evaluation object P, its related Factor Set $U = \{u_1, u_2, \cdots, u_n\}$, and a Rating Set $V = \{v_1, v_2, \cdots, v_m\}$ for each of the factors, after performing fuzzy evaluation on the Rating Set of each factor in U based on the membership function, this paper could obtain a Fuzzy Evaluation Matrix with $m \times n$ elements:

$$R = \begin{bmatrix} R_1 \\ R_2 \\ \vdots \\ R_4 \end{bmatrix} = \begin{bmatrix} r_{11} & r_{21} & \cdots & r_{m1} \\ r_{12} & r_{22} & \cdots & r_{m2} \\ \vdots & \vdots & \ddots & \vdots \\ r_{1n} & r_{2n} & \cdots & r_{mn} \end{bmatrix}. \tag{1}$$

The element r_{ij} in the above matrix represents the Fuzzy Membership Degree of factor u_i with respect to the Rating element v_i, that is, a fuzzy relationship of U to V, thereby determining the Fuzzy Evaluation Matrix of the evaluation object P.

(2) Determine the Weight of Indicators by the Entropy Method

In the Fuzzy Comprehensive Evaluation Matrix, the Fuzzy Comprehensive Evaluation Vector needs to be obtained by weighted summation. This paper has used the Entropy Method to determine the weight of different indicators of the same rating. Since the calculation of the Entropy Method uses the proportion of a certain indicator of each rating to the sum of indicators with the same nature, no standardization is needed. The specific calculation steps are as follows:

A. Calculate the proportion of the jth indicator preferred by the ith program (P_{ij}):

$$P_{ij} = \frac{r_{ij}}{\sum_{i=1}^{m} r_{ij}}, \quad (i = 1, 2, \cdots, m; j = 1, 2, \cdots, n), \tag{2}$$

B. Calculate the entropy value of the jth indicator (e_j):

$$e_j = -k * \sum_{i=1}^{m} P_{ij} ln P_{ij}, \tag{3}$$

where ln represents the natural logarithm, and the constant k is related to the m of the Rating Set. Generally, let $k = \frac{1}{\ln m}$, and it would have $0 \le e_j \le 1$.

C. Calculate the variation coefficient of the jth indicator (g_j):

$$g_j = 1 - e_j. \tag{4}$$

The above formula shows that, for the jth indicator, the smaller the Entropy Value (e_j) is, the larger the variation coefficient (g_j) becomes.

(4) Calculate the weight of the jth indicator:

$$w_j = \frac{g_j}{\sum_{j=1}^{n} g_j}. \tag{5}$$

(3) Calculate the Fuzzy Comprehensive Evaluation Score

After calculating the weights of different indicators, this paper could obtain the Fuzzy Evaluation Set B by matrix and vector algorithm based on the Fuzzy Evaluation Matrix and Weight Vector:

$$B = W * R^T = \begin{bmatrix} r_{11} & r_{21} & \cdots & r_{m1} \\ r_{12} & r_{22} & \cdots & r_{m2} \\ \vdots & \vdots & \ddots & \vdots \\ r_{1n} & r_{2n} & \cdots & r_{mn} \end{bmatrix} * [w_1, w_2, \cdots, w_n]^T = [b_1, b_2, \cdots, b_m]. \tag{6}$$

Finally, this paper could obtain $M = max(b_1, b_2, \cdots, b_m)$ based on the principle of maximum membership degree, whose value represents the Fuzzy Comprehensive Evaluation Score of the evaluation object (please refer to Appendix A for the MATLAB Algorithm for Fuzzy Optimization Model Based on Entropy Weight Method). The higher the score, the better the city's air quality is; the lower the score, the worse the city's air quality is.

One disadvantage of the above method, especially the entropy weight method, is that the index values are all required to be greater than zero. If the value of some indicators has an outlier of zero or less than zero, the calculation result of the entropy method will be invalidated [51,52]. Since the data of the six pollutant indicators involved in this study are all greater than zero, this deficiency of the entropy weight method is avoided.

3. Results

The data used in this paper are from the monthly air quality and pollutant monitoring data of cities in Zhejiang Province published by China's National Environmental Monitoring Center [53], and the Data Center of China's Ministry of Environmental Protection [54]. The study period is from January 2015 to April 2018, covering the monthly average concentration data of the six major air pollutants of $PM_{2.5}$, PM_{10}, CO, NO_2, O_3 and SO_2 in China's current air quality standards [26,33]. The data used in this study is the monthly average concentration data of the six pollutants in the target cities (calculated according to the daily data of each city, including 1215 days). In the study of this paper, according to the availability and completeness of the data, the selected period is from January 2015 to April 2018 (partial city data is missing in 2014, so it is not included in the calculation range). The data quoted here are the raw data of the day obtained by the official observation points, so we calculated the monthly average concentration data of the six pollutants in the target cities based on the original daily data, and then used the method in Part 2 to achieve the evaluation outcomes.

Based on the Fuzzy Optimization Model with Entropy Weight Method introduced in Part 2.1 and the above air pollutant data, this paper has calculated the air quality evaluation scores of the 11 cities in Zhejiang Province from January 2015 to April 2018 (as shown in Figure 3 below and Tables A1–A4 in Appendix B):

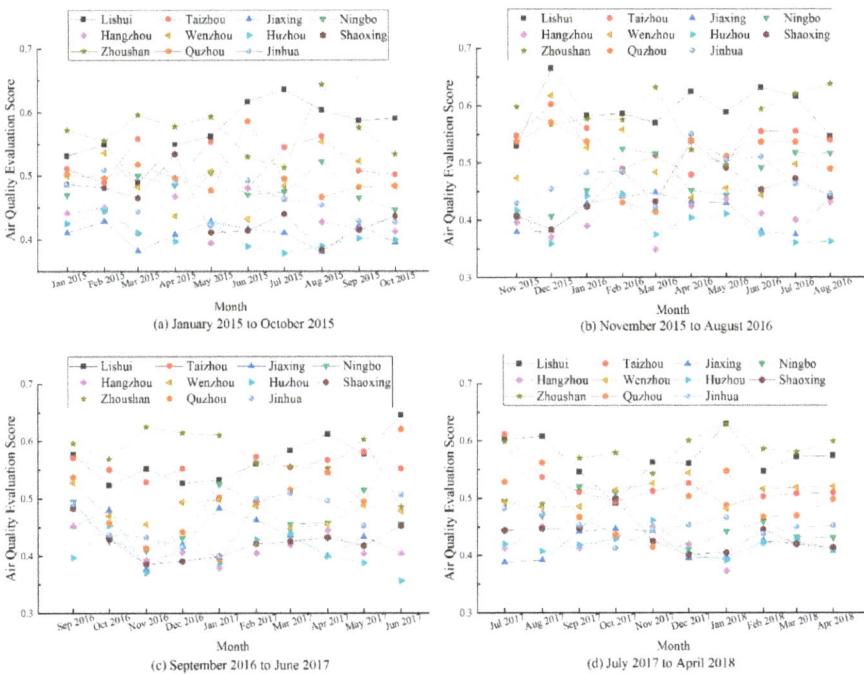

Figure 3. Air quality evaluation score of cities in Zhejiang Province: (**a**) January 2015 to October 2015; (**b**) November 2015 to August 2016; (**c**) September 2016 to June 2017; (**d**) July 2017 to April 2018.

4. Discussion

According to the above model and calculation results, the higher the score, the better the city's air quality is; the lower the score, the worse the city's air quality is. Therefore, the below characteristics in the air quality scores of cities in Zhejiang Province from January 2015 to April 2018 have been found:

(1) Overall evaluation results and trends. According to official statistics, since January 2013, the Ministry of Environmental Protection of China has started air quality monitoring and evaluation

on key regions (such as the Beijing–Tianjin–Hebei region, the Yangtze River Delta region, and the Pearl River Delta region) and the so-called "74 cities" (including municipalities, provincial capitals, and other key cities) with air quality reports published monthly. According to the air quality reports published during the study period, the air quality of cities in Zhejiang Province was generally at the middle level [55], which fully supports the calculation results in this paper. During the study period, the air quality evaluation scores of cities in Zhejiang range from 0.3 to 0.7. Although the evaluation scores of some cities (such as Jiaxing, Huzhou and Hangzhou) have always been low (never exceeded 0.5), the lowest air quality score is still above 0.34 (0.3491 of Hangzhou in March 2016). Therefore, the air pollution of Zhejiang Province is not as enduring and severe as that in areas such as Beijing, Tianjin and Hebei [56–58].

(2) Seasonal trends. Although the differences between the quarters were not large during the study period, the air pollution in cities in Zhejiang Province still showed a certain seasonal trend—the air quality in autumn and winter was generally relatively poor, while in spring and summer is generally relatively good. This trend has been confirmed in relevant researches [59–62]. The reason is that, although Zhejiang is located in southern China, with higher temperature in winter than that in the northern regions such as Beijing, Tianjin and Hebei, the demand for heating in winter still exists, so the increase in energy consumption has increased air pollution to a considerable extent.

Compared with the above research, we calculated the air quality evaluation values of all prefecture-level cities in Zhejiang Province based on the six kinds of atmospheric pollutants, which overcame the limitation of only 1–2 pollutants and several cities of Zhejiang. Moreover, we calculated the weight of different pollutant indicators by using the fuzzy optimization model and based on the entropy weight method, to enrich existing research methods of seasonal changes literature.

(3) The air quality in Zhoushan and the southern cities of Zhejiang is obviously improved. As the first prefecture-level city in China that is formed by an archipelago, Zhoushan consists of large numbers of islands and is surrounded by sea, and thus enjoys unique geographical advantages in maintaining good air quality. In 2017, the air quality of Zhoushan ranked third among all cities in China [63]. According to the calculation results of this paper, the air quality evaluation score of Zhoushan has always ranked top in Zhejiang Province, and its air quality score has improved by 4.67% at the last study period when compared with that of the beginning period. At the same time, the air quality of Lishui and Wenzhou in southern Zhejiang has improved by 8.04% and 4.22% respectively during the study period (please refer to Table 1 below).

Table 1. The air quality assessment results for key months in Zhoushan and the southern cities of Zhejiang.

	January 2015	April 2018	Improvement Ratio at the End of the Study Period (April 2018) Compared to the Beginning of the Study Period (January 2015)
Lishui	0.5312	0.5739	8.04%
Wenzhou	0.4999	0.5210	4.22%
Zhoushan	0.5722	0.5989	4.67%

Apart from geographical reasons, these three cities have formulated tailored air pollution prevention and control measures based on their own characteristics and the "Regulations on the Prevention and Control of Air Pollution in Zhejiang Province" officially passed by the Standing Committee of Zhejiang Provincial People's Congress on 1 July 2016 [64]:

- Lishui City has formulated a "Detailed Code of Practice for Air Pollution Prevention and Control Campaigns", clearly specifying air pollution prevention and control work in terms of six perspectives: industrial structure, energy structure, mobile pollution sources (motor vehicles), industrial waste gas, urban smoke and dust, and rural waste gas [65].

- Based on the characteristics of its industries, Wenzhou is working to build a low-carbon city to control air pollution by developing a recycling economy. Wenzhou is promoting industrial transformation and upgrade by cultivating large clustering industries and large enterprise headquarters, vigorously develops the marine economy, and aims to build an industrial structure with the modern service industry as the main part, supported by an advanced manufacturing industry, and with an urban modern agriculture in synergetic development. By the end of 2017, the energy consumption per unit of industrial added value of Wenzhou has reduced by 17% compared with that in 2012. The recycled proportion of main non-ferrous metals and steel was over 40%, and more than 70% of the industrial parks of the above provincial level have completed recycling upgrade and reconstruction [66].
- Zhoushan strictly controls coal consumption: it has set a control target regarding total coal consumption, and determines detailed responsibility and accountability of specific enterprises and equipment that consume coal with monthly monitoring. In addition, Zhoushan also works hard to develop clean energy by taking advantage of its favorable geographical location (surrounded by sea), such as building onshore wind farms and independent power supply systems on islands in order to utilize ocean energy [67].

(4) The air quality of the northern cities of Zhejiang is relatively poor. During the study period, the air quality of the four cities in northern Zhejiang (Jiaxing, Hangzhou, Huzhou, and Shaoxing) has been ranked at the bottom. The air quality score of Shaoxing even declined by 14.85% at the ending period when compared with that of the beginning period, and Hangzhou's air quality score also decreased by 6.69% (please refer to Table 2 below).

Table 2. The air quality assessment results for key months in the northern cities of Zhejiang.

	January 2015	April 2018	Improvement Ratio at the End of the Study Period (April 2018) Compared to the Beginning of the Study Period (January 2015)
Jiaxing	0.4104	0.4069	−0.85%
Hangzhou	0.4412	0.4117	−6.69%
Huzhou	0.4252	0.4103	−3.50%
Shaoxing	0.4869	0.4146	−14.85%

The main reason behind is that these four cities have always been the most densely populated and economically developed region in Zhejiang Province; their industries have generated large amounts of air pollutants. Moreover, the four cities are geographically located in the inland areas of northern Zhejiang, and are heavily affected by the air pollutants spread from northern China [68,69], unlike coastal cities (such as Zhoushan) where the air pollutants can be easily dispersed [70,71]:

- Shaoxing is located in the intersection of the hills of western Zhejiang, the mountains of eastern Zhejiang and the northern Zhejiang plain, surrounded by mountains and with a terrain high in the south and low in the north [72]. Once the air pollutants gathered over the city, it is more difficult for them to dissipate. Meanwhile, Shaoxing has been relying on the heavily polluting printing and dyeing industry as a pillar industry for many years, with this industry's production capacity taking over 60% of the total production capacity in Zhejiang Province, and accounting for one-third of the national production capacity of the printing and dyeing industry in China [73].
- As for Hangzhou, the provincial capital of Zhejiang, its permanent population reached 9.47 million at the end of 2017, and its total population may have exceeded 10 million if taking the migrant population into consideration; its annual industrial investment was 86.1 billion yuan in 2017 and its total number of motor vehicles reached 2.79 million at the end of 2017 [74], resulting in large amounts of air pollutant emissions of all kinds that have brought tremendous pressures to the environment.

Therefore, although these cities have adopted a series of pollution prevention measures, the air quality of these cities still ranked bottom among cities in Zhejiang during the study period.

(5) The air quality of the four cities in central Zhejiang (Jinhua, Ningbo, Quzhou and Taizhou) has declined at the last study period when compared with that of the beginning period. Among them, Jinhua and Ningbo have seen a large decline in air quality, which are 6.80% and 8.15%, respectively (please refer to Table 3 below).

Table 3. The air quality assessment results for key months in the central cities of Zhejiang.

	January 2015	April 2018	Improvement Ratio at the End of the Study Period (April 2018) Compared to the Beginning of the Study Period (January 2015)
Jinhua	0.4853	0.4523	−6.80%
Ningbo	0.4697	0.4314	−8.15%
Quzhou	0.5023	0.4981	−0.84%
Taizhou	0.5109	0.5099	−0.20%

- In Jinhua City, industrial emissions are the main source of air pollutants responsible for 67.31% of the SO_2, 34.42% of the NO_x, 30.39% of the CO, 53.02% of the PM_{10} and 50.95% of the $PM_{2.5}$. Among the industrial pollution sources, the building materials manufacturing industry and the textile printing and dyeing industry account for the largest proportion [75].
- Ningbo is also short in natural resources and lacks the energy resources needed to fuel economic growth, with more than 90% of its energy imported from other regions in which industrial consumption accounts for over 75% [76]. Meanwhile, the heavy usage of fossil energy in industrial production is also an important reason for decline in air quality of Ningbo. Therefore, although Ningbo have adopted a series of measures to control air pollution, given that their industrial structure and energy structure have not fundamentally changed, their air quality has not improved during the study period and has even deteriorated along with rapid economic development [77].
- The same thing is observed in Quzhou and Taizhou too. Although the decline in their air quality is not as big as in Ningbo and Jinhua (Quzhou 0.84% and Taizhou 0.20%), the fact that there is no obvious improvement in their air quality when comparing that in the last period with that of the beginning period has indicated that the air pollution control campaign still has a long way to go for these cities.

In summary, the main reasons for the above air quality status during the study period in cities in Zhejiang Province are:

(1) The industrial structure is not reasonable and industrial upgrading is not yet in place.
(2) Coal-based energy consumption structure has not changed, and energy efficiency is not ideal.
(3) The laws and regulations on air pollution control need to be further improved.

Compared with similar literature, this paper has made innovations and contributions in the following two aspects:

(1) By constructing the six major pollutants defined by China's air quality standards, we have evaluated the air quality of Zhejiang Province more comprehensively than the existing research. In similar studies of air quality in Zhejiang Province, scholars often use particulate pollutants as research objects [27,28,78]. Then, few research works contain the six major pollutants.

(2) In terms of research methods, the paper has introduced the Entropy Weight Method based on the traditional Fuzzy Optimization Model to construct a Fuzzy Optimization Measurement and Evaluation Model, in order to calculate the weight more accurately. Comparing the current literature on the evaluation of air quality in Zhejiang and the Yangtze River Delta region (such as Shanghai and Jiangsu) [79–83], we have innovated the traditional evaluation model and tried to make the evaluation results more reliable.

5. Conclusions

This paper has constructed an air quality evaluation system incorporating the six major air pollutants covered in China's current national air quality standards (SO_2, NO_2, CO, PM_{10}, $PM_{2.5}$ and O_3), and attempts to obtain a comprehensive evaluation of the air quality of cities in Zhejiang by analyzing the monthly data of the six pollutants in the 11 cities of Zhejiang Province from January 2015 to April 2018. In terms of the research method, this paper has introduced the Entropy Weight Method to the traditional Fuzzy Optimization Model to construct a Fuzzy Optimization Measurement and Evaluation Model for the air quality of Zhejiang Province. The conclusions of this paper are:

(1) During the study period, the air pollution in cities in Zhejiang Province still showed a certain seasonal trend. Moreover, the air quality scores of cities in Zhejiang Province range from 0.3 to 0.7, which indicates the air pollution in Zhejiang is not as severe as in Beijing, Tianjin, Hebei and other regions.

(2) The air quality of northern Zhejiang cities (such as Jiaxing, Huzhou, Shaoxing and Hangzhou) is worse than that of the southern cities. The air quality of Shaoxing has dropped by 14.85% at the last study period when compared with that of the beginning period, and Hangzhou, the provincial capital of Zhejiang, has also seen a decrease of 6.69% in air quality. Apart from geographical factors, the industrial structure of these cities is also an important reason for their poor air quality.

(3) The air quality of Lishui, Zhoushan and Wenzhou has improved by 8.04%, 4.67% and 4.22%, respectively, at the last study period when compared with that of the beginning period. In addition to geographical reasons, these three cities have formulated tailored air pollution prevention and control measures based on their own characteristics and the "Regulations on the Prevention and Control of Air Pollution in Zhejiang Province".

(4) The air quality of the four cities in central Zhejiang (Ningbo, Jinhua, Quzhou and Taizhou) has declined at the last study period when compared with that of the beginning period given that their industrial structure and energy structure have not fundamentally changed during the study period. Among them, Ningbo and Jinhua have seen a large decline in air quality, which are 8.15% and 6.80%, respectively.

The main feature of this study is the comprehensive evaluation of air quality in Zhejiang Province using the data of six air pollutants. Compared with the existing Chinese air pollution research literature, on the one hand, this paper comprehensively evaluates the impact of six kinds of pollutants instead of 1–2 kinds of pollutants according to Chinese national standards. On the other hand, we try to make innovation in the research methods of air pollution by using the Fuzzy Optimization Model Based on Entropy Weight Method, in order to enrich the literature on air pollution and sustainable development.

Based on the findings above, this paper has provided the following policy recommendations for further enhancing air pollution control in the cities of Zhejiang Province:

(1) Adjust the industrial structure as soon as possible to promote industrial upgrading. In the cities of Zhejiang Province, especially in Shaoxing and Ningbo, industry still accounts for a large proportion of the Gross Domestic Product (GDP). In 2017, Shaoxing's secondary industry output accounted for 48.8% of its total GDP, while its service industry output accounted for 47.2% of its GDP [84]; Ningbo's secondary industry output accounted for 51.8% of its total GDP, while its service industry output accounted for 45.0% of its total GDP [85]. These cities need to accelerate the development of their service industry and promote the transition of its industrial structure towards a low carbon and environmental friendly structure. The key in reducing air pollution in these cities is to optimize the industrial structure and promote industrial upgrade. In order to promote the transformation and upgrade of heavy pollution industrial enterprises, these cities must abandon the old production model and processing mode that sacrifice energy and environment, improve their GDP structure, and lower their level of air pollution. At the same time, it is necessary to vigorously develop advanced manufacturing industries, improve the industrial capacity with help from the advancement of information technologies, promote environmentally friendly industries to form related industrial clusters, and control the emissions of industrial waste gas from their sources. In addition, given the current industrial waste gas emission

Sustainability **2019**, *11*, 2143

levels of Zhejiang Province, it is also necessary to speed up the construction of eco-industrial parks, attract funds and talents, and reduce the cost of sewage disposal and pollution treatment in order to facilitate resource saving, industrial upgrading, and pollution/emission reduction.

(2) Increase investment in science and technology, change the current energy consumption structure that heavily relies on coal, and improve the efficiency of energy utilization. Currently, the cities in Zhejiang Province still rely on coal as their main energy source for production. In 2016, the province's total coal consumption reached 139.48 million tons [14], which increased by 0.89% compared with that in year 2015 [86]. Therefore, cities in Zhejiang Province need to enhance the investment in science and technology, change the energy structure over-relying on coal as soon as possible, eliminate coal-fired equipment with heavy pollution, reduce emissions resulted from coal burning, and improve their overall air quality and environment quality. On the one hand, they should continuously reduce the pollution emissions of coal-fired equipment; on the other hand, they need to continuously increase the proportion of clean energy and renewable energy in energy consumption, such as by using subsidies and incentives to encourage enterprises and residents to use clean energy, in order to fundamentally change the current coal-based energy consumption structure. At the same time, they should work to increase the output per unit of energy consumption by applying the latest scientific and technological achievements in the energy field to industrial production and daily life, reducing the energy consumption per unit of regional GDP with help from scientific and technological advancement in order to effectively control air pollution.

(3) Further improve laws and regulations on air pollution control, and enhance the legislation and law enforcement work on environmental protection. All cities in Zhejiang Province need to strengthen their legislation on air pollution control, especially the legislation of laws and regulations on industrial waste gas emission. The cities need to scientifically classify the enterprises by pollutant emissions, monitor and control the pollution emissions based on a grading standard and have clear rules and standards to follow. In law enforcement work, the cities need to suspend the production of enterprises that fail to meet national and provincial standards and order them to rectify, strictly supervise and control the approval and construction of high-pollution projects, and completely eradicate illegal pollution emission. It is necessary to strictly restrict the emissions of enterprises by legal provisions, investigate the enterprises that have violated the laws, and hold their responsible persons accountable by law. Furthermore, it is necessary to strengthen environmental protection supervision work, determine and clarify the accountability in environmental protection among governments at all levels, and actively cooperate with environmental protection departments to perform supervision/investigation and law enforcement work. At the same time, the cities should encourage their residents to participate in environmental supervision and management, and accept social supervision in order to eventually establish a long-term system for air pollution control with continuous improvement.

Author Contributions: Conceptualization, X.W. and Z.Y.; Methodology, X.W.; Validation, Z.Y.; Formal analysis, X.W. and Z.Y.; Resources, X.W. and Z.Y.; Investigation, X.W. and Z.Y.; Data Curation, Z.Y.; Writing—original draft preparation, X.W. and Z.Y.; Writing—review and editing, X.W. and Z.Y.; Visualization, X.W.; Supervision, X.W.

Funding: This research received no external funding.

Conflicts of Interest: The authors declare no conflict of interest.

Appendix A MATLAB Algorithm for Fuzzy Optimization Model Based on Entropy Weight Method

Algorithm: Fuzzy Optimization Model Algorithm

```
function[B]=fuzzy_zhpj(model,A,R)
B=[];
[m,s1]=size(A);
[s2,n]=size(R);
if(s1~=s2)
    disp('The column of A is not equal to the row of R');
else
  if(model==1)
      for(i=1:m)
        for(j=1:n)
            B(i,j)=0;
            for(k=1:s1)
                x=0;
                if(A(i,k)<R(k,j))
                    x=A(i,k);
                else
                    x=R(k,j);
                end
                if(B(i,j)<x)
                    B(i,j)=x;
                end
            end
        end
      end
  elseif(model==2)
      for(i=1:m)
        for(j=1:n)
            B(i,j)=0;
            for(k=1:s1)
                x=A(i,k)*R(k,j);
                if(B(i,j)<x)
                    B(i,j)=x;
                end
            end
        end
      end
  elseif(model==3)
        for(i=1:m)
          for(j=1:n)
            B(i,j)=0;
            for(k=1:s1)
                B(i,j)=B(i,j)+A(i,k)*R(k,j);
            end
          end
        end
```

```
    else if(model==4)
        for(i=1:m)
            for(j=1:n)
                B(i,j)=0;
                for(k=1:s1)
                    x=0;
                    x=min(A(i,k),R(k,j));
                    B(i,j)=B(i,j)+x;
                end
                B(i,j)=min(B(i,j),1);
            end
        end
    elseif(model==5)
        C=[];
        C=sum(R);
        for(j=1:n)
            for(i=1:s2)
                R(i,j)=R(i,j)/C(j);
            end
        end
        for(i=1:m)
            for(j=1:n)
                B(i,j)=0;
                for(k=1:s1)
                    x=0;
                    x=min(A(i,k),R(k,j));
                    B(i,j)=B(i,j)+x;
                end
            end
        end
    else
        disp('improper assignment of model');
    end
end
end
```

Appendix B Air Quality Evaluation Score of Cities in Zhejiang Province (January 2015 to April 2018)

Table A1. Air quality evaluation score of cities in Zhejiang Province (January 2015 to October 2015).

	January 2015	February 2015	March 2015	April 2015	May 2015	June 2015	July 2015	August 2015	September 2015	October 2015
Hangzhou	0.4412	0.4498	0.4095	0.4673	0.3943	0.4803	0.4642	0.4274	0.4165	0.4127
Huzhou	0.4252	0.4450	0.4110	0.3972	0.4239	0.3892	0.3783	0.3895	0.4015	0.3994
Jiaxing	0.4104	0.4285	0.3821	0.4073	0.4281	0.4167	0.4101	0.3807	0.4210	0.3955
Jinhua	0.4853	0.5090	0.4434	0.4891	0.4218	0.4923	0.4625	0.4537	0.4290	0.4268
Lishui	0.5312	0.5491	0.4890	0.5497	0.5623	0.6167	0.6359	0.6039	0.5872	0.5906
Ningbo	0.4697	0.4453	0.5002	0.4850	0.5051	0.4703	0.4747	0.5221	0.4654	0.4466
Quzhou	0.5023	0.4958	0.5172	0.4962	0.4766	0.5859	0.4952	0.4663	0.4820	0.4838
Shaoxing	0.4869	0.4811	0.4651	0.5340	0.4111	0.4137	0.4401	0.3834	0.4149	0.4366
Taizhou	0.5109	0.4895	0.5587	0.4929	0.5529	0.4802	0.5447	0.5624	0.5076	0.5016
Wenzhou	0.4999	0.5358	0.4820	0.4367	0.5066	0.4318	0.4837	0.5537	0.5228	0.4828
Zhoushan	0.5722	0.5555	0.5959	0.5781	0.5933	0.5294	0.5127	0.6437	0.5758	0.5338

Table A2. Air quality evaluation score of cities in Zhejiang Province (November 2015 to August 2016).

	November 2015	December 2015	January 2016	February 2016	March 2016	April 2016	May 2016	June 2016	July 2016	August 2016
Hangzhou	0.3966	0.3710	0.3902	0.4437	0.3491	0.4250	0.4368	0.4120	0.4010	0.4321
Huzhou	0.4173	0.3594	0.4429	0.4461	0.3754	0.4048	0.4114	0.3763	0.3609	0.3633
Jiaxing	0.3796	0.3819	0.4290	0.4417	0.4485	0.4320	0.4302	0.3807	0.3750	0.4413
Jinhua	0.4300	0.4543	0.4833	0.4856	0.4227	0.5505	0.5048	0.5107	0.4630	0.4463
Lishui	0.5290	0.6652	0.5824	0.5862	0.5702	0.6248	0.5887	0.6318	0.6159	0.5464
Ningbo	0.4031	0.4070	0.4524	0.5240	0.5162	0.4520	0.4436	0.4919	0.5173	0.5163
Quzhou	0.5369	0.5715	0.5369	0.4306	0.4145	0.5395	0.5077	0.5365	0.5364	0.4894
Shaoxing	0.4079	0.3843	0.4243	0.4854	0.4328	0.5381	0.4915	0.4536	0.4727	0.4410
Taizhou	0.5480	0.6026	0.5607	0.4896	0.5127	0.4792	0.5112	0.5555	0.5559	0.5394
Wenzhou	0.4736	0.6182	0.5263	0.5582	0.4840	0.4388	0.4558	0.4438	0.4974	0.4894
Zhoushan	0.5980	0.5672	0.5769	0.5754	0.6321	0.5227	0.4964	0.5939	0.6206	0.6375

Table A3. Air quality evaluation score of cities in Zhejiang Province (September 2016 to June 2017).

	September 2016	October 2016	November 2016	December 2016	January 2017	February 2017	March 2017	April 2017	May 2017	January 2017
Hangzhou	0.4519	0.4324	0.3930	0.4066	0.3790	0.4053	0.4198	0.4448	0.4048	0.4046
Huzhou	0.3973	0.4532	0.3705	0.4159	0.3868	0.4285	0.4393	0.3994	0.3874	0.3562
Jiaxing	0.4517	0.4798	0.3765	0.4209	0.4833	0.4623	0.4361	0.4000	0.4338	0.4052
Jinhua	0.4910	0.4372	0.4330	0.4154	0.3983	0.4994	0.5093	0.4960	0.4527	0.5069
Lishui	0.5768	0.5227	0.5519	0.5271	0.5329	0.5606	0.5839	0.6125	0.5776	0.6460
Ningbo	0.4947	0.4268	0.4094	0.4317	0.5254	0.4926	0.4555	0.4585	0.5147	0.4558
Quzhou	0.5368	0.4575	0.4136	0.4421	0.3928	0.4919	0.5153	0.5448	0.4945	0.6207
Shaoxing	0.4831	0.4309	0.3856	0.3909	0.3993	0.4214	0.4258	0.4325	0.4183	0.4528
Taizhou	0.5710	0.5500	0.5287	0.5521	0.5019	0.5721	0.5544	0.5673	0.5813	0.5523
Wenzhou	0.5277	0.4701	0.4554	0.4940	0.4979	0.4864	0.4490	0.4549	0.4880	0.4781
Zhoushan	0.5961	0.5685	0.6248	0.6144	0.6103	0.5612	0.5555	0.5527	0.6030	0.4855

Table A4. Air quality evaluation score of cities in Zhejiang Province (July 2017 to April 2018).

	July 2017	August 2017	September 2017	October 2017	November 2017	December 2017	January 2018	February 2018	March 2018	April 2018
Hangzhou	0.4126	0.4507	0.4128	0.4328	0.4504	0.4202	0.3729	0.4218	0.4232	0.4117
Huzhou	0.4207	0.4078	0.4189	0.4294	0.4619	0.4043	0.3914	0.4233	0.4285	0.4103
Jiaxing	0.3884	0.3918	0.4426	0.4472	0.4427	0.3954	0.3969	0.4259	0.4207	0.4069
Jinhua	0.4826	0.4730	0.4530	0.4131	0.4449	0.4535	0.4665	0.4382	0.4496	0.4523
Lishui	0.6025	0.6076	0.5461	0.4913	0.5623	0.5604	0.6296	0.5470	0.5723	0.5739
Ningbo	0.4949	0.4691	0.5208	0.5101	0.4259	0.4110	0.4422	0.4598	0.4330	0.4314
Quzhou	0.5285	0.5619	0.4675	0.4365	0.4150	0.5031	0.5472	0.4677	0.4696	0.4981
Shaoxing	0.4444	0.4474	0.4480	0.4994	0.4249	0.4007	0.4052	0.4459	0.4203	0.4146
Taizhou	0.6114	0.5367	0.5113	0.4940	0.5119	0.5258	0.4875	0.5026	0.5079	0.5099
Wenzhou	0.4934	0.4845	0.4858	0.5146	0.5259	0.5442	0.4817	0.5162	0.5188	0.5210
Zhoushan	0.5984	0.4899	0.5695	0.5788	0.5423	0.6000	0.6289	0.5857	0.5798	0.5989

References

1. Cojocaru, C.; Cocârță, D.M.; Istrate, I.A.; Crețescu, I. Graphical Methodology of Global Pollution Index for the Environmental Impact Assessment Using Two Environmental Components. *Sustainability* **2017**, *9*, 593. [CrossRef]
2. Yang, W.; Li, L. Efficiency evaluation of industrial waste gas control in China: A study based on data envelopment analysis (DEA) model. *J. Clean. Prod.* **2018**, *179*, 1–11. [CrossRef]
3. Tsinganos, K.; Gerasopoulos, E.; Keramitsoglou, I.; Pirrone, N.; the ERA-PLANET Team. ERA-PLANET, a European Network for Observing Our Changing Planet. *Sustainability* **2017**, *9*, 1040. [CrossRef]
4. Li, L.; Yang, W. Total Factor Efficiency Study on China's Industrial Coal Input and Wastewater Control with Dual Target Variables. *Sustainability* **2018**, *10*, 2121. [CrossRef]
5. Yuan, J.; Ou, X.; Wang, G. Establishing a Framework to Evaluate the Effect of Energy Countermeasures Tackling Climate Change and Air Pollution: The Example of China. *Sustainability* **2017**, *9*, 1555. [CrossRef]
6. Gao, G.; Zeng, X.; An, C.; Yu, L. A Sustainable Industry-Environment Model for the Identification of Urban Environmental Risk to Confront Air Pollution in Beijing, China. *Sustainability* **2018**, *10*, 962. [CrossRef]
7. Liu, C.; Côté, R. A Framework for Integrating Ecosystem Services into China's Circular Economy: The Case of Eco-Industrial Parks. *Sustainability* **2017**, *9*, 1510. [CrossRef]

8. Yang, W.; Li, L. Efficiency Evaluation and Policy Analysis of Industrial Wastewater Control in China. *Energies* **2017**, *10*, 1201. [CrossRef]
9. Xu, Y.; Yang, W.; Wang, J. Air quality early-warning system for cities in China. *Atmos. Environ.* **2017**, *148*, 239–257. [CrossRef]
10. Wang, C.; Bi, J.; Rikkert, M.G.M.O. Early warning signals for critical transitions in cardiopulmonary health, related to air pollution in an urban Chinese population. *Environ. Int.* **2018**, *121*, 240–249. [CrossRef]
11. Liu, X.; Gao, X. A New Study on Air Quality Standards: Air Quality Measurement and Evaluation for Jiangsu Province Based on Six Major Air Pollutants. *Sustainability* **2018**, *10*, 3561. [CrossRef]
12. Li, C.; Zhu, Z. Research and application of a novel hybrid air quality early-warning system: A case study in China. *Sci. Total Environ.* **2018**, *626*, 1421–1438. [CrossRef]
13. Yang, Y.; Yang, W. Does Whistleblowing Work for Air Pollution Control in China? A Study Based on Three-party Evolutionary Game Model under Incomplete Information. *Sustainability* **2019**, *11*, 324. [CrossRef]
14. National Bureau of Statistics of China. *China Statistical Yearbook, 2016*; China Statistic Press: Beijing, China, 2017.
15. Rohde, R.A.; Muller, R.A. Air pollution in China: Mapping of concentrations and sources. *PLoS ONE* **2015**, *10*, e0135749. [CrossRef]
16. The World Bank Group; IHME. *The Cost of Air Pollution: Strengthening the Economic Case for Action*; World Bank Group: Seattle, WA, USA, 2016.
17. Lai, J.; Yang, Z.; Feng, J. The Conversion of Dynamic Structure and Mechanism in Zhejiang's Economy Growth. *Zhejiang Soc. Sci.* **2016**, *4*, 144–147.
18. He, W.; Zhu, Q. Industrial Structure Upgrade, Economic Development and Tax Growth: An Empirical Analysis Based on VAR Models in Zhejiang Province and Jiangsu Province. *Public Financ. Res. J.* **2018**, *2*, 52–63.
19. National Bureau of Statistics of China. *China Statistical Yearbook, 2000*; China Statistic Press: Beijing, China, 2001.
20. Statistics Bureau of Zhejiang Province. *Statistical Communique of Zhejiang National Economic and Social Development in 2017*; Statistics Bureau of Zhejiang Province: Hangzhou, China, 2018.
21. Yang, W.; Li, L. Analysis of Total Factor Efficiency of Water Resource and Energy in China: A Study Based on DEA-SBM Model. *Sustainability* **2017**, *9*, 1316. [CrossRef]
22. The People's Government of Zhejiang Province. Geographical Profile of Zhejiang Province. Available online: http://www.zj.gov.cn/col/col1544746/index.html (accessed on 1 March 2019).
23. The World Bank. The Land Area of Countries. Available online: http://api.worldbank.org/v2/en/indicator/AG.LND.TOTL.K2?downloadformat=excel (accessed on 1 March 2019).
24. Ni, Z.; Luo, K.; Zhang, J.; Feng, R.; Zheng, H.; Zhu, H.; Wang, J.; Fan, J.; Gao, X.; Cen, K. Assessment of winter air pollution episodes using long-range transport modeling in Hangzhou, China, during World Internet Conference, 2015. *Environ. Pollut.* **2018**, *236*, 550–561. [CrossRef] [PubMed]
25. Feng, R.; Wang, Q.; Huang, C.; Liang, J.; Luo, K.; Fan, J.; Cen, K. Investigation on air pollution control strategy in Hangzhou for post-G20/pre-Asian-games period (2018–2020). *Atmos. Pollut. Res.* **2018**. [CrossRef]
26. Ministry of Environmental Protection of the People's Republic of China. *Technical Regulation on Ambient Air Quality Index (on Trial): HJ 633-2012*; China Environmental Science Press: Beijing, China, 2012.
27. Wu, H.; Yang, C.; Chen, J.; Yang, S.; Lu, T.; Lin, X. Effects of Green space landscape patterns on particulate matter in Zhejiang Province, China. *Atmos. Pollut. Res.* **2018**. [CrossRef]
28. Xu, J.; Xu, M.; Snape, C.; He, J.; Behera, S.N.; Xu, H.; Ji, D.; Wang, C.; Yu, H.; Xiao, H.; et al. Temporal and spatial variation in major ion chemistry and source identification of secondary inorganic aerosols in Northern Zhejiang Province, China. *Chemosphere* **2017**, *179*, 316–330. [CrossRef]
29. Fu, Q.; Mo, Z.; Lyu, D.; Zhang, L.; Qin, Z.; Tang, Q.; Yin, H.; Xu, P.; Wu, L.; Lou, X.; et al. Air pollution and outpatient visits for conjunctivitis: A case-crossover study in Hangzhou, China. *Environ. Pollut.* **2017**, *231*, 1344–1350. [CrossRef]
30. Xu, J.; Xu, H.; Xiao, H.; Tong, L.; Snape, C.E.; Wang, C.; He, J. Aerosol composition and sources during high and low pollution periods in Ningbo, China. *Atmos. Res.* **2016**, *178–179*, 559–569. [CrossRef]
31. Sun, J.; Wang, Q.; Zhuang, S.; Zhang, A. Occurrence of polybrominated diphenyl ethers in indoor air and dust in Hangzhou, China: Level, role of electric appliances, and human exposure. *Environ. Pollut.* **2016**, *218*, 942–949. [CrossRef]

32. Mestl, H.E.S.; Edwards, R. Global burden of disease as a result of indoor air pollution in Shaanxi, Hubei and Zhejiang, China. *Sci. Total Environ.* **2011**, *409*, 1391–1398. [CrossRef] [PubMed]

33. Ministry of Environmental Protection of the People's Republic of China. *Ambient Air Quality Standards: GB3095-2012*; China Environmental Science Press: Beijing, China, 2012.

34. Hopwood, B.; Mellor, M.; O'Brien, G. Sustainable development: mapping different approaches. *Sustain. Dev.* **2005**, *13*, 38–52. [CrossRef]

35. Ding, G.K.C. Sustainable construction—The role of environmental assessment tools. *J. Environ. Manag.* **2008**, *86*, 451–464. [CrossRef]

36. Moldan, B.; Janoušková, S.; Hák, T. How to understand and measure environmental sustainability: Indicators and targets. *Ecol. Indic.* **2012**, *17*, 4–13. [CrossRef]

37. Basiago, A.D. Economic, social, and environmental sustainability in development theory and urban planning practice. *Environmentalist* **1998**, *19*, 145–161. [CrossRef]

38. Hueting, R.; Reijnders, L. Broad sustainability contra sustainability: the proper construction of sustainability indicators. *Ecol. Econ.* **2004**, *50*, 249–260. [CrossRef]

39. Chen, L.; Wang, Y.; Lai, F.; Feng, F. An investment analysis for China's sustainable development based on inverse data envelopment analysis. *J. Clean. Prod.* **2017**, *142*, 1638–1649. [CrossRef]

40. Zhang, P.; Yuan, H.; Tian, X. Sustainable development in China: Trends, patterns, and determinants of the "Five Modernizations" in Chinese cities. *J. Clean. Prod.* **2019**, *214*, 685–695. [CrossRef]

41. Lamata, M.T.; Pelta, D.; Verdegay, J.L. Optimisation problems as decision problems: The case of fuzzy optimisation problems. *Inf. Sci. (NY)* **2018**, *460–461*, 377–388. [CrossRef]

42. Milan, S.G.; Roozbahani, A.; Banihabib, M.E. Fuzzy Optimization Model and Fuzzy Inference System for Conjunctive Use of Surface and Groundwater Resources. *J. Hydrol.* **2018**. [CrossRef]

43. Xie, Y.L.; Xia, D.X.; Ji, L.; Huang, G.H. An inexact stochastic-fuzzy optimization model for agricultural water allocation and land resources utilization management under considering effective rainfall. *Ecol. Indic.* **2018**, *92*, 301–311. [CrossRef]

44. Liagkouras, K.; Metaxiotis, K. Multi-period mean–variance fuzzy portfolio optimization model with transaction costs. *Eng. Appl. Artif. Intell.* **2018**, *67*, 260–269. [CrossRef]

45. Mi, Z.; Pan, S.; Yu, H.; Wei, Y. Potential impacts of industrial structure on energy consumption and CO2 emission: a case study of Beijing. *J. Clean. Prod.* **2015**, *103*, 455–462. [CrossRef]

46. Mi, Z.; Wei, Y.; Wang, B.; Meng, J.; Liu, Z.; Shan, Y.; Liu, J.; Guan, D. Socioeconomic impact assessment of China's CO_2 emissions peak prior to 2030. *J. Clean. Prod.* **2017**, *142*, 2227–2236. [CrossRef]

47. Wu, G.; Duan, K.; Zuo, J.; Zhao, X.; Tang, D. Integrated Sustainability Assessment of Public Rental Housing Community Based on a Hybrid Method of AHP-Entropy Weight and Cloud Model. *Sustainability* **2017**, *9*, 603.

48. Tufano, D.; Sotoudeh, Z. Exploring the entropy concept for coupled oscillators. *Int. J. Eng. Sci.* **2017**, *112*, 18–31. [CrossRef]

49. Seitz, W.; Kirwan, A.D. Incomparability, entropy, and mixing dynamics. *Phys. A Stat. Mech. Its Appl.* **2018**, *506*, 880–887. [CrossRef]

50. Suh, D.H. An Entropy Approach to Regional Differences in Carbon Dioxide Emissions: Implications for Ethanol Usage. *Sustainability* **2018**, *10*, 243. [CrossRef]

51. Dos Santos, L.F.S.; Neves, L.A.; Rozendo, G.B.; Ribeiro, M.G.; do Nascimento, M.Z.; Tosta, T.A.A. Multidimensional and fuzzy sample entropy (SampEnMF) for quantifying H&E histological images of colorectal cancer. *Comput. Biol. Med.* **2018**, *103*, 148–160. [CrossRef]

52. Du, Y.; Wang, S.; Wang, Y. Group fuzzy comprehensive evaluation method under ignorance. *Expert Syst. Appl.* **2019**. [CrossRef]

53. China's National Environmental Monitoring Center. The City Air Quality Publishing Platform. Available online: http://106.37.208.233:20035/ (accessed on 16 February 2019).

54. Data Center of China's Ministry of Ecology and Environment. Concentration of Main Pollutants in Cities of China, 2015–2018. Available online: http://datacenter.mep.gov.cn/websjzx/queryIndex.vm (accessed on 16 February 2019).

55. Ministry of Environmental Protection of the People's Republic of China. *Air Quality Report on the 74 Key Monitored Cities in China, 2015–2018*; Ministry of Environmental Protection of the People's Republic of China: Beijing, China, 2018.

56. Wang, J.; Zhao, Q.; Zhu, Z.; Qi, L.; Wang, J.X.L.; He, J. Interannual variation in the number and severity of autumnal haze days in the Beijing–Tianjin–Hebei region and associated atmospheric circulation anomalies. *Dyn. Atmos. Ocean.* **2018**, *84*, 1–9. [CrossRef]

57. Liang, Y.; Niu, D.; Zhou, W.; Fan, Y. Decomposition Analysis of Carbon Emissions from Energy Consumption in Beijing–Tianjin–Hebei, China: A Weighted-Combination Model Based on Logarithmic Mean Divisia Index and Shapley Value. *Sustainability* **2018**, *10*, 2535. [CrossRef]

58. Yuan, G.; Yang, W. Evaluating China's Air Pollution Control Policy with Extended AQI Indicator System: Example of the Beijing–Tianjin–Hebei Region. *Sustainability* **2019**, *11*, 939. [CrossRef]

59. Niu, Y.; Gu, J.; Pu, J.; Li, W.; Wu, J. The Long-Term Variation of Haze Weather in Urban Areas of Zhejiang. *J. Trop. Meteorol.* **2010**, *26*, 807–812.

60. Yu, K.; Chen, L.; Zhang, J.; Zhu, C.; Lin, C.; Hu, X.; Chen, D. Variation Characteristics of Atmospheric Mixing Layer Height in Zhejiang. *Meteorol. Sci. Technol.* **2017**, *45*, 735–744.

61. Li, H.; Yin, H.; Yang, Y.; Qiu, J.; Zhang, X. Air Diffusion Ability in Huzhou: Variation Characteristics and Causes. *Chin. Agric. Sci. Bull.* **2018**, *34*, 121–125.

62. Yang, W.; Yuan, G.; Han, J. Is China's air pollution control policy effective? Evidence from Yangtze River Delta cities. *J. Clean. Prod.* **2019**, *220*, 110–133. [CrossRef]

63. Ministry of Environmental Protection of the People's Republic of China Air Quality Status in Key Regions and 74 Cities in December and January–December 2017. Available online: http://www.zhb.gov.cn/gkml/hbb/qt/201801/t20180118_429903.htm (accessed on 28 August 2018).

64. Standing Committee of Zhejiang Provincial People's Congress. Regulations on the Prevention and Control of Air Pollution in Zhejiang Province. Available online: http://www.zj.gov.cn/art/2016/6/15/art_14433_278191.html (accessed on 28 August 2018).

65. Lishui Municipal Government. *Air Pollution Prevention and Control Action Regulations of Lishui*; Lishui Municipal Government: Lishui, China, 2017.

66. Wenzhou Municipal Government. *Air Pollution Prevention and Control Action Regulations of Wenzhou*; Wenzhou Municipal Government: Wenzhou, China, 2017.

67. Zhoushan Municipal Government. *Report on the Implementation of the "One Law One Case" of Air Pollution Control in Zhoushan Municipal Government*; Zhoushan Municipal Government: Zhoushan, China, 2016.

68. Chen, A.; Chen, L. Discussion of the Space-Time Distribution Characteristics of Haze Pollution in Some Cities in Zhejiang Province and the Prevention and Control Strategies. *Contam. Control Air-Cond. Technol.* **2017**, *1*, 67–71.

69. Shen, Y.; Sheng, Y.; Xu, B.; Chen, L.; Shen, L. On Multi-dimensional Monitoring Network of Atmospheric Pollution in Zhejiang Province. *Environ. Sustain. Dev.* **2018**, *4*, 52–54.

70. Shi, K.; Liu, C.; Wu, S. Self-organized Evolution of Trans-boundary PM10 Pollution in Zhoushan City. *Acta Sci. Circumstantiae* **2014**, *34*, 1125–1132.

71. Fang, L.; Fu, X.; Xie, L.; Liao, W.; Yu, J. Numerical Simulation of Air Pollution Transport in Zhoushan Island. *Res. Environ. Sci.* **2014**, *27*, 1087–1094.

72. Shaoxing Municipal Government. Natural Geography of Shaoxing. Available online: http://www.sx.gov.cn/col/col1461898/index.html (accessed on 28 August 2018).

73. Shaoxing Investigation Team of National Bureau of Statistics of China. Research on the Role Positioning and Transformation Development of Shaoxing Printing and Dyeing Industry. *Stat. Theory Pract.* **2017**, *1*, 31–34.

74. Bureau of Statistics of Hangzhou. Statistical Communique of 2017 National Economic and Social Development of Hangzhou. Available online: http://tjj.hangzhou.gov.cn/content-getOuterNewsDetail.action?newsMainSearch.id=78193c3e-6229-11e8-97a6-d89d676397bf (accessed on 28 August 2018).

75. Xiang, C.; Wang, Y.; Zhou, H.; Bao, S.; Zhang, M.; Wu, M. Development of the Anthropogenic Air Pollutants Emission Inventory in Jinhua. *Environ. Sci. Technol.* **2017**, *40*, 229–237.

76. Jiang, L.; Xu, X.; Yang, H. Analysis and Countermeasures on Fog and Haze in Ningbo. *Envion. Sci. Manag.* **2015**, *40*, 43–46.

77. Yang, W.; Li, L. Energy Efficiency, Ownership Structure, and Sustainable Development: Evidence from China. *Sustainability* **2017**, *9*, 912. [CrossRef]

78. Hu, K.; Guo, Y.; Hu, D.; Du, R.; Yang, X.; Zhong, J.; Fei, F.; Chen, F.; Chen, G.; Zhao, Q.; et al. Mortality burden attributable to PM1 in Zhejiang province, China. *Environ. Int.* **2018**, *121*, 515–522. [CrossRef]

Sustainability **2019**, *11*, 2143

79. Han, T.; Yao, L.; Liu, L.; Xian, A.; Chen, H.; Dong, W.; Chen, J. Baosteel emission control significantly benefited air quality in Shanghai. *J. Environ. Sci.* **2018**, *71*, 127–135. [CrossRef]

80. Li, L.; Zhao, Q.; Zhang, J.; Li, H.; Liu, Q.; Li, C.; Chen, F.; Qiao, Y.; Han, J. Bottom-up emission inventories of multiple air pollutants from open straw burning: A case study of Jiangsu province, Eastern China. *Atmos. Pollut. Res.* **2019**, *10*, 501–507. [CrossRef]

81. Guo, H.; Chen, M. Short-term effect of air pollution on asthma patient visits in Shanghai area and assessment of economic costs. *Ecotoxicol. Environ. Saf.* **2018**, *161*, 184–189. [CrossRef]

82. Ji, X.; Meng, X.; Liu, C.; Chen, R.; Ge, Y.; Kan, L.; Fu, Q.; Li, W.; Tse, L.A.; Kan, H. Nitrogen dioxide air pollution and preterm birth in Shanghai, China. *Environ. Res.* **2019**, *169*, 79–85. [CrossRef]

83. Wang, Q.; Wang, Y.; Zhou, P.; Wei, H. Whole process decomposition of energy-related SO_2 in Jiangsu Province, China. *Appl. Energy* **2017**, *194*, 679–687. [CrossRef]

84. Bureau of Statistics of Shaoxing. Statistical Communique of 2017 National Economic and Social Development of Shaoxing. Available online: http://www.sx.gov.cn/art/2018/4/26/art_1462885_18295578.html (accessed on 28 August 2018).

85. Bureau of Statistics of Ningbo. Statistical Communique of 2017 National Economic and Social Development of Ningbo. Available online: http://tjj.ningbo.gov.cn/read/20180206/30511.aspx (accessed on 28 August 2018).

86. National Bureau of Statistics of China. *China Statistical Yearbook, 2015*; China Statistic Press: Beijing, China, 2016.

sustainability

MDPI

Article

Primary Pollutants and Air Quality Analysis for Urban Air in China: Evidence from Shanghai

Ying Yan [1,2], Yuangang Li [3,*], Maohua Sun [4] and Zhenhua Wu [5]

[1] Business School, University of Shanghai for Science and Technology, Shanghai 200093, China; firstsnowy@outlook.com

[2] School of International Affairs and Public Administration, Shanghai University of Political Science and Law, Shanghai 201701, China

[3] Dalian University of Foreign Languages, Dalian 116044, China

[4] School of Foreign Language, Dalian Jiaotong University, Dalian 116028, China; maohua.sun@outlook.com

[5] School of Information Management and Engineering, Shanghai University of Finance and Economics, Shanghai 200433, China; wu.zhenhua@mail.shufe.edu.cn

* Correspondence: yuangang.li@outlook.com; Tel.: +86-411-8280-3121

Received: 17 February 2019; Accepted: 15 April 2019; Published: 17 April 2019

Abstract: In recent years, China's urban air pollution has caused widespread concern in the academic world. As one of China's economic and financial centers and one of the most densely populated cities, Shanghai ranks among the top in China in terms of per capita energy consumption per unit area. Based on the Shanghai Energy Statistical Yearbook and Shanghai Air Pollution Statistics, we have systematically analyzed Shanghai's atmospheric pollutants from three aspects: Primary pollutants, pollutants changing trends, and fine particulate matter. The comprehensive pollution index analysis method, the grey correlation analysis method, and the Euclid approach degree method are used to evaluate and analyze the air quality in Shanghai. The results have shown that Shanghai's primary pollutants are $PM_{2.5}$ and O_3, and the most serious air pollution happens during the first half of the year, particularly in the winter. This is because it is the peak period of industrial energy use, and residential heating will also lead to an increase in energy consumption. Furthermore, by studying the particulate pollutants of $PM_{2.5}$ and PM_{10}, we clearly disclosed the linear correlation between $PM_{2.5}$ and PM_{10} concentrations in Shanghai which varies seasonally.

Keywords: primary pollutants; air quality; comprehensive pollution index analysis; grey correlation analysis; Euclid approach degree method

---

1. Introduction

In recent years, with China's rapid economic development, consumption of fossil energy has also grown rapidly, and its air quality, especially in cities, has deteriorated drastically, causing a significant negative impact on people's health as well as climate change [1–4]. It has been realized that the scope and severity of urban air pollution are affected by the nature of air pollutants and pollution sources [5], weather conditions [6–8], as well as properties of the land surface [9–11]. These factors are influenced by natural factors (such as air pressure [12], temperature [13], wind direction and speed [14], etc.), but human factors (such as industrial waste gas emissions [15], domestic coal combustion [16,17], automobile exhaust emissions [18], etc.) have a greater impact on the urban air quality. At the same time, human activities also affect natural factors to a certain extent, and a considerable part of the human factors come from the unreasonable consumption of primary energy and secondary energy. The energy consumption structure is closely connected to the industrial structure [19,20]. The current industrial structure with high consumption and low output has further resulted in the deterioration of air quality.

Shanghai is China's largest industrial city and an energy-consuming city, with a per capita energy consumption and unit area energy consumption much higher than the national average. Its total energy consumption has increased from 106.71 million tons of standard coal in 2010 to 117.12 million tons of standard coal in 2016 [21], while the total energy consumption nationwide was 4.36 billion tons of standard coal in 2016 [22], which means Shanghai's total energy consumption accounted for 3% of the total energy consumption of 338 cities in China. What comes with such high energy consumption density is the deterioration of Shanghai's urban environment. According to the data in the Shanghai Environmental Condition Bulletin, the number of days with good air quality was only 275 in 2017, with an Air Quality Index (AQI) good rate of 75.3% [23]. The requirement of continuous economic growth, the increasing consumption of energy and resources, and the continuous deterioration of air quality have brought tremendous pressure and severe challenges to the sustained and stable development of Shanghai's economy and society.

In order to meet the requirements of air quality under the new circumstances, in 2012, China issued a new national ambient air quality standard (GB 3095-2012), which clarified the calculation method of AQI [24]:

First, calculate the Individual Air Quality Index of certain pollutant ($IAQI_P$):

$$IAQI_P = \frac{IAQI_{Hi} - IAQI_{Lo}}{BP_{Hi} - BP_{Lo}} (C_P - BP_{Lo}) + IAQI_{Lo} \tag{1}$$

In the equation above, C_P represents the mass concentration of pollutant P; BP_{Hi} is the higher threshold of pollutant concentration near C_P corresponding to the specified IAQI (Individual Air Quality Index) regulated by government policy; BP_{Lo} is the lower threshold of pollutant concentration near C_P regulated by the government; $IAQI_{Hi}$ is the corresponding IAQI to BP_{Hi}; while $IAQI_{Lo}$ is the corresponding IAQI to BP_{Lo}.

Then, take the largest number from all $IAQI_P$ to calculate the AQI:

$$AQI = max\{IAQI_1, IAQI_2, IAQI_3, \cdots, IAQI_n\} \tag{2}$$

In 2013, the first year the new ambient air quality standard was implemented, the air quality monitoring and evaluation work of Shanghai started to follow the new standards including the Ambient Air Quality Standards (GB3095-2012) and the Technical Regulation on Ambient Air Quality Index (HJ 633-2012) [25]. This was a great opportunity to study the impact of Shanghai's energy consumption structure on its air quality, accelerate the optimization of Shanghai's energy consumption structure, and build an energy-saving society, which is of great significance to the construction of an international city.

Currently, studies on related fields mainly focus on three aspects. The first is the analysis of fine particle pollution and its impact on atmospheric visibility in cities. The second is the concentration feature and chemical composition of air pollutants. The third is the description of emission factors of air pollutants.

Li et al. (2019) studied the meteorological conditions of the severe haze weather that frequently occurred in North China and concluded two main reasons for the decrease in visibility [26]. The first is the influence of meteorological conditions such as atmospheric currents, and the second is the change in the average astigmatism coefficient caused by the absorption and scattering of light due to fine particles and major air pollutants [26]. Golly et al. [27] (2019) conducted experiments on the chemical characterization of $PM_{2.5}$ particles in five rural areas of France, and conducted chemical analysis on the samples every 6 days, including their organic carbon (OC), elemental carbon (EC), ion species, etc. The results showed that wood combustion had made high contributions to the organic carbon (OC), and in some rural areas, the contribution rate of wood combustion to OC could be as high as 90% in winters; the contribution of terrestrial protozoa organic components was also significant in summers and autumns, with a monthly $PM_{2.5}$ contribution rate of 4.5–9.5% [27]. Ryu et al. (2019)

studied the PM (Particulate Matter) removal effect of plant evapotranspiration by using the PM removal performance of five plants and the relative humidity (RH) in a closed chamber as control parameters. The results showed that under effective transpiration, honeysuckle had higher efficiency for aerosol $PM_{2.5}$ removal [28].

At the same time, relevant departments of different countries have also formulated different emission inventories in response to air pollution. The U.S. Environmental Protection Agency (EPA) has established the emission inventory for pollutants through direct measurements of power plants stacks, which provides emission measurements that have an error of less than 2% [29]. The establishment of this emission inventory has provided valuable guidance to the study of the impact of energy consumption on the atmospheric environment. The European Environment Agency (EEA) has established an emission inventory for 30 countries and regions including France and Germany, which covers 8 pollutants (NO_x, SO_2, CO, NH_3, CH_4, N_2O, CO_2, NMVOC) [30]. The study of the emission inventory in Asia started relatively late. Ohara et al. established an emission inventory of Asia from 1980–2020, in which the pollutants mainly come from energy consumption such as the combustion of fossil fuel and biomass fuel for industrial, power, transportation and civil use [31]. This is a relatively comprehensive emission inventory for Asia so far. Meanwhile, Korea and Japan are expected to have their own emission inventory [32–35].

In current studies, there is a lack of systematic and quantitative research on the migration characteristics of urban air pollutants under the influence of energy consumption and estimation of pollutants produced by energy consumption. Therefore, it is important to analyze the characteristics of urban air pollution by relating to the energy consumption needs of Shanghai as a mega-city in its economic and social development, in order to improve its air quality as well as the life quality of its residents. This paper has adopted the Comprehensive Pollution Index Method, the Improved Grey Relational Degree Method, and the Euclid Approach Degree Method to evaluate the air quality of Shanghai, and systematically analyzed the changing pattern and correlation of fine particle pollutants ($PM_{2.5}$ and PM_{10}) in Shanghai, in order to achieve innovations as following:

(1) By introducing the pollution index analysis method, the grey correlation analysis method, and the Euclid approach degree method comprehensively, we hope to overcome their respective deficiencies and make new additions to existing research methods.

(2) By further discussing the changing pattern and correlation of the fine particle pollutants ($PM_{2.5}$ and PM_{10}), we hope to provide new evidence of the interrelationship between major atmospheric pollutants in China.

In the following parts of this paper: Section 2 introduces the backgrounds and methods of this paper and introduces three study methods. Section 3 uses the three methods to calculate and evaluate the air quality of Shanghai from 1 November 2017 to 31 October 2018. Based on the above assessment, Section 4 further discusses the changing pattern and correlation of the fine particle pollutants ($PM_{2.5}$ and PM_{10}) in Shanghai during the study period. Finally, Section 5 provides conclusions of this paper.

2. Materials and Methods

2.1. Introduction of China's AQI System

In order to meet the public's increasing requirement of air quality, and objectively reflect the air pollution situation in China at the same time, in the first half of 2012, the Ministry of Environmental Protection of China issued the Technical Regulation on Ambient Air Quality Index (HJ 633-2012) to replace the previous Air Pollution Index (API). The pollutants covered by this new standard increased to 6 items (SO_2, NO_2, PM_{10}, $PM_{2.5}$, O_3, and CO). The AQI is divided into six levels, which represent superior air quality, good air quality, mild pollution, moderate pollution, heavy pollution, and severe pollution respectively from the highest to the lowest level. The corresponding Air Quality Indexes are: Level I—0–50, Level II—50–100, Level III—101–150, Level IV—151–200, Level V—201–300, and Level VI—above 300 [25]. See Table 1 for details of each level.

Table 1. Air Quality Index Range and corresponding impact.

Air Quality Index Range	Air Quality Level	Air Quality Category	Representative Color	Impacts on Human Health and Recommended Actions
0–50	Level I	Superior	Green	The air quality is satisfactory. There is basically no air pollution, and no impact on human activities.
51–100	Level II	Good	Yellow	The air quality is acceptable. There are certain air pollutants that may cause health issues to a small number of people who should reduce outdoor activities.
101–150	Level III	Mild Pollution	Orange	Symptoms in susceptible people would intensify, and healthy people would show irritation symptoms. Elderly people and children should avoid long hours of high-intensity outdoor exercises.
151–200	Level IV	Moderate Pollution	Red	Symptoms in susceptible people would further intensify, and the breathing of healthy people would be affected. Elderly people and children should avoid outdoor sports.
201–300	Level V	Heavy Pollution	Purple	Ordinary people would show symptoms. Elderly people and children should avoid outdoor sports. The general population should reduce outdoor activities.
>300	Level VI	Severe Pollution	Maroon	Obvious and strong symptoms would appear, and all groups of people should avoid outdoor activities.

2.2. Overview of Shanghai Air Quality

The main air pollutants in Shanghai include SO_2, NO_2, PM_{10}, $PM_{2.5}$, O_3, and CO. According to the data released by the Shanghai Environmental Hotline, the main pollutants published before 2012 include SO_2, NO_2, and PM_{10}. Since 2012, $PM_{2.5}$, O_3, and CO have been added to the published main pollutants [36].

Taking 2017 as an example, according to the AQI evaluation, the number of days with superior and good air quality in Shanghai was 275, which was 1 day less than that in 2016. The good AQI rate was 75.3%, which was 0.1% point lower than that of 2016. Overall speaking, there were 58 days with superior air quality, 217 days with good air quality, 71 days with mild pollution, 17 days with moderate air pollution, and 2 days with heavy pollution. The number of days with heavy pollution was the same with that in 2016. In those 90 days with air pollution, there were 52 days in which ozone (O_3) was the primary air pollutant (the maximum IAQI air pollutant when AQI is greater than 50 [25]), accounting for 57.8% of the pollution days; there were 23 days in which fine particles ($PM_{2.5}$) was the primary air pollutant, accounting for 25.6% of the pollution days; there were 12 days in which nitrogen dioxide (NO_2) was the primary air pollutant, accounting for 13.3% of the pollution days; there were 2 days in which inhalable particles (PM_{10}) was the primary air pollutant (due to the transportation of sand dust), accounting for 2.2% of the pollution days; there was 1 day in which $PM_{2.5}$ and NO_2 were the primary air pollutants, accounting for 1.1% of the pollution days [23].

In 2017, the annual average concentration of $PM_{2.5}$ in Shanghai was 39 μg/m³, which exceeded the Level II national air quality standard of 4 μg/m³ and decreased by 13.3% and 37.1% respectively compared with that of 2016 and the base year 2013. In 2017, the annual average concentration of PM_{10} in Shanghai was 55 μg/m³, which met the Level II national air quality standard, and decreased by 6.8% compared with that of 2016. In 2017, the annual average concentration of SO_2 in Shanghai was 12 μg/m³, which met the Level I national air quality standard, and decreased by 20.0% compared with that of 2016. In 2017, the annual average concentration of NO_2 in Shanghai was 44 μg/m^3, which exceeded the Level II national air quality standard of 4 μg/m^3, and increased by 2.3% compared with that of 2016. In 2017, the 90th percentile of the daily maximum 8-h average concentration of O_3 in Shanghai was 181 μg/m^3, which exceeded the Level II national air quality standard of 21 μg/m^3, which increased by 10.4% compared with that of 2016. In 2017, the daily average concentration of CO in Shanghai ranged from 0.4–1.8 mg/m^3, which met the Level II national air quality standard. The annual average concentration of CO in Shanghai was 0.76 mg/m^3 in 2017, which decreased by 3.8% compared with that of 2016 [23].

Figure 1 below shows the good rate of overall air quality (the ratio of air quality rated as Level I or II in Table 1) in Shanghai from 2013–2017 [23,37–40]. It can be seen from the figure that since 2013, Shanghai's air quality has shown an improvement trend, despite a slight decline in 2015, which was mainly due to the fine particle pollution (PM$_{2.5}$) during autumn and winter, and ozone (O_3) pollution during summer.

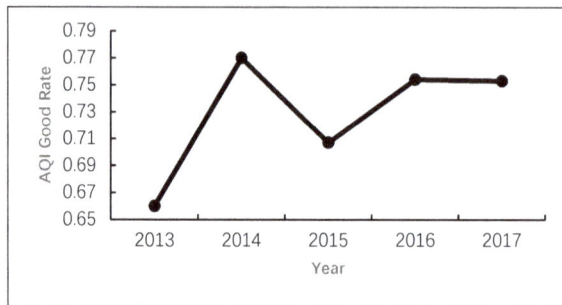

Figure 1. Shanghai Air Quality (AQI) Good Rate from 2013–2017.

2.3. Overview of Shanghai Climate

The climate of Shanghai is a typical subtropical maritime monsoon climate, mild and humid, with four distinct seasons. The spring in Shanghai is warm but often has sudden cold currents. The summer is hot with frequent heavy rains. The autumn is cool with dry weather. The winter is cold and accompanies fog and haze weather.

The Shanghai Meteorological Department began to accelerate the construction of automatic weather stations in 2002. Up to now, there are more than 200 automatic weather stations that have been built and used effectively [41]. The main observations indices of those stations include temperature, rainfall, air pressure, wind, visibility and dew point, etc. This paper selects 67 automatic stations with temperature observation records starting from 1 January 2006 and analyzes the climate data during the study period [42]. We found that Shanghai's climate has the following distinct features:

(1) The climate in Shanghai is with a monthly average relative humidity of over 75%, and the annual precipitation is 1100 millimeters, which helps to relieve air pollution to some extent. So, Shanghai is a city with stable humidity. This will not cause the time difference of its PM$_{2.5}$.

(2) The average annual temperature is 16.7 °C. The average highest temperature in July and August is 28 °C and the extreme temperature in summer is 40.2 °C. The average lowest temperature in January is 4 °C, and the extreme temperature in winter is −12.1 °C [23].

(3) The northeast wind and the northwest wind are the dominant winds in winter, while the southeast wind and the southwest wind are the dominant winds in the summer. Because the east side of Shanghai is facing the sea, the easterly wind brings the clean air from the sea, while the westerly wind facilitates the spread of air pollutants from neighboring regions to Shanghai [43,44]. Shanghai is a city with many winds all year round. The average wind speed is relatively stable. At the same

time, Shanghai is located in the plain, and there will be no conduction effect of pollutants due to wind direction problems.

2.4. Three Air Quality Assessment Methods

2.4.1. Comprehensive Pollution Index Method

In terms of the Comprehensive Pollution Index Method, the first step is to analyze the pollution load of the main air pollutants. The formula of the pollution load coefficient is as follows:

$$f_i = \frac{P_i}{P}, P_i = \frac{C_i}{S_i}, P = \sum_{i=1}^{k} P_i \tag{3}$$

where:

C_i is the annual average concentration of the ith pollutant in the atmosphere;
S_i is the evaluation criteria of the ith pollutant in the atmosphere;
P_i is the sub-index of the ith pollutant;
f_i is the pollution load coefficient of the ith pollutant.

Then calculate the Comprehensive Pollution Index I according to Equation (2):

$$I = \sqrt{max\left(\frac{c_1}{s_1}, \frac{c_2}{s_2}, \dots, \frac{c_k}{s_k}\right) \frac{1}{k \sum_{i=1}^{k} \frac{c_i}{s_i}}} \tag{4}$$

where:

c_k is the observed concentration value of a pollutant;
s_k is the corresponding evaluation criteria in Level II national air quality standard for the pollutant;
I is the Comprehensive Pollution Index.

China issued the new Ambient Air Quality Standards (GB3095-2012) in 2012, which reclassified the atmospheric functional zones from the original three categories into two categories. Nature reserves, tourist attractions and other areas that require special protection belong to the first category, referred to as the Category I Zone, which applies to the Level I concentration limit. Commercial areas, industrial parks, and rural areas belong to the second category, referred to as the Category II Zone, which applies to the Level II concentration limit [27]. See Table 2 for the concentration limits of different functional zones.

The air quality can be evaluated by comparing the calculated Comprehensive Pollution Index I with the thresholds in the Air Quality Index scale. Table 3 below has provided the Air Pollution Grading System.

The Comprehensive Pollution Index Method determines the air quality level based on the calculated pollution index value, and considers the average level of various pollutants and the damage level of a single pollutant in the calculation, which is simple and easy to conduct. However, the main disadvantage of this method is that when the value of the pollution index is exactly at the threshold between two air quality levels, it would be arbitrary to determine the air quality only based on one cut-off value and this would diminish the credibility of the evaluation result. Meanwhile, the calculation result depends on the ratio of the observed highest pollutant concentration value to the corresponding standard value, which would result in a higher Comprehensive Pollution Index if the observed value of a certain pollutant is relatively higher [45,46].

Table 2. Basic air pollutants assessment standards.

Pollutant	Average Value	Concentration Threshold		Unit of Measurement
		Level I	Level II	
Sulfur Dioxide (SO$_2$)	Annual Average	20	60	$\mu g/m^3$
	24-h Average	50	150	
	Hourly Average	150	500	
Nitrogen Dioxide (NO$_2$)	Annual Average	40	40	$\mu g/m^3$
	24-h Average	80	80	
	Hourly Average	200	200	
Particulate Matter (PM$_{10}$)	Annual Average	40	70	$\mu g/m^3$
	24-h Average	50	150	
Particulate Matter (PM$_{2.5}$)	Annual Average	15	35	$\mu g/m^3$
	24-h Average	35	75	
Ozone (O$_3$)	24-h Average	100	160	$\mu g/m^3$
	Hourly Average	160	200	
Carbonic Oxide (CO)	24-h Average	4	4	mg/m^3
	Hourly Average	10	10	

Table 3. Air pollution grading system.

Air Quality Level	Clean	Mild Pollution	Moderate Pollution	Heavy Pollution	Severe Pollution
I	<0.6	0.6–1	1–1.9	1.9–2.8	>2.8
Pollution Level	Safe	Standard	Alert	Warning	Emergency

2.4.2. The Improved Grey Relational Degree Method

Let the reference sequence be $\{X_{i(k)}\}$. Compare the two sequences of $\{X_{j(k)}\}$ and $\{X_{i(k)}\} = \{X_{i(1)}, X_{i(2)}, \ldots X_{i(n)}\}$, $K = 1, 2, \ldots n$. The correlation coefficient $(\xi_{ij(k)})$ of $\{X_{i(k)}\}$ and $\{X_{j(k)}\}$ at point K (reflecting the correlation of the comparison sequence and the reference sequence at a certain point) can be defined by:

$$\xi_{ij(k)} = \frac{\underset{j}{min}\;\underset{k}{min}\;\Delta_{ij(k)} + \rho\;\underset{j}{max}\;\underset{k}{max}\;\Delta_{ij(k)}}{\Delta_{ij(k)} + \rho\;\underset{j}{max}\;\underset{k}{max}\;\Delta_{ij(k)}} \tag{5}$$

where:

$\Delta_{ij(k)} = |X_{i(k)} - X_{j(k)}|$ is the difference in the absolute value of $\{X_{i(k)}\}$ and $\{X_{j(k)}\}$ at point K;

$\underset{j}{min}\;\underset{k}{min}\;\Delta_{ij(k)}$ is the minimum differnce between two levels;

$\underset{j}{max}\;\underset{k}{max}\;\Delta_{ij(k)}$ is the maximum difference between two levels;

ρ is the distinguishing coefficient, which takes a value between 0 and 1. After comparing the ρ values in the related literature [47,48], we set $\rho = 0.5$ to avoid the influence of the extreme values on the calculation results.

Integrate the correlation coefficients at different points $(K = 1, 2 \ldots n)$ to obtain the overall correlation of the comparison sequence $\{X_{j(k)}\}$ and the reference sequence $\{X_{i(k)}\}$, as shown in the following equation:

$$\gamma_{i,j} = \frac{1}{n}\sum_{k=1}^{n}\xi_{ij(k)} \tag{6}$$

If $\{X_{1(k)}\}, \{X_{2(k)}\}, \ldots, \{X_{j(k)}\}$ are N known comparison sequences, and $\{X_{i(k)}\}$ is a known reference series, there would be:

$$\gamma_{i,s}{}^* = max\{\gamma_{i,j}\}, 1 \leq j \leq n \tag{7}$$

At this time, the reference sequence $\{X_{i(k)}\}$ would have the best correlation with the comparison sequence $\{X_{j(k)}\}$.

Obviously, if $\{X_{i(k)}\}$ represents the sequence made up by the observed mass concentration values of different pollutants, $\{X_{j(k)}\}$ represents the sequence made up by the evaluation standards of a certain level of different pollutants. Because of the good correlation between $\{X_{i(k)}\}$ and $\{X_{j(k)}\}$, it is most appropriate to evaluate the air quality of Sample Point i as the corresponding Level S.

Furthermore, normalize the data. Let $S(k,j) = s_{Kj}/s_{Ij}$ be the equal-standard grading standard, and $I(i,j) = c_{ij}/s_{Ij}$ be the Air Pollution Index to be evaluated. The weight of the pollutant $w(i,j)$ can be written as follows:

$$w(i,j) = \frac{log100I(i,j)}{\sum_{i=1}^{k} log100I(i,j)} \tag{8}$$

where:

S_{Kj} is the Graded Index of the jth pollution indicator on Level K;
S_{Ij} is the Graded Index of the jth pollution indicator on Level I;
C_{ij} is the observed value of the jth pollution indicator in the ith monitoring point.

Based on this, the correlation between the quality of air samples to be evaluated and the standard air quality of different levels can be calculated by:

$$G(i,k) = \sum_{i=1}^{k} W(i,j) \cdot \gamma_{i,s}{}^* \tag{9}$$

The traditional Grey Relational Degree Method is relatively simple in calculation. However, when the pollution factors are significantly different, the average value with equal weights would understate the pollution factor with high concentration while overstate the pollution factor with low concentration, which would differ from the actual pollution condition [49–51]. The improved method above determines different air quality levels based on the observed concentration of different pollution factors, and calculates the weights and correlation coefficients of each pollution factor accordingly. When evaluating the air quality, the Improved Grey Relational Degree Method not only enhances the weights of the pollution factors with high concentration, but also takes into account the combined effects of different pollution factors on air quality, so that basically no information is lost during the evaluation process. It also comprehensively considers the effects of different pollutant weights and the interactions between different pollutants, thus improving the accuracy of the evaluation result.

2.4.3. Euclid Approach Degree Method

First, determine the characteristic value $\lambda(K,j)$ of the pollution level, as shown in the following equation:

$$\lambda(K,j) = \begin{cases} s_{Kj}/2, K=1 \\ [s_{Kj} + s_{K-1j}]/2, K=2 \end{cases} \tag{10}$$

where S_{Kj} represents the grading index of the jth pollutant of Level K.
Then determine the index weight $w(i,j)$ of different pollutants:

$$w(i,j) = \frac{(\lambda_{Kj} + x_{ij})/\overline{\lambda_J}}{\sum_{i=1}^{n}(\lambda_{Kj} + x_{ij})/\overline{\lambda_J}} \tag{11}$$

where:

x_{ij} represents the observed concentration value of the *j*th pollution indicator at the *i*th monitoring point;
$\overline{\lambda_j}$ represents the the mean value of the characteristic values of different levels of the *j*th pollution indicator;
λ_{kj} represents the characteristic value of Level II of the *j*th pollution indicator.

Normalize the observed result by:

$$x(i,j) = x_{ij} / \lambda_{kj}$$

Calculate the Proximity Degree $\eta(i,j)$ of the air sample to be evaluated:

$$\eta(i,j) = \sqrt{\sum_{i=1}^{n} W(i,j)[x(i,j) - \lambda(i,j)]^2}$$

Then, determine the respective air quality level of each monitoring point based on the principle of minimum proximity degree.

For evaluation purpose, the Euclid Approach Degree Method needs to establish two membership functions of the observed value and the standard level. All valid data observed have been taken into consideration in the modeling and calculation process. Therefore, there won't be any information loss during the evaluation process and the actual condition of the environment could be comprehensively reflected [52–54].

3. Results

The study period of this paper is from 1 November 2017 to 31 October 2018. The air pollutants as the study object include SO_2, NO_2, PM_{10}, $PM_{2.5}$, O_3 and CO. The seasons are determined based on the months: Spring (March, April, May), summer (June, July, August), autumn (September, October, November), and winter (December, January, February). See Table 4 for the average concentration levels of various air pollutants in different seasons.

Table 4. Average Concentration of Main Air Pollutants by Season ($\mu g/m^3$).

Pollutant / Season	$PM_{2.5}$	PM_{10}	SO_2	NO_2	O_3	CO
Winter	52.00	66.67	14.33	58.00	83	0.82
Spring	42.33	62.67	12.00	44.33	97	0.76
Summer	24.33	36.67	8.00	25.33	112	0.68
Autumn	31.67	52.67	9.67	46.00	103	0.53

The Air Pollution Grading Indexes of different seasons obtained through the Comprehensive Pollution Grading Method are shown in Table 5.

Table 5. Comprehensive Pollution Grading Index by Season.

Season	Winter	Spring	Summer	Autumn
Comprehensive Pollution Index	1.24	1.02	0.59	0.92

It can be seen from Table 5 that from 1 November 2017 to 31 October 2018, the average air quality in Shanghai during winter (December, January, February) was heavy pollution; the average air quality during spring (March, April, May) was moderate pollution; the average air quality during summer (June, July, August) was clean; while the average air quality during autumn (September, October,

November) was mild pollution. The results above indicate that the air quality of Shanghai was still not ideal and further pollution control measures are needed.

Table 6 has shown the Air Pollution Grading Index by season obtained through the Improved Grey Relational Degree Method and the Euclid Approach Degree Method introduced in Part 2.4 (please refer to Appendix A for the MATLAB algorithm—MATLAB 2017b, MathWorks, Natick, USA).

Table 6. Air Pollution Grading Index by season obtained by the improved grey relational degree method and the Euclid approach degree method.

Method Season	Improved Grey Relational Degree Method	Euclid Approach Degree Method
Winter	II	II
Spring	II	II
Summer	I	I
Autumn	I	II

It can be seen that the evaluation results obtained through the Improved Grey Relational Degree Method and the Euclid Approach Degree Method are consistent except for autumn. The air quality in winter has met the Level II national air quality standard stipulated in GB3095-2012; the air quality in spring has also met the Level II standard; while the air quality in summer has reached the Level I national air quality standard.

It can be seen from the calculation results obtained by the three evaluation methods above that there is some concern in the air quality of Shanghai, especially during winters when the air pollution is most severe.

4. Discussion

Based on the above calculation results, this paper further analyzes the fine particle pollution of Shanghai during the study period. The particulate pollutants in the atmosphere can be categorized into total suspended particulates (TSP), PM_{10}, and $PM_{2.5}$ based on the particle size [55–57]. TSP generally refers to the particulate matters floating in the air with a particle size of less than 100 μm, including solid particles and liquid particles [58,59]. PM_{10} refers to particulate matters with a particle size of 10 μm or less. Most PM_{10} could reach the throat or even further in the respiratory tract [60,61]. $PM_{2.5}$ refers to particulate matters with a particle size of below 2.5 μm. Most $PM_{2.5}$ can settle in the respiratory tract, and a small number of $PM_{2.5}$ could even reach the pulmonary alveoli which are very difficult to get rid of and extremely harmful to the human body [62–64]. In recent research, $PM_{2.5}$ and PM_{10} have been the focus of air pollution control in China [65–68]. According to studies at home and abroad, there exist certain correlations between PM_{10} and $PM_{2.5}$ [69–72]. In order to fully understand the relationship between $PM_{2.5}$ and other major pollutants in Shanghai, we have calculated the ratio of $PM_{2.5}/PM_{10}$, $PM_{2.5}/SO_2$, $PM_{2.5}/NO_2$, $PM_{2.5}/CO$, and $PM_{2.5}/O_3$, according to the 2017 Shanghai Environmental Bulletin [26] and the Shanghai Air Quality Monthly Report from January–October 2018 [73]. The results showed that the variation range of $PM_{2.5}/PM_{10}$ was [0.4–0.7], while the ratio of $PM_{2.5}/SO_2$, $PM_{2.5}/NO_2$, $PM_{2.5}/CO$, and $PM_{2.5}/O_3$ was low (see Figure 2).

Hence, we will focus on the correlation between $PM_{2.5}$ and PM_{10} concentration in Shanghai. The ratio of $PM_{2.5}/PM_{10}$ in Shanghai ranged from 0.50–0.91 during the study period [23,73]. The monthly ratios are shown in Figure 3 below, which was highest in January and lowest in August. Overall speaking, the ratios were volatile, with an average value of 0.68. Among the 90 pollution days in 2017 as published in 2017 Shanghai Environmental Bulletin, there are 25.6% of the days in which fine particles ($PM_{2.5}$) was the primary air pollutant [23].

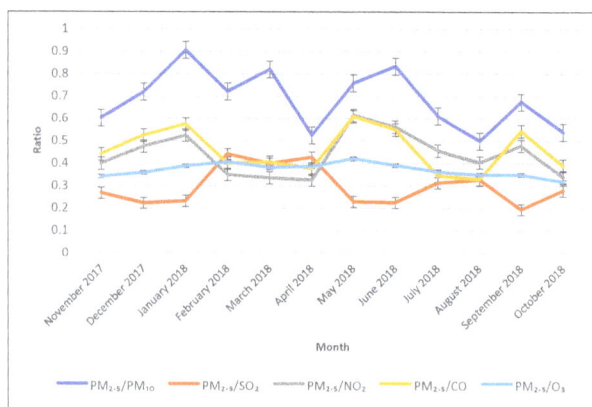

Figure 2. Ratio of $PM_{2.5}/PM_{10}$, $PM_{2.5}/SO_2$, $PM_{2.5}/NO_2$, $PM_{2.5}/CO$, and $PM_{2.5}/O_3$ by month.

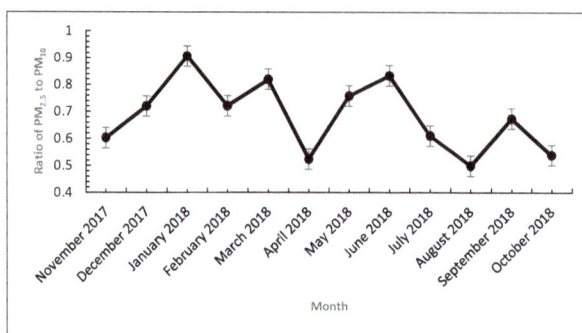

Figure 3. Ratio of $PM_{2.5}$ to PM_{10} by month.

It can be seen from the seasonal change of the $PM_{2.5}/PM_{10}$ ratio in Figure 4 that the seasonal trend of this ratio is: winter > spring > summer > autumn. Meanwhile, this ratio in winter is 1.3 times of that in summer. According to the relevant literature, we found that it is due to the increased energy consumption in winter heating, less rainfall and more fog weathers in winters, which do not facilitate the movement of fine particles and results in less sedimentation. In springs, the increased wind frequency and air flow, especially the northwest wind would bring coarse particulate pollution to Shanghai. In summers, the high temperature and rising hot air do not facilitate the sedimentation of fine particles. In autumns, the cool weather and air flow help to spread and subside fine particles, and therefore the degree of fine particle pollution is lower [74–78].

Through the quarterly linear regression analysis of $PM_{2.5}$ and PM_{10} in Shanghai from 1 November 2017 to 31 October 2018, this paper has found a significant linear relationship between $PM_{2.5}$ and PM_{10}.

As shown in Figure 5a, although the linear correlation between $PM_{2.5}$ and PM_{10} varies from season to season, there is still a strong correlation between $PM_{2.5}$ and PM_{10} concentrations, which is the strongest during winters and summers. In winter, the correlation coefficient reached $R^2 = 0.9655$, while in summer, the correlation coefficient $R^2 = 0.9112$. The corresponding regression equations are $y = 0.9333x - 5.2223$ and $y = 0.7734x - 2.614$, respectively. Taking winter as an example, the t-test on the three-month data of winter provided a confidence interval of [53.4543, 68.4943] with 95% confidence, and a significance probability of 0, which is less than 0.05. Therefore, we can say there is a significant linear relationship between $PM_{2.5}$ and PM_{10} concentration in winter. In spring and autumn, there is also a linear relationship between $PM_{2.5}$ and PM_{10} concentration, but the correlation coefficient is smaller. The regression equation in spring is $y = 0.5524x - 11.206$, with a correlation coefficient of $R^2 = 0.7379$; while the regression equation in autumn is $y = 0.5731x + 5.597$, with a correlation coefficient of $R^2 = 0.7282$.

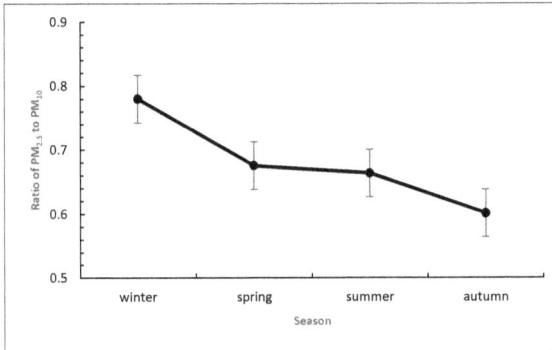

Figure 4. Ratio of $PM_{2.5}$ to PM_{10} by Season.

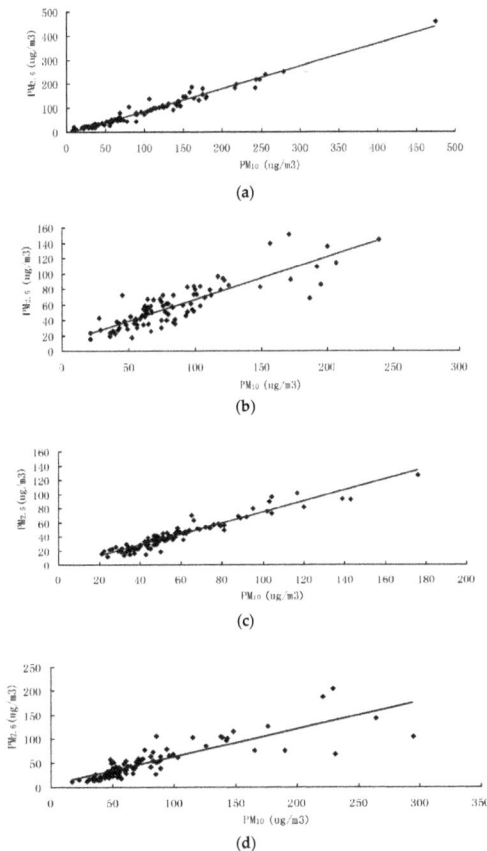

(a)

(b)

(c)

(d)

Figure 5. The Regression Curve of $PM_{2.5}$ and PM_{10} Mass Concentration in Shanghai from November 2017–October 2018: (**a**) in winter, (**b**) in spring, (**c**) in summer, (**d**) in autumn.

The regression fitting results above show that there is a significant linear relationship between $PM_{2.5}$ and PM_{10} concentration in winters and summers, while their linear correlation is less significant during spring and autumn, which is mainly due to temperature reasons. The cold weather in winters and hot weather in summers of Shanghai do not facilitate the spread of particle pollutants. The particles tend to float in the air, showing a significant linear correlation. On the other hand, in springs and autumns, the temperature is moderate with frequent and strong monsoon which helps to increase air flow and facilitate the diffusion and sedimentation of particle pollutants. Fine particles and coarse particles respond differently to these climate factors. Therefore, the linear correlation between the mass concentration of particulate matters $PM_{2.5}$ and PM_{10} is less significant in springs and autumns.

Although air quality has shown improvement in the past decade, there are numerous challenges in the coming years. With the construction of the Yangtze River Delta urban agglomeration and the Yangtze River Economic Belt during the 13th Five-Year Plan period, there will be strong economic growth as well as a continuous increase in air pollutant emissions in neighboring cities and other provinces and cities at the upper and middle region of the Yangtze River. If we cannot establish an effective and coordinated regional air pollution prevention and control mechanism, it would greatly affect Shanghai's air quality. Moreover, since the parameters we used are derived from official data from Shanghai [21,23,36–40,73], and the aforementioned research methods have been widely recognized in the academic world, the research design of this paper has exportability under the premise of using other reliable data sources.

5. Conclusions

This paper has evaluated the quarterly air quality of Shanghai by using the Comprehensive Pollution Index Method, the Improved Grey Relational Degree Method, and the Euclid Approach Degree Method and based on the technical norms of China's current AQI and analysis of Shanghai's overall climate. By analysis on the air pollutants (SO_2, NO_2, PM_{10} and $PM_{2.5}$) in Shanghai from 1 November 2017 to 31 October 2018, this paper has reached the following conclusions:

(1) The air quality of Shanghai has moderate pollution in winters and springs, clean in summers, and mild pollution in autumns. The evaluation results obtained by the Improved Grey Relational Degree Method and the Euclid Approach Degree Method are basically consistent. The air quality in winter has met the Level II national air quality standard in GB3095-2012; the air quality in spring has also met the Level II standard; while the air quality in summer has reached the Level I national air quality standard. These results are consistent between the Improved Grey Relational Degree Method and the Euclid Approach Degree Method. However, in autumn, the air quality evaluation result according to the Improved Grey Relational Degree Method is Level I, while the evaluation result according to the Euclid Approach Degree Method is Level II. Therefore, there exist some concern in the air quality of Shanghai, especially during winters when the air pollution is most severe.

(2) The air pollutants in Shanghai have shown a seasonal pattern of high concentration in winters and low concentration in summers; meanwhile, the pollutant concentration is higher in the first half of the year than in the second half. This is because the first half of the year is the peak period of industrial energy consumption, and both industrial and residential heating needs in winter would inevitably cause increase in energy consumption such as the coal [79], which would undoubtedly increase the concentration of air pollutants.

(3) By analyzing the particle pollutants of $PM_{2.5}$ and PM_{10}, this paper has found that the linear correlation between the two varies with the seasons, which is most significant during winters and summers.

Author Contributions: Conceptualization, Y.Y. and Y.L.; Methodology, Y.Y.; Software, M.S.; Validation, Z.W.; Formal Analysis, Y.L.; Investigation, Y.Y.; Resources, Y.Y.; Data Curation, Y.L; Writing-Original Draft Preparation, M.S.; Writing-Review & Editing, Y.Y. and Y.L.; Visualization, Z.W.; Project Administration, Y.Y.

Funding: This research received no external funding.

Conflicts of Interest: The authors declare no conflict of interest.

Appendix A

The MATLAB algorithm for calculating the distance between observations and standard values

```
clc;
close;
clear all;
format short;
% raw data
X = [];% input variable, a column of standard values (in Table 2), a column of observations (in Table 4)
n1 = size(x,1);
for i = 1:n1
x(i,:) = x(i,:)/x(i,1);
end
data = x;
consult = data(6:n1,:);
m1 = size(consult,1);
compare = data(1:5,:);
m2 = size(compare,1);
for i = 1:m1
for j = 1:m2
t(j,:) = compare(j,:)-consult(i,:);
end
min_min = min(min(abs(t')));
max_max = max(max(abs(t')));
resolution = 0.5;
coefficient = (min_min+resolution*max_max)./(abs(t)+resolution*max_max);
corr_degree = sum(coefficient')/size(coefficient,2);
r(i,:) = corr_degree;
end
```

References

1. Zhang, D.; Liu, J.; Li, B. Tackling Air Pollution in China—What do We Learn from the Great Smog of 1950s in LONDON. *Sustainability* **2014**, *6*, 5322–5338. [CrossRef]
2. Yang, W.; Li, L. Energy efficiency, ownership structure, and sustainable development: Evidence from China. *Sustainability* **2017**, *9*, 912. [CrossRef]
3. Yang, Y.; Yang, W. Does Whistleblowing Work for Air Pollution Control in China? A Study Based on Three-party Evolutionary Game Model under Incomplete Information. *Sustainability* **2019**, *11*, 324. [CrossRef]
4. Yuan, G.; Yang, W. Evaluating China's Air Pollution Control Policy with Extended AQI Indicator System: Example of the Beijing-Tianjin-Hebei Region. *Sustainability* **2019**, *11*, 939. [CrossRef]
5. Li, H.; Tan, X.; Guo, J.; Zhu, K.; Huang, C. Study on an Implementation Scheme of Synergistic Emission Reduction of CO_2 and Air Pollutants in China's Steel Industry. *Sustainability* **2019**, *11*, 352. [CrossRef]
6. Bloss, W. Measurement of Air Pollutants. In *Reference Module in Earth Systems and Environmental Sciences*; Elsevier: Amsterdam, The Netherlands, 2018; ISBN 978-0-12-409548-9.
7. Park, J.H.; Lee, S.H.; Yun, S.J.; Ryu, S.; Choi, S.W.; Kim, H.J.; Kang, T.K.; Oh, S.C.; Cho, S.J. Air pollutants and atmospheric pressure increased risk of ED visit for spontaneous pneumothorax. *Am. J. Emerg. Med.* **2018**, *36*, 2249–2253. [CrossRef] [PubMed]
8. Filonchyk, M.; Yan, H.; Li, X. Temporal and spatial variation of particulate matter and its correlation with other criteria of air pollutants in Lanzhou, China, in spring-summer periods. *Atmos. Pollut. Res.* **2018**, *9*, 1100–1110. [CrossRef]

9. Dirgawati, M.; Heyworth, J.S.; Wheeler, A.J.; McCaul, K.A.; Blake, D.; Boeyen, J.; Cope, M.; Yeap, B.B.; Nieuwenhuijsen, M.; Brunekreef, B.; et al. Development of Land Use Regression models for particulate matter and associated components in a low air pollutant concentration airshed. *Atmos. Environ.* **2016**, *144*, 69–78. [CrossRef]

10. Son, Y.; Osornio-Vargas, Á.R.; O'Neill, M.S.; Hystad, P.; Texcalac-Sangrador, J.L.; Ohman-Strickland, P.; Meng, Q.; Schwander, S. Land use regression models to assess air pollution exposure in Mexico City using finer spatial and temporal input parameters. *Sci. Total Environ.* **2018**, *639*, 40–48. [CrossRef]

11. Tao, H.; Xing, J.; Zhou, H.; Chang, X.; Li, G.; Chen, L.; Li, J. Impacts of land use and land cover change on regional meteorology and air quality over the Beijing-Tianjin-Hebei region, China. *Atmos. Environ.* **2018**, *189*, 9–21. [CrossRef]

12. Pérez, J.F.; Sabatino, S.; Galia, A.; Rodrigo, M.A.; Llanos, J.; Sáez, C.; Scialdone, O. Effect of air pressure on the electro-Fenton process at carbon felt electrodes. *Electrochim. Acta* **2018**, *273*, 447–453. [CrossRef]

13. Kalisa, E.; Fadlallah, S.; Amani, M.; Nahayo, L.; Habiyaremye, G. Temperature and air pollution relationship during heatwaves in Birmingham, UK. *Sustain. Cities Soc.* **2018**, *43*, 111–120. [CrossRef]

14. Yu, Y.; Kwok, K.C.S.; Liu, X.P.; Zhang, Y. Air pollutant dispersion around high-rise buildings under different angles of wind incidence. *J. Wind Eng. Ind. Aerodyn.* **2017**, *167*, 51–61. [CrossRef]

15. Yang, W.; Li, L. Efficiency evaluation of industrial waste gas control in China: A study based on data envelopment analysis (DEA) model. *J. Clean. Prod.* **2018**, *179*, 1–11. [CrossRef]

16. Cheng, M.; Zhi, G.; Tang, W.; Liu, S.; Dang, H.; Guo, Z.; Du, J.; Du, X.; Zhang, W.; Zhang, Y.; et al. Air pollutant emission from the underestimated households' coal consumption source in China. *Sci. Total Environ.* **2017**, *580*, 641–650. [PubMed]

17. Cai, P.; Nie, W.; Chen, D.; Yang, S.; Liu, Z. Effect of air flowrate on pollutant dispersion pattern of coal dust particles at fully mechanized mining face based on numerical simulation. *Fuel* **2019**, *239*, 623–635. [CrossRef]

18. Liu, B.; Wu, S.; Shen, L.; Zhao, T.; Wei, Y.; Tang, X.; Long, C.; Zhou, Y.; He, D.; Lin, T.; et al. Spermatogenesis dysfunction induced by PM2.5 from automobile exhaust via the ROS-mediated MAPK signaling pathway. *Ecotoxicol. Environ. Saf.* **2019**, *167*, 161–168. [CrossRef]

19. Yang, W.; Li, L. Analysis of Total Factor Efficiency of Water Resource and Energy in China: A Study Based on DEA-SBM Model. *Sustainability* **2017**, *9*, 1316.

20. Li, L.; Yang, W. Total Factor Efficiency Study on China's Industrial Coal Input and Wastewater Control with Dual Target Variables. *Sustainability* **2018**, *10*, 2121. [CrossRef]

21. Shanghai Municipal Statistics Bureau. *Shanghai Statistical Yearbook 2017*; China Statistics Press: Shanghai, China, 2018.

22. National Bureau of Statistics of the People's Republic of China. *China Statistical Yearbook 2017*; China Statistics Press: Beijing, China, 2018.

23. Shanghai Municipal Environmental Protection Bureau. *2017 Shanghai Environmental Bulletin*; Shanghai Municipal Environmental Protection Bureau: Shanghai, China, 2018.

24. Ministry of Environmental Protection of the People's Republic of China. *Ambient Air Quality Standards: GB3095-2012*; China Environmental Science Press: Beijing, China, 2012.

25. Ministry of Environmental Protection of the People's Republic of China. *Technical Regulation on Ambient Air Quality Index (on Trial): HJ 633-2012*; China Environmental Science Press: Beijing, China, 2012.

26. Li, X.; Gao, Z.; Li, Y.; Gao, C.Y.; Ren, J.; Zhang, X. Meteorological conditions for severe foggy haze episodes over north China in 2016–2017 winter. *Atmos. Environ.* **2019**, *199*, 284–298. [CrossRef]

27. Golly, B.; Waked, A.; Weber, S.; Samake, A.; Jacob, V.; Conil, S.; Rangognio, J.; Chrétien, E.; Vagnot, M.P.; Robic, P.Y.; et al. Organic markers and OC source apportionment for seasonal variations of PM2.5 at 5 rural sites in France. *Atmos. Environ.* **2019**, *198*, 142–157. [CrossRef]

28. Ryu, J.; Kim, J.J.; Byeon, H.; Go, T.; Lee, S.J. Removal of fine particulate matter (PM2.5) via atmospheric humidity caused by evapotranspiration. *Environ. Pollut.* **2019**, *245*, 253–259. [CrossRef] [PubMed]

29. United States Environmental Protection Agency. *Air Emissions Inventories*; United States Environmental Protection Agency: Washington, DC, USA, 2018.

30. European Environment Agency. *EMEP/EEA Air Pollutant Emission Inventory Guidebook 2016*; European Environment Agency: Copenhagen, Denmark, 2016.

31. Ohara, T.; Akimoto, H.; Kurokawa, J.I.; Horii, N.; Yamaji, K.; Yan, X.; Hayasaka, T. An Asian emission inventory of anthropogenic emission sources for the period 1980–2020. *Atmos. Chem. Phys.* **2007**, *7*, 4419–4444.

32. Kim, J.-H.; Park, J.-M.; Lee, S.-B.; Pudasainee, D.; Seo, Y.-C. Anthropogenic mercury emission inventory with emission factors and total emission in Korea. *Atmos. Environ.* **2010**, *44*, 2714–2721. [CrossRef]

33. Kim, N.K.; Kim, Y.P.; Morino, Y.; Kurokawa, J.; Ohara, T. Verification of NOx emission inventory over South Korea using sectoral activity data and satellite observation of NO2 vertical column densities. *Atmos. Environ.* **2013**, *77*, 496–508. [CrossRef]

34. Kannari, A.; Streets, D.G.; Tonooka, Y.; Murano, K.; Baba, T. MICS-Asia II: An inter-comparison study of emission inventories for the Japan region. *Atmos. Environ.* **2008**, *42*, 3584–3591. [CrossRef]

35. Li, Y.; Qian, X.; Zhang, L.; Dong, L. Exploring spatial explicit greenhouse gas inventories: Location-based accounting approach and implications in Japan. *J. Clean. Prod.* **2017**, *167*, 702–712. [CrossRef]

36. Shanghai Municipal Environmental Protection Bureau. *2012 Shanghai Environmental Bulletin*; Shanghai Municipal Environmental Protection Bureau: Shanghai, China, 2013.

37. Shanghai Municipal Environmental Protection Bureau. *2013 Shanghai Environmental Bulletin*; Shanghai Municipal Environmental Protection Bureau: Shanghai, China, 2014.

38. Shanghai Municipal Environmental Protection Bureau. *2014 Shanghai Environmental Bulletin*; Shanghai Municipal Environmental Protection Bureau: Shanghai, China, 2015.

39. Shanghai Municipal Environmental Protection Bureau. *2015 Shanghai Environmental Bulletin*; Shanghai Municipal Environmental Protection Bureau: Shanghai, China, 2016.

40. Shanghai Municipal Environmental Protection Bureau. *2016 Shanghai Environmental Bulletin*; Shanghai Municipal Environmental Protection Bureau: Shanghai, China, 2017.

41. Shen, Z.; Liang, P.; He, J. Analysis on the climatic characteristics of the fine structure of the urban heat island in Shanghai. *Trans. Atmos. Sci.* **2017**, *40*, 369–378.

42. Meteorological Data Center of China Meteorological Administration Observation Data of Shanghai Ground Meteorology. Available online: https://data.cma.cn/ (accessed on 17 March 2019).

43. Guo, H.; Chen, M. Short-term effect of air pollution on asthma patient visits in Shanghai area and assessment of economic costs. *Ecotoxicol. Environ. Saf.* **2018**, *161*, 184–189. [CrossRef] [PubMed]

44. Ji, X.; Meng, X.; Liu, C.; Chen, R.; Ge, Y.; Kan, L.; Fu, Q.; Li, W.; Tse, L.A.; Kan, H. Nitrogen dioxide air pollution and preterm birth in Shanghai, China. *Environ. Res.* **2019**, *169*, 79–85. [CrossRef] [PubMed]

45. Xiaoyu, L.; Jun, Y.; Pengfei, S. Structure and Application of a New Comprehensive Environmental Pollution Index. *Energy Procedia* **2011**, *5*, 1049–1054. [CrossRef]

46. Gąsiorek, M.; Kowalska, J.; Mazurek, R.; Pająk, M. Comprehensive assessment of heavy metal pollution in topsoil of historical urban park on an example of the Planty Park in Krakow (Poland). *Chemosphere* **2017**, *179*, 148–158. [CrossRef] [PubMed]

47. Zhu, C.; Li, N. Study on Grey Clustering Model of Indoor Air Quality Indicators. *Procedia Eng.* **2017**, *205*, 2815–2822. [CrossRef]

48. Yunlong, W.; Kai, L.; Guan, G.; Yanyun, Y.; Fei, L. Evaluation method for Green jack-up drilling platform design scheme based on improved grey correlation analysis. *Appl. Ocean Res.* **2019**, *85*, 119–127. [CrossRef]

49. Yan, F.; Qiao, D.; Qian, B.; Ma, L.; Xing, X.; Zhang, Y.; Wang, X. Improvement of CCME WQI using grey relational method. *J. Hydrol.* **2016**, *543*, 316–323. [CrossRef]

50. Sun, G.; Guan, X.; Yi, X.; Zhou, Z. Grey relational analysis between hesitant fuzzy sets with applications to pattern recognition. *Expert Syst. Appl.* **2018**, *92*, 521–532. [CrossRef]

51. Li, X.; Wang, Z.; Zhang, L.; Zou, C.; Dorrell, D.D. State-of-health estimation for Li-ion batteries by combing the incremental capacity analysis method with grey relational analysis. *J. Power Sources* **2019**, *410–411*, 106–114. [CrossRef]

52. Du, X.; Yang, Z.; Li, G.; Jia, Y.; Chen, C.; Gui, L. Remaining useful life assessment of machine tools based on AHP method and Euclid approach degree. In Proceedings of the International Conference on System Reliability and Science (ICSRS), Paris, France, 15–18 November 2016; pp. 46–52.

53. Wang, S.; Zhang, T.; Cheng, L.; Li, J.; Tang, H.; Guo, M. Comprehensive performance of compound fabrics in terms of electromagnetic shielding and wearability based on the Euclid approach degree of fuzzy matter elements. *J. Text. Inst.* **2017**, *108*, 341–346.

54. Niu, D.; Song, Z.; Wang, M.; Xiao, X. Improved TOPSIS method for power distribution network investment decision-making based on benefit evaluation indicator system. *Int. J. Energy Sect. Manag.* **2017**, *11*, 595–608.

55. Gysel, N.; Welch, W.A.; Chen, C.-L.; Dixit, P.; Cocker, D.R.; Karavalakis, G. Particulate matter emissions and gaseous air toxic pollutants from commercial meat cooking operations. *J. Environ. Sci.* **2018**, *65*, 162–170. [CrossRef]

56. Feng, Y.; Li, Y.; Cui, L. Critical review of condensable particulate matter. *Fuel* **2018**, *224*, 801–813. [CrossRef]

57. Brodny, J.; Tutak, M. Analysis of the diversity in emissions of selected gaseous and particulate pollutants in the European Union countries. *J. Environ. Manag.* **2019**, *231*, 582–595. [CrossRef]

58. Ren, G.; Chen, Z.; Feng, J.; Ji, W.; Zhang, J.; Zheng, K.; Yu, Z.; Zeng, X. Organophosphate esters in total suspended particulates of an urban city in East China. *Chemosphere* **2016**, *164*, 75–83. [CrossRef] [PubMed]

59. Gonçalves, C.; Figueiredo, B.R.; Alves, C.A.; Cardoso, A.A.; da Silva, R.; Kanzawa, S.H.; Vicente, A.M. Chemical characterisation of total suspended particulate matter from a remote area in Amazonia. *Atmos. Res.* **2016**, *182*, 102–113. [CrossRef]

60. Feng, W.; Li, H.; Wang, S.; Van Halm Lutterodt, N.; An, J.; Liu, Y.; Liu, M.; Wang, X.; Guo, X. Short-term PM10 and emergency department admissions for selective cardiovascular and respiratory diseases in Beijing, China. *Sci. Total Environ.* **2019**, *657*, 213–221. [CrossRef] [PubMed]

61. Marchetti, S.; Longhin, E.; Bengalli, R.; Avino, P.; Stabile, L.; Buonanno, G.; Colombo, A.; Camatini, M.; Mantecca, P. In vitro lung toxicity of indoor PM10 from a stove fueled with different biomasses. *Sci. Total Environ.* **2019**, *649*, 1422–1433. [CrossRef] [PubMed]

62. Martins, N.R.; da Graça, G.C. Impact of PM2.5 in indoor urban environments: A review. *Sustain. Cities Soc.* **2018**, *42*, 259–275. [CrossRef]

63. Xu, M.; Ge, C.; Qin, Y.; Gu, T.; Lou, D.; Li, Q.; Hu, L.; Feng, J.; Huang, P.; Tan, J. Prolonged PM2.5 exposure elevates risk of oxidative stress-driven nonalcoholic fatty liver disease by triggering increase of dyslipidemia. *Free Radic. Biol. Med.* **2019**, *130*, 542–556. [CrossRef]

64. Qiu, Y.; Wang, G.; Zhou, F.; Hao, J.; Tian, L.; Guan, L.; Geng, X.; Ding, Y.; Wu, H.; Zhang, K. PM2.5 induces liver fibrosis via triggering ROS-mediated mitophagy. *Ecotoxicol. Environ. Saf.* **2019**, *167*, 178–187. [CrossRef]

65. Duan, J.; Chen, Y.; Fang, W.; Su, Z. Characteristics and Relationship of PM, PM10, PM2.5 Concentration in a Polluted City in Northern China. *Procedia Eng.* **2015**, *102*, 1150–1155. [CrossRef]

66. Fang, X.; Bi, X.; Xu, H.; Wu, J.; Zhang, Y.; Feng, Y. Source apportionment of ambient PM10 and PM2.5 in Haikou, China. *Atmos. Res.* **2017**, *190*, 1–9. [CrossRef]

67. Pan, S.; Du, S.; Wang, X.; Zhang, X.; Xia, L.; Liu, J.; Pei, F.; Wei, Y. Analysis and interpretation of the particulate matter (PM10 and PM2.5) concentrations at the subway stations in Beijing, China. *Sustain. Cities Soc.* **2019**, *45*, 366–377. [CrossRef]

68. Xue, H.; Liu, G.; Zhang, H.; Hu, R.; Wang, X. Similarities and differences in PM10 and PM2.5 concentrations, chemical compositions and sources in Hefei City, China. *Chemosphere* **2019**, *220*, 760–765. [CrossRef] [PubMed]

69. Chu, H.; Huang, B.; Lin, C. Modeling the spatio-temporal heterogeneity in the PM10-PM2.5 relationship. *Atmos. Environ.* **2015**, *102*, 176–182. [CrossRef]

70. gon Ryou, H.; Heo, J.; Kim, S.-Y. Source apportionment of PM10 and PM2.5 air pollution, and possible impacts of study characteristics in South Korea. *Environ. Pollut.* **2018**, *240*, 963–972. [CrossRef]

71. Sahanavin, N.; Prueksasit, T.; Tantrakarnapa, K. Relationship between PM10 and PM2.5 levels in high-traffic area determined using path analysis and linear regression. *J. Environ. Sci.* **2018**, *69*, 105–114. [CrossRef]

72. Gao, Y.; Ji, H. Microscopic morphology and seasonal variation of health effect arising from heavy metals in PM2.5 and PM10: One-year measurement in a densely populated area of urban Beijing. *Atmos. Res.* **2018**, *212*, 213–226. [CrossRef]

73. Shanghai Municipal Bureau of Ecology and Enviroment. 2018 Shanghai Air Quality Monthly Report. Available online: http://www.sepb.gov.cn/fa/cms/shhj2143/shhj5157/index.shtml (accessed on 10 January 2018).

74. Hu, Q.; Fu, H.; Wang, Z.; Kong, L.; Chen, M.; Chen, J. The variation of characteristics of individual particles during the haze evolution in the urban Shanghai atmosphere. *Atmos. Res.* **2016**, *181*, 95–105. [CrossRef]

75. Chen, L.; Ma, J.; Zhen, X.; Cao, Y. Variation characteristics and meteorological influencing factors of air pollution in Shanghai. *J. Meteorol. Environ.* **2017**, *33*, 59–67. [CrossRef]

76. Gao, S.; Tian, R.; Guo, B.; Zhang, L.; Ma, X. Characteristics of PM2.5 Concentration and its Relations with Meteorological Factors in Typical Cities of the Yangtze River Delta. *Sci. Technol. Eng.* **2018**, *18*, 142–155.

77. Liu, Y.; Yu, Y.; Liu, M.; Lu, M.; Ge, R.; Li, S.; Liu, X.; Dong, W.; Qadeer, A. Characterization and source identification of PM2.5-bound polycyclic aromatic hydrocarbons (PAHs) in different seasons from Shanghai, China. *Sci. Total Environ.* **2018**, *644*, 725–735. [CrossRef] [PubMed]

78. Yang, W.; Yuan, G.; Han, J. Is China's air pollution control policy effective? Evidence from Yangtze River Delta cities. *J. Clean. Prod.* **2019**, *220*, 110–133. [CrossRef]

79. Yang, W.; Li, L. Efficiency Evaluation and Policy Analysis of Industrial Wastewater Control in China. *Energies* **2017**, *10*, 1201. [CrossRef]

sustainability

MDPI

Article

Study on Development Sustainability of Atmospheric Environment in Northeast China by Rough Set and Entropy Weight Method

Yuangang Li [1,†], Maohua Sun [2,†], Guanghui Yuan [3,*,†], Qi Zhou [3,†] and Jinyue Liu [4,†]

[1] Business School, Dalian University of Foreign Languages, Dalian 116044, China
[2] School of Foreign Language, Dalian Jiaotong University, Dalian 116028, China
[3] Fintech Research Institute & School of Information Management and Engineering, Shanghai University of Finance and Economics, Shanghai 200433, China
[4] Business School, University of Shanghai for Science and Technology, Shanghai 200093, China
*   Correspondence: guanghuiyuan@outlook.com; Tel.: +86-21-6590-4028
†   All the authors contributed equally to this work.

Received: 25 May 2019; Accepted: 8 July 2019; Published: 11 July 2019

Abstract: In order to evaluate the atmospheric environment sustainability in the provinces of Northeast China, this paper has constructed a comprehensive evaluation model based on the rough set and entropy weight methods. This paper first constructs a Pressure-State-Response (PSR) model with a pressure layer, state layer and response layer, as well as an atmospheric environment evaluation system consisting of 17 indicators. Then, this paper obtains the weight of different indicators by using the rough set method and conducts equal-width discrete analysis and clustering analysis by using SPSS software. This paper has found that different discrete methods will end up with different reduction sets and multiple indicators sharing the same weight. Therefore, this paper has further introduced the entropy weight method based on the weight solution determined by rough sets and solved the attribute reduction sets of different layers by using the Rosetta software. Finally, this paper has further proved the rationality of this evaluation model for atmospheric environment sustainability by comparing the results with those of the entropy weight method alone and those of the rough set method alone. The results show that the sustainability level of the atmospheric environment in Northeast China provinces has first improved, and then worsened, with the atmospheric environment sustainability level reaching the highest level of 0.9275 in 2014, while dropping to the lowest level of 0.6027 in 2017. Therefore, future efforts should focus on reducing the pressure layer and expanding the response layer. Based on analysis of the above evaluation results, this paper has further offered recommendations and solutions for the improvement of atmospheric environment sustainability in the three provinces of Northeast China.

Keywords: PSR Model; rough set; entropy weight method; attribute reduction

---

1. Introduction

Sustainable development refers to "meeting the needs of contemporary people while at the same time sustaining the ability of future generations to meet their needs" [1]. In 1992, the Conference on Global Environment and Development in Rio de Janeiro adopted the Agenda of the Century [2]. After that, China also adopted The 21st Century Agenda of China. Both agendas have set sustainable development that "meets the current needs and pursuits of human without damaging the needs and pursuits of the future" as the goal of future economic development [2,3].

Sustainable development refers to development that meets the needs of the present without compromising the ability of future generations to meet their needs [4–6]. The atmosphere has significant implications to people's lives, which makes the topic of atmospheric environment sustainability one of the hottest discussions of today [7–9]. As a crucial part of human activities, the atmosphere and its quality directly impact the life and daily production of human beings. Since the origin of species, humans cannot survive without the air. The issue of air quality and safety is one of the key issues to humankind [10–12]. The Los Angeles photochemical smog episode in the 1940s, the serious sulfur dioxide pollution in Donora, Pennsylvania in October 1948, and the London smog incidents in December 1952, etc., all have warned us of the consequences of air pollution and unsustainable development [13–15].

As the world's second largest economy, China plays an increasingly important role in global economic development [16–18]. However, China is also a huge energy consumer [19–21]. China's energy consumption accounts for about 23% of world total consumption, and its coal consumption accounts for 59% of China's energy consumption, with an annual consumption of 4.64 billion tons. In 2018, China's consumption of coal, crude oil, natural gas and electricity increased by 1.0%, 6.5%, 17.7% and 8.5%, respectively [22].

Since the reform and opening up, some regions of China have paid too much attention to economic interests and neglected the protection of natural resources and the ecological environment, resulting in ecological imbalances and serious pollution, especially in provinces and cities that focus on heavy industry, where air pollution has become a common thing and has seriously damaged the health of the local population [23–25]. Such issues have not only brought dilemma to social development, but also developed new issues, in which ecological challenges are the most alarming [26–28].

With their vast land, rich fossil fuels resources such as coal, oil and natural gas, as well as their leading industrial foundation in the country, the three Northeastern provinces of China (Liaoning, Jilin and Heilongjiang) used to be the fastest-growing regions in modern China, known as the "Cradle of the Nation's Industry" [29,30]. However, in recent years, the three Northeastern provinces have experienced four key issues of high energy consumption, high resource dependence, high environmental degradation, and a high ratio of brain drain, and are now facing unprecedented energy and environmental crises. Such a phenomenon is called the "Northeast Phenomenon" [31–33].

Currently, the air pollution problem in Northeast China is also quite serious [34–36]. Although total emissions of SO_2 and NO_X declined during the period between 2011 and 2017, SO_2 and NO_X emissions are still 849,477.79 tons and 1,270,006.47 tons in 2017, respectively (please refer to Figure 1) [37]. In this context, it is of great practical and theoretical significance to evaluate the development sustainability of the atmospheric environment in the Northeast region and explore effective air pollution control measures. It is necessary to study how to effectively find the weak points in atmospheric environment protection and take targeted improvement measures in provinces with lagging economic development to achieve sustainable development under the background of China's rapid economic growth.

When studying the sustainable development of the atmospheric environment, academic circles use the rough set method and the entropy weight method, respectively. For example, Lai et al. studied low carbon technology integration management using the rough set method. Based on a questionnaire survey and exploratory factor analysis results on the selected indexes, they implemented a rough set method to identify the weight of all the indexes. Their results showed that the constructed evaluation framework can properly reflect the integrity, and the rough set evaluation could well reflect the overall performance of low carbon project evaluation [38]. Xue et al. developed a fuzzy rough set algorithm to identify the spatial variability, driving forces, and uncertainties of the net ecosystem exchange of carbon between the temperate forests and the atmosphere. Their results showed the advantages of the new rough set algorithm and explained the most important variables for net ecosystem exchange in the northeastern United States [39]. Zhao et al. established a fuzzy comprehensive model based on entropy technology for air quality assessment. By improving the computing factors' weights with Entropy Weight Method, they used the new model to assess the air quality of Fuxin city, China. The results coincided with the objective air quality condition of Fuxin city greatly, which proved the effectiveness of the entropy weight method [40]. Chen et al. developed a hybrid approach, combining a land use regression model and the entropy weight method to estimate the PM2.5 concentrations on a national scale in China. They proved that the hybrid model could potentially provide more valid predictions than a commonly-used model. With $R^2 = 0.82$ and root mean square error of 4.6 µg/m^3 [41]. Liu et al. estimated the relationship between urbanization and atmospheric environment security in Jinan City from 1996 to 2004 on the basis of the theory of environmental Kuznets curves. Employing the entropy method to determine the index weight, they constructed a comprehensive index system for urbanization and atmospheric environment security. They determined the main factors that influence the system to provide a basis for creating scientific urban development strategies and atmospheric environment protection measures [42].

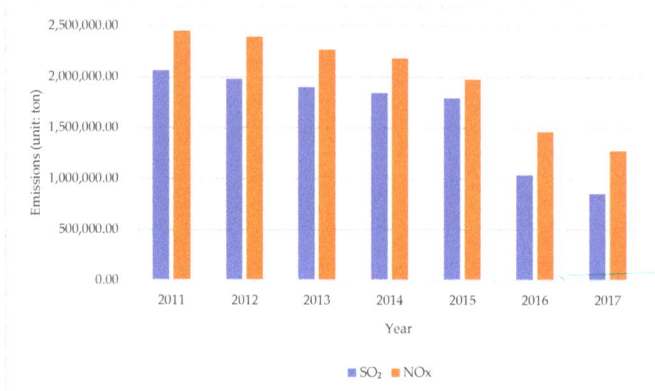

Figure 1. SO$_2$ and NOx emissions in Northeast China from 2011 to 2017.

However, it is rare to combine the two methods to learn from each other. Based on rough set and entropy weight theory, this paper has determined the comprehensive weight by taking both the weight calculated by the attribute reduction set and the entropy weight into consideration, and that has improved the credibility and feasibility of the indicator weights. This paper has also compared experimental results by using different discrete methods, and found that the ranking of attribute values is basically consistent both before and after introducing the entropy weight, thus solving the problem of inconsistent results under different discrete methods and experimental methods. After comparing the results by this new method and by the method of determining the objective weights of the attributes based on the ideal structure optimization model, this paper has found that the rankings are basically the same, which further proves the rationality of this method. Finally, this paper has graded the indicator values into four levels (Excellent, Good, Medium, Poor) according to the actual level of atmospheric environment sustainability as well as previous studies on indicator grading. By determining the levels, this paper could obtain the level of sustainable development of each region based on a comprehensive indicator value.

2. Materials and Methods

2.1. Modeling Basis

Different from the traditional development concept solely based on economic development, sustainable development means balancing economic development with environmental protection and that the two things cannot be separated. When evaluating whether ecosystems are healthy, researchers often use the conceptual PSR Model based on the logic of "Pressure-State-Response" [43–46]. By analyzing the causal relationship inherent in the system, this model discovers the causal chain in the system, and then takes targeted adjustment measures to achieve system sustainability. When analyzing the relationship between humans and nature, this model believes that because of the pressure brought by human activities to the natural environment, the quantity of natural resources and some of the original properties of nature are changed, thus calling for society to take countermeasures through various environmental, economic and governance strategies. This process repeats over and over again, constituting the relationship between human beings and the natural environment. Whether the indicator system is reasonably constructed or not determines the accuracy of the evaluation results [47,48]. Therefore, it is necessary to select generalized and representative indicators as the evaluation indicators from a wide range of potential indicators. The following principles need to be followed: scientific, complete, principal components, and independent [49,50]. Regarding the principal components, this means that when selecting evaluation indicators, if the information given is sufficient, we should select as few indicators as possible and make sure that all indicators are representative.

2.1.1. Establishment of the Indicator Layer for the PSR Model

Pressure indicators refer to the environmental impact of human economic and social activities, such as the damage and disturbance to air quality caused by economic growth, social development, and emissions by various industries during their daily operation in the Northeast region. Therefore, the pressure layer refers to the collection of related indicators that cause damage and disturbance to the air quality and its sustainable development. [51]

State indicators refer to the environmental conditions and environmental changes within specific time periods, such as the consumption of coal, crude oil, and natural gas that affect the sustainable development of the Northeast region. Therefore, the state layer refers to the collection of economic indicators and energy consumption indicators generated by human life in economic activities. This layer includes indicators that are responsive assessments of human activities that can effectively describe the process of regional sustainable development [52].

Response indicators refer to how the society and individuals act to mitigate, stop, and prevent the negative impacts of human activities on the environment, as well as the remedial measures for ecological changes that have already occurred and hindered human survival and development, such as the waste treatment measures in the Northeast region. Therefore, the response layer refers to the collection of measures that society and individuals actively take action to mitigate, prevent, restore and prevent the negative impacts of human activities on the environment and remedy the environmental damage already caused [53].

Relieving pressure is at the core of the system; the state is the basis for the system to decide whether to respond; while response is the main way to achieve sustainable development. The three aspects of pressure, state and response interact with each other to form an organic feedback loop. Effective response behaviors would help to maintain the air quality of Northeast China at a good level. Otherwise, inappropriate response behaviors would cause the air quality of Northeast China to fall into a vicious circle.

2.1.2. Construction of the Indicator System for the PSR Model

Through empirical research, US economists Grossman and Krueger studied the relationship between the quality of the ecological environment and GDP per capita. They found that environmental pollution showed an upward trend with the growth of GDP per capita when the relative income was low, reached the peak at certain turning point when the countries entered the high-income stage, and then gradually declined with the growth of GDP per capita [54].

According to the above logic of causality, this paper has allocated various factors into the pressure layer, the state layer and the response layer respectively. Drawing on existing research [51,55,56] and combining the characteristics of the three provinces, we selected 17 indicators including core and supplementary indicators based on the data of air quality and GDP per capita of the Northeastern provinces from 2009 to 2017 to construct the PSR Model [57]. This indicator system is complete and independent with a principal component and accurately reflects the sustainability level of the atmospheric environment in Northeast China, as shown in Table 1.

Table 1. The indicators in the Pressure-State-Response (PSR) Model for Northeast China atmospheric environment sustainability measurement and their standardized values from 2009 to 2017.

The Element Layer	The Indicator Layer	2009	2010	2011	2012	2013	2014	2015	2016	2017
The Pressure Layer	SO_2	0.060	0.061	0.000	0.015	0.063	0.095	0.126	0.352	0.437
	Nitric Oxide	0.004	0.177	0.000	0.004	0.041	0.068	0.177	0.311	0.477
	Smoke (Dust)	0.111	0.093	0.174	0.119	0.090	0.000	0.188	0.437	0.493
	Oil Reserves	0.000	0.000	0.060	0.080	0.132	0.168	0.192	0.218	0.218
	Natural Gas Reserves	0.084	0.000	0.034	0.053	0.073	0.079	0.098	0.110	0.079
	Coal Reserves	0.000	0.109	0.951	0.966	1.000	0.903	0.971	0.882	1.000
The State Layer	Regional GDP (Hundred Million RMB)	0.539	0.651	0.791	0.861	0.909	0.946	0.948	0.967	1.000
	Value Added of the Secondary Industry (Hundred Million RMB)	0.671	0.831	0.987	1.000	0.968	0.918	0.793	0.727	0.671
	Value Added of the Service Industry (Hundred Million RMB)	0.377	0.453	0.552	0.623	0.690	0.775	0.861	0.936	1.000
	Industrial Value Added (Hundred Million RMB)	0.675	0.844	0.999	1.000	0.971	0.912	0.772	0.694	0.633
	GDP per capita (RMB)	0.535	0.646	0.783	0.852	0.899	0.936	0.941	0.965	1.000
	Coal Consumption (Ten Thousand Tons)	0.786	0.869	0.939	0.994	0.944	0.967	0.956	0.999	1.000
	Crude Oil Consumption (Ten Thousand Tons)	0.933	0.952	0.996	0.980	0.962	0.969	0.960	1.000	0.973
	Natural Gas Consumption (Hundred Million Cubic Meters)	0.012	0.000	0.135	0.464	0.598	0.686	0.727	1.000	0.792
The Response Layer	Investment in Industrial Pollution Control (Ten Thousand RMB)	0.479	0.238	0.487	0.189	1.000	0.858	0.934	0.840	0.440
	Investment in Waste Gas Control (Ten Thousand RMB)	0.271	0.091	0.422	0.149	1.000	0.871	0.730	0.835	0.373
	Local Fiscal Expenditure on Environmental Protection (Hundred Million RMB)	0.000	0.223	0.248	0.341	0.423	0.391	0.719	0.405	1.000

2.2. Evaluation Model on Atmospheric Environment Sustainability

2.2.1. Basic Concept of the Model

First, this paper determines the weight of each layer by using the rough set and entropy weight methods. Because the two methods are both objective methods for determining weights, the combination of the two has taken both the importance of each attribute to decision-making and the influence of the information quantity within each attribute on decision-making into consideration, thus achieving more precise weights [58–61]. Secondly, by assuming the weights of element layers, this paper has constructed a comprehensive evaluation model on atmospheric environment sustainability to obtain the sustainability indicator value. At last, this paper has ranked the objects under assessment based on the indicator values, and respectively determined each object's influence on the sustainability level.

2.2.2. Evaluation Steps

1. Standardize the raw data. Due to different properties of the evaluation indicators, the original data are in different dimensions and orders of magnitude. In order to eliminate the influence of difference in dimensions and ensure the comparability of data as well as the feasibility of the decision-making results, it is necessary to properly standardize the original data. Suppose there are N evaluation objects, M evaluation indicators, the evaluation value of the indicators is expressed as $m'_{ij}(i \in N, J \in M)$, and the standardized indicator value is written as m^i_{ij}. It can be concluded from the relevant literature that for benefit attributes, the larger the indicator value, the better; while for cost attributes, the smaller the indicator value, the better [62,63]. In this paper, the indicators in the state layer and the response layer are benefit attributes, while the indicators in the pressure layer are cost attributes. Then we have:

$$m_{ij} = \frac{m'_{ij} - \min_{i \in N}\{m'_{ij}\}}{\min_{i \in N}\{m'_{ij}\} - \max_{i \in N}\{m'_{ij}\}} \tag{1}$$

$$m_{ij} = \frac{\max_{i \in N}\{m'_{ij}\} - m'_{ij}}{\max_{i \in N}\{m'_{ij}\} - \min_{i \in N}\{m'_{ij}\}} \tag{2}$$

2. Determine the Decision Weights. When rough sets are used to process data, the general approach is to determine the indicator weights based on the importance of various attributes [64,65]. However, this method is flawed. The reduction set of attributes refers to the smallest set of attributes whose number is greater than zero while division is the same with the original data set. The intersection of all reduction sets is called the core attribute. It can be found by data analysis that when the core attribute does not exist, the indicator weights based on the importance of various attributes all equal to zero, which is in contradiction with the fact. When the core attribute does not exist, the attributes in all the reduction sets are relatively necessary attributes. It can be learned from the literature that the importance of each attribute can be determined by the ratio of the number of occurrences of the attribute in the reduction sets to the total number of reduction sets [66,67].

Let H be the total number of reduction sets. H is the number of reduction sets that contain the relatively necessary attribute L_j, then the weight of L_j is $X^*_{L_j} = g/G$, which can be normalized as:

$$X_{L_J} = \left(X^1_{L_J} + X^2_{L_J}\right)/2 \tag{3}$$

Although the above method can determine the weights of relatively necessary attributes, on the one hand, because the data processed by this method is discrete, different attribute discretization methods would result in different reduction sets, thus obtaining inconsistent attribute weights; on the other hand, it happens that some attributes would have the same weights as calculated by Formula (3). Therefore, it can be seen that this method is inadequate to properly calculate the weights of relatively necessary attributes.

In order to make up for the shortcomings of the above two methods, this paper has integrated the entropy weight method with the attribute reduction set method for weight calculation. The entropy weight method is an objective weighting method based on normalization matrix calculation and is not suitable for discrete data [68,69]. The entropy weight method analyzes the influence of indicator variation on the weight [70,71], while the attribute reduction set method examines the dependence of decision attributes on conditional attributes [72–74]. By combining the weights obtained by both methods, this paper has comprehensively considered the importance of each attribute to decision-making and the influence of information quantity within each attribute on decision-making, thus determining the weight of attributes based on two aspects and making up for the shortcomings of the attribute reduction set method in weight determination.

According to existing literature, the entropy weight of indicator L_j is [75]:

$$X_{L_j}^2 = \left(1 + s\sum_{i=1}^{N} Int_{i_j}\right) \bigg/ \left(M + s\sum_{j=1}^{M}\sum_{i=1}^{N} t_{i_j} Int_{i_j}\right) \tag{4}$$

in which $s = 1/lnN$, $t_{i_j} = m_{i_j} / \sum\limits_{i=1}^{N} m_{i_j}$. Assume that when $t_{i_j} = 0$, $h_{i_j} lnh_{i_j} = 0$.

By combining the weight obtained from the attribute reduction set method ($X_{L_j}^1$) and the entropy weight ($X_{L_j}^2$), this paper has obtained the new weight for indicator L_j (X_{L_j}), which comprehensively considers the importance of the attribute itself and its information quantity to decision-making, and therefore makes the weighting more reasonable.

3. Calculate the Comprehensive Evaluation Value. Suppose the importance weights of the pressure layer, the state layer and the response layer in sustainable development are ω, ψ, ξ, respectively. Then, the comprehensive value can be expressed as:

$$Z_i = \omega \sum_{j=1}^{M_1} m_{i_j} X_{L_j} + \psi \sum_{j=M_1+1}^{M_2} m_{i_j} X_{L_j} + \xi \sum_{j=M_2+1}^{M_3} m_{L_j} X_{L_j} \tag{5}$$

where M_1, M_2, and M_3 respectively stand for the number of indicators in the pressure layer, the state layer and the response layer; ω, ψ, ξ are determined by the weighting method based on standard deviation, i.e., by Formula (7) [15]; e_{i_j} is the indicator value of the pressure layer, the state layer and the response layer; $\bar{e}_j$ is the mean value of the indicator values.

$$X_j = \sqrt{\sum_{i=1}^{N} (e_{i_j} - \bar{e}_j)^2 / (N-1)}, X_j = X_j / \sum_{j=1}^{3} X_j \tag{6}$$

Based on previous research results on the development sustainability level and the actual situation of sustainable development of the atmospheric environment [76,77], this paper has graded sustainability indicator values into four levels (Excellent, Good, Medium, Poor). When the sustainability indicator value is no lower than 0.9 ($Z \geq 0.9$), its level of sustainable development is Excellent, indicating a high level of sustainability of the atmospheric environment in various aspects including the economic, environmental and social aspects. When $0.75 < Z < 0.9$, its level of sustainable development is Good; when $0.6 < Z \leq 0.75$, its level of sustainable development is Medium, which indicates that the sustainability of the atmospheric environment in this region has improved but this improvement is not prominent. When $Z \leq 0.6$, its level of sustainable development is Poor, indicating a low level of sustainability of the atmospheric environment in this region which requires significant improvement efforts.

3. Results

3.1. Standardize the Evaluation Indicator

By standardizing the air quality data of the three Northeastern provinces from 2009 to 2017 according to Formulas (1) and (2), this paper has obtained the standardized values of various indicators (as shown in Table 1).

This paper has adopted the rough set method to process discrete data and used SPSS software to perform equal-width discrete analysis and clustering analysis on the data [20]. The equal-width discrete algorithm is a typical unsupervised discretization method, which equally divides the standardized data between [0,1] into four intervals: (0.9, 1.0), (0.65, 0.9), (0.5, 0.65), and (0, 0.5). Four discrete values are selected for these intervals: 4, 3, 2, and 1, corresponding to Excellent, Good, Medium, and Poor, respectively. Clustering analysis is used to obtain decision attributes, which are divided into four categories according to the conditional attributes, and the decision information system is eventually determined as shown in Table 2 below:

Table 2. Decision table of the PSR Model.

The Element Layer	The Indicator Layer	2011	2012	2013	2014	2015	2016	2017
The Pressure Layer	SO$_2$	1	1	1	1	1	2	2
	Nitric Oxide	1	1	1	1	1	2	2
	Smoke (Dust)	1	1	1	1	1	2	2
The State Layer	Oil Reserves	1	1	1	1	1	1	1
	Natural Gas Reserves	1	1	1	1	1	1	1
	Coal Reserves	4	4	4	4	4	4	4
	Regional GDP (Hundred Million RMB)	4	4	4	4	4	4	4
	Value Added of the Secondary Industry (Hundred Million RMB)	4	4	4	4	4	3	3
	Value Added of the Service Industry (Hundred Million RMB)	3	3	3	4	4	4	4
	Industrial Value Added (Hundred Million RMB)	4	4	4	4	4	3	3
	GDP per capita (RMB)	4	4	4	4	4	4	4
	Coal Consumption (Ten Thousand Tons)	4	4	4	4	4	4	4
	Crude Oil Consumption (Ten Thousand Tons)	4	4	4	4	4	4	4
	Natural Gas Consumption (Hundred Million Cubic Meters)	1	2	3	3	3	4	4
The Response Layer	Investment in Industrial Pollution Control (Ten Thousand RMB)	2	1	4	4	4	4	2
	Investment in Waste Gas Control (Ten Thousand RMB)	2	1	4	4	3	4	2
	Local Fiscal Expenditure on Environmental Protection (Hundred Million RMB)	1	2	2	2	3	2	4

Table 2 shows the change in the discrete values of the indicators in the three Northeastern provinces from 2011 to 2017. It can be seen that although the values of the pressure layer elements (SO$_2$, Nitric Oxide, Smoke/Dust) have shown an improvement trend, they still belong to the Medium Level. Therefore, it can be concluded that the air quality of Northeast China is still quite poor, and it is necessary to look for a sustainable development plan for the Northeast region regarding the atmospheric environment. The reserves of oil, natural gas and coal resources are generally stable, of which oil and natural gas are relatively scarce, while coal reserves are relatively abundant. On the other hand, it can be seen by analysis of the elements of the state layer that the Northeast region has not only maintained excellent regional GDP, but the industrial structure of the three Northeast provinces is also undergoing certain changes. For example, the added value of the secondary industry has decreased compared with that of the service industry; the growth of the industry has slowed down but the regional GDP per capita has still remained at an excellent level. Overall, the consumption of coal and oil is still quite high, and the consumption of natural gas is growing. The investment in industrial pollution control and waste gas control has first increased but later decreased, while the local fiscal expenditure on environmental protection has shown an increasing trend. Based on the above table, we could clearly understand the changes in the relevant factors that influence the air quality sustainability of the Northeast region as well as their values in different time periods. It can be seen that the three Northeast provinces have made efforts to improve their air quality in recent years and have achieved certain results. However, there is still a long way to go before these provinces have accomplished a true transformation and upgrade.

3.2. Determine the Weights

Attribute reduction refers to selecting the minimum condition subset while ensuring an unchanged correlation coefficient as the decision system so as to determine the condition attributes in the decision rule. By using Table 2 and the Rosetta software, this paper has obtained the attribute reduction set of each layer.

The Attribute Reduction Set of the Pressure Layer:

Heilongjiang: $\{a_2, a_6\}$, $\{a_1, a_6\}$, $\{a_3, a_6\}$;
Jilin: $\{a_1, a_3, a_6\}$, $\{a_2, a_3, a_6\}$;
Liaoning: $\{a_1, a_3, a_5, a_6\}$, $\{a_2, a_3, a_5, a_6\}$.

The Attribute Reduction Set of the State Layer:

Heilongjiang: $\{b_3, b_4, b_5, b_8\}$, $\{b_1, b_3, b_4, b_8\}$, $\{b_2, b_3, b_5, b_8\}$, $\{b_1, b_2, b_3, b_8\}$;
Jilin: $\{b_3, b_4, b_8\}$, $\{b_2, b_3, b_5, b_8\}$, $\{b_1, b_2, b_3, b_8\}$;
Liaoning: $\{b_2, b_3, b_8\}$, $\{b_3, b_4, b_8\}$.

The Attribute Reduction Set of the Response Layer:

Heilongjiang: $\{c_1, c_3\}$, $\{c_2, c_3\}$;
Jilin: $\{c_1, c_2, c_3\}$;
Liaoning: $\{c_1, c_2, c_3\}$.

Based on the attribute reduction set of various indicator layers and Formula (3), this paper has obtained the relatively necessary attribute weight $X_{L_j}^1$; based on Formulas (4) and (5) and Table 1, this paper has obtained the entropy weight of each indicator layer $X_{L_j}^2$ and the new weight X_{L_j}. The calculation results are shown in Table 3. Different from subjective methods that rely on expert experience and lack objectivity and objective methods with poor explanatory power, the method in this paper uses the rough set method to explore the internal relationship within the experimental data so that the weight obtained by this method could demonstrate the information quantity within each attribute. In this way, it can be ensured that the indicators selected could reflect most of the original information, thus achieving more effective and objective results. Moreover, during data analysis, this paper has performed equal-width discrete analysis, thus preventing the potential problem of inconsistent weighting under different discretization methods. By combining the entropy weight method with the attribute reduction set method, this paper has made up for the shortcomings of the reduction set method and obtained more reasonable weighting results.

Table 3. The weights of various indicators.

The Element Layer	Indicator	Entropy-Based Weight X_{Lj}^1	Rough Weight X_{Lj}^2	Average Weight X_{Lj}
The Pressure Layer	a1	0.2409	0.1667	0.2038
	a2	0.2862	0.1667	0.2264
	a3	0.1533	0.1667	0.1600
	a4	0.1525	0.0000	0.0763
	a5	0.0753	0.0000	0.0376
	a6	0.0919	0.5000	0.2959
The State Layer	b1	0.0375	0.1250	0.0813
	b2	0.0256	0.1250	0.0753
	b3	0.0995	0.2500	0.1747
	b4	0.0311	0.1250	0.0780
	b5	0.0380	0.1250	0.0815
	b6	0.0059	0.0000	0.0029
	b7	0.0005	0.0000	0.0002
	b8	0.7619	0.2500	0.5060
The Response Layer	c1	0.2158	0.2500	0.2329
	c2	0.3568	0.2500	0.3034
	c3	0.4274	0.5000	0.4637

3.3. Calculate the Comprehensive Evaluation Value

Based on Table 1, Table 3, and Formula (6), this paper has obtained the attribute values and sustainability levels of each indicator layer as shown in Table 4 below. From Table 4, it can be seen that the sustainability levels from 2009 to 2017 are ranked as: $Z_{2014} > Z_{2013} > Z_{2015} > Z_{2011} > Z_{2012} > Z_{2017} > Z_{2009} > Z_{2010} > Z_{2016}$.

Table 4. The comprehensive management and sustainability level obtained by the PSR Model.

Year	p Value	S Value	R Value	Sustainable Value Z	Level
2009	0.1677	0.0023	0.0273	0.5810	Poor
2010	0.1549	0.0028	0.0189	0.5455	Poor
2011	0.1672	0.0034	0.0155	0.7287	Medium
2012	0.1576	0.0037	0.0151	0.6777	Medium
2013	0.1541	0.0041	0.0397	0.8009	Good
2014	0.1719	0.0043	0.0617	0.9275	Excellent
2015	0.1581	0.0043	0.0332	0.7723	Good
2016	0.0997	0.0037	0.0298	0.5459	Poor
2017	0.0850	0.0039	0.0266	0.6027	Medium

4. Discussion

By comparing the above method with the entropy weight method and the rough set method alone, we can obtain the rationality of the method adopted in this paper, i.e., introducing the entropy weight method based on the rough set method to construct a sustainable development model. The calculation results of the three different methods for Northeast China as a whole are shown in Table 5 below (Please refer to Appendix A for the calculation results of each province in Northeast China.):

Table 5. Comparison of comprehensive values calculated by different methods for Northeast China.

The Element Layer	Year	The Entropy Weight Method	The Rough Set Method	Rough Set + Entropy Weight Method
The Pressure Layer	2009	0.7090	0.6882	0.7092
	2010	0.5728	0.7016	0.6364
	2011	0.6807	0.7199	0.7239
	2012	0.5878	0.6831	0.6473
	2013	0.6439	0.6915	0.6624
	2014	0.6928	0.7776	0.7033
	2015	0.6724	0.7164	0.6486
	2016	0.4083	0.4420	0.4418
	2017	0.3944	0.3380	0.3233
The State Layer	2009	0.3063	0.3587	0.2879
	2010	0.3195	0.2727	0.2727
	2011	0.4249	0.4063	0.4061
	2012	0.4159	0.4477	0.4270
	2013	0.4996	0.4999	0.4497
	2014	0.4640	0.4757	0.4574
	2015	0.4876	0.5377	0.4659
	2016	0.3623	0.4488	0.4274
	2017	0.4944	0.5168	0.4596
The Response Layer	2009	0.5036	0.4075	0.4039
	2010	0.2560	0.2846	0.2557
	2011	0.3749	0.2587	0.2562
	2012	0.2300	0.2195	0.2389
	2013	0.7124	0.6638	0.6034
	2014	0.8988	0.7609	0.8225
	2015	0.5120	0.4447	0.4475
	2016	0.5429	0.5613	0.4860
	2017	0.3794	0.3243	0.3577
The Sustainable Layer	2009	0.3349	0.4542	0.4180
	2010	0.3781	0.5273	0.4273
	2011	0.3898	0.4698	0.5072
	2012	0.4127	0.5105	0.5143
	2013	0.3490	0.5667	0.5460
	2014	0.5182	0.6392	0.6799
	2015	0.4124	0.5746	0.5630
	2016	0.5036	0.4915	0.4454
	2017	0.4663	0.4455	0.4544

Different discrete methods would lead to different ranking results, especially the ranking of sustainability indicators. By the entropy weight method, the year of 2009 had the lowest sustainability value of 0.3349; while by the rough set method, the year of 2017 had the lowest sustainability value of 0.4455, which indicates large differences in the sustainability evaluation result by different weighting methods, and that the choice of weighting methods directly affects the accuracy of the evaluation. However, if the entropy weight method is further introduced based on the rough set method, the result obtained is consistent with the result obtained through the original Method, as shown in Figure 2 below. According to the result, in Northeast China, the same year 2009 had the lowest sustainability level for atmospheric environment, with an indicator value of 0.4180; while the year 2014 has the best sustainability level with an indicator value of 0.6799. This proves that by introducing the entropy weight method, this paper has solved the issue of inconsistent results by different weighting methods. The rough set method categorizes the data based on attributes and examines the degree of approximation by finding the upper and lower approximations and finding the positive domain. In the assessment of sustainable development of the atmospheric environment in the Northeast region, the rough set method can be used to draw rough conclusions. However, because different weighting methods would lead to different results, this would have a great impact on the accuracy of the data as well as the final conclusion. The entropy weight method is an objective weighting method which only depends on the discreteness of the data. The entropy value can be used to determine the discrete degree of the indicator, and thereby obtain the weight of the indicator in overall comprehensive evaluation. Thanks to this feature of the entropy weight method, the accuracy of the evaluation result can be effectively improved so as to make up for the flaw of the rough set method that different weighting methods would lead to different conclusions [78–80].

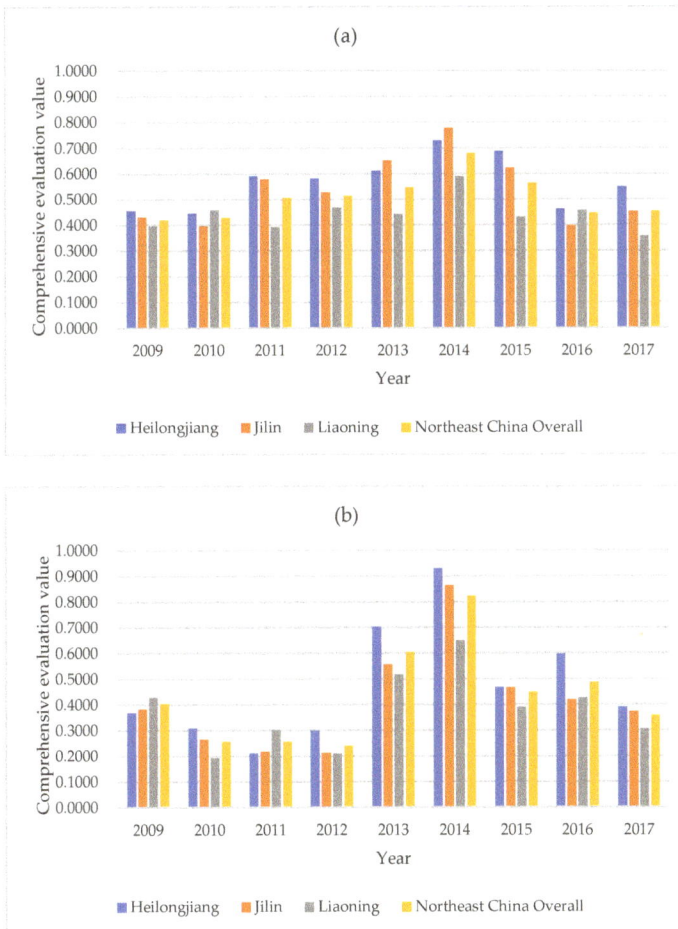

Figure 2. Calculation results of the rough set plus entropy weight method for Northeast China: (**a**) Results of the sustainable layer; (**b**) Results of the response layer.

From Table 5, we could learn the sustainability level of the atmospheric environment in Northeast China: the state layer and the response layer have both shown an upward trend, while the pressure layer has shown a downtrend. Now we can perform a targeted analysis based on the different trends of various indicator layers:

1. Pressure Indicators: it can be seen from the comprehensive value of the pressure indicators that 2014 and 2015 have the best performance, but there has been a downtrend in recent years, with a record low in 2017, which is probably due to increasing demands for resources with the increase of population pressure. The Northeast region is the industrial heartland of China. Back in 2016, the National People's Congress and Chinese People's Political Consultative Conference reaffirmed the policy of "revitalizing the old industrial base in Northeast China", which probably had led to a growing demand for resources in this region. In addition, the Northeast region faces severe cold during winters. The strong demand for heating and large consumption of coal would cause the environmental quality to further deteriorate.

2. State Indicators: the indicators of the state layer have generally shown an upward trend, indicating an overall improvement in resource utilization rate in the Northeast region with growing environmental awareness. In addition, the state layer indicators have reflected the fact that in the wave of globalization, the industrial development of the Northeast region has had a great impact on the atmospheric environment sustainability. As the cradle of China's core industry, while it continues to develop its industry, the Northeast region should also pay attention to the reuse of resources and increase investment in clean energy development.

3. Response Indicators: the indicators of the response layer have generally shown a downward trend. The investment reached its peak in 2014, and according to various indicators, the sustainability level of the Northeast provinces also reached its best in 2014, indicating that enterprises and local government are working towards a correct direction in terms of environmental protection investment. The Northeast provinces could see more achievements if they increase their investment in environmental protection.

4. Sustainability Indicators: there have been large fluctuations in the sustainability indicators, which are mainly affected by the air quality protection investment by enterprises and the local government, the promotion of clean energy such as natural gas, and the environmental awareness of people. This indicates that achieving sustainable development is the common responsibility of everyone in our society. Individuals, enterprises and the government should all establish a sense of responsibility and strengthen their understanding of sustainable development concepts, thus promoting the sustainable development of the Northeastern provinces.

It can also be found that Heilongjiang Province has done the best in the sustainable development of the atmospheric environment during the research period. On the whole, the main reasons are:

(1) Heilongjiang strictly abides by the state's laws, regulations and action guidelines on air pollution control, and has formulated a series of sustainable development policies for atmospheric environment with local characteristics and specificity. It has issued and released "Implementation Rules of Air Pollution Prevention Action Plan in Heilongjiang" [81], "Special Action Plan for Air Pollution Prevention and Control in Heilongjiang Province, 2016–2018" [82], and "Air Pollution Prevention and Control Regulations of Heilongjiang Province" [83], etc. These regulations have played positive roles in reducing air pollution emissions and maintaining the sustainable development of the atmospheric environment in Heilongjiang.

(2) In accordance with national and local regulations, Heilongjiang has established a strict environmental accountability mechanism to give political sanctions and economic penalties to units and individuals that are not effective in preventing and controlling air pollution, which urges government officials and enterprises to attach great importance to the sustainable development of the atmospheric environment. In early 2018, the provincial government reproached 65 civil servants, including eight cadres at the departmental level and 39 at the sectional level, who were are directly responsible for the poor effectiveness of air pollution prevention and control in October 2017. Among

them, 45 were given Party and administrative discipline, and 17 were conducted admonition talks. At the same time, the Provincial Agricultural Commission was ordered to make a written inspection to the provincial government [84].

(3) Heilongjiang has focused on controlling major air pollution sources such as coal-fired heating and biomass burning emissions in winter and adopted new technologies for pollution source grid monitoring and straw resource utilization. By the end of 2015, the comprehensive utilization rate of straw in the province exceeded 80%, which effectively reduced atmospheric pollution [82].

5. Conclusions

This paper has adopted a new method that can determine the weight more objectively. First, this paper has constructed a PSR Model for the sustainability evaluation of the atmospheric environment of Northeast China as well as an air quality evaluation system consisting of 17 indicators. After this paper obtained the weights by using the rough set method, it further introduced the entropy weight method to make up for the flaws of the rough set method, and effectively solved the issue of large differences in evaluation results by different weighting methods under the rough set method, thus improving the accuracy of the results. This paper obtained the data on changes in the atmospheric environment of the three Northeast provinces from 2009 to 2017. Finally, by consolidating the data of the three provinces, this paper obtained the evaluation result of the atmospheric environment sustainability in Northeast China, which showed that the sustainability level of the atmospheric environment in Northeast China provinces first improved, and then worsened, with the atmospheric environment sustainability level reaching the highest level of 0.9275 in 2014, while dropping to the lowest level of 0.6027 in 2017. In Northeast China, Heilongjiang Province has done the best in the sustainable development of the atmospheric environment. From 2009 to 2017, the sustainability level of Northeast China's atmospheric environment has risen at first but then declined and achieved the best level in 2014.

The research results of this paper have good applicability to the sustainable development of China's atmospheric environment. On the one hand, this paper has obtained a more complete evaluation on the actual sustainability level of the atmospheric environment based on the existing data. The main feature of this study is that it uses the rough set method to calculate the weight of each condition attribute and introduces the entropy weight method to obtain an objective evaluation of the data based on discrete values. On the other hand, this paper has combined the entropy weight method with the rough set method to make up for the flaws of the rough set method and effectively makes use of the advantages of the two methods. The two methods supplement each other and lead to more accurate conclusions regarding the development sustainability of the atmospheric environment. This paper uses the entropy value to solve the problem of inconsistent data under different weighting methods and has innovated a model for atmospheric sustainable development studies as well as enriched the literature on atmospheric environment and sustainable development. Therefore, compared with the previous research, the advantages and applicability of this paper are mainly reflected in:

1. The analysis and calculation of the pressure layer, state layer and response layer. This paper reflects the various factors affecting the sustainable development of atmospheric environment more comprehensively, and provides a more scientific and complete basis for the future sustainable development policies.

2. As far as the research method is concerned, the entropy weight method is used to make up for the shortcomings of the rough set method. The problem of difference in evaluation results caused by using rough set method alone is effectively solved (see Table 5), so that calculation accuracy is optimized.

3. China is vast in territory, and the sources and effects of air pollution are not exactly the same everywhere. By comparing and analyzing the impacts of various factors affecting the sustainable development of the atmospheric environment on different provinces in the same region, this paper could help different regions of China (such as the Yangtze River Delta, Pearl River Delta, etc.) to analyze their own atmospheric environmental impact factors more effectively.

Based on the evaluation results, this paper has proposed the following suggestions in order for the atmospheric environment of the Northeast region to maintain sustainable development:

1. Develop relevant technologies to improve energy efficiency and reduce waste and pollutant emissions. Meanwhile, establish an incentive program and punishment policy for technology-based enterprises to encourage the research and development as well as application of high-quality technology.

2. Increase investment in industrial pollution control, establish a systematized industrial pollution and waste treatment system, and process different types of pollutants by category. Establish relevant laws and regulations, strictly limit and monitor the pollution emission quota of different production units and enhance law enforcement efforts to crack down on illegal pollution behavior.

3. Further adjust and optimize the industrial structure by lifting the proportion of the service industry in economic development and advancing the transformation and upgrading of the secondary industry; gradually phase out heavily polluting enterprises.

4. The local government of Northeast provinces should also actively encourage and advocate for people to use clean energy and reduce the use of traditional energy. The local government should further enhance the public's environmental awareness [85], and establish an air quality management system in accordance with the concepts of sustainable development with the participation of the public in order to truly realize the "revitalization of the old industrial base in Northeast China".

Author Contributions: Conceptualization, Y.L. and G.Y.; Methodology, M.S.; Software, G.Y.; Validation, Y.L., G.Y. and Q.Z.; Formal Analysis, M.S.; Investigation, Q.Z.; Resources, J.L.; Data Curation, M.S.; Writing-Original Draft Preparation, Y.L.; Writing-Review and Editing, M.S.; Visualization, G.Y.; Supervision, J.L.; Project Administration, Y.L.

Funding: This research received no external funding.

Conflicts of Interest: The authors declare no conflict of interest.

Appendix A Calculation Results by the Three Different Methods for Each Province in Northeast China

Table A1. Comparison of comprehensive values calculated by different methods for Heilongjiang.

The Element Layer	Year	The Entropy Weight Method	The Rough Set Method	Rough Set + Entropy Weight Method
The Pressure Layer	2009	0.8541	0.7277	0.8312
	2010	0.7441	0.7259	0.8248
	2011	0.8253	0.7698	0.8810
	2012	0.7355	0.6947	0.8000
	2013	0.7490	0.7143	0.8162
	2014	0.8288	0.7736	0.8737
	2015	0.8480	0.6049	0.6996
	2016	0.4321	0.4541	0.5118
	2017	0.4223	0.2848	0.3320
The State Layer	2009	0.2802	0.3164	0.3423
	2010	0.2978	0.2491	0.2798
	2011	0.3399	0.4173	0.4551
	2012	0.4399	0.4096	0.4523
	2013	0.4805	0.4349	0.4811
	2014	0.5015	0.5482	0.5972
	2015	0.4858	0.4646	0.5147
	2016	0.4005	0.4534	0.5164
	2017	0.5748	0.4524	0.5162

Table A1. *Cont.*

The Element Layer	Year	The Entropy Weight Method	The Rough Set Method	Rough Set + Entropy Weight Method
The Response Layer	2009	0.5419	0.3788	0.3680
	2010	0.2241	0.3162	0.3089
	2011	0.2706	0.2277	0.2108
	2012	0.2022	0.3053	0.2977
	2013	0.7917	0.7251	0.7017
	2014	1.0193	0.9406	0.9290
	2015	0.5652	0.4886	0.4666
	2016	0.5460	0.6186	0.5958
	2017	0.3421	0.3951	0.3896
The Sustainable Layer	2009	0.5670	0.2794	0.4556
	2010	0.5161	0.3241	0.4454
	2011	0.5873	0.4156	0.5912
	2012	0.5710	0.4463	0.5815
	2013	0.5674	0.3967	0.6110
	2014	0.7953	0.5237	0.7278
	2015	0.6302	0.4894	0.6869
	2016	0.4433	0.4326	0.4624
	2017	0.5015	0.5720	0.5493

Table A2. Comparison of comprehensive values calculated by different methods for Jilin.

The Element Layer	Year	The Entropy Weight Method	The Rough Set Method	Rough Set + Entropy Weight Method
The Pressure Layer	2009	0.5787	0.8304	0.7045
	2010	0.5291	0.7724	0.6507
	2011	0.5639	0.8408	0.7024
	2012	0.5307	0.7932	0.6620
	2013	0.5203	0.7740	0.6471
	2014	0.6030	0.8412	0.7221
	2015	0.5501	0.7781	0.6641
	2016	0.3501	0.4873	0.4187
	2017	0.3017	0.4121	0.3569
The State Layer	2009	0.2179	0.2761	0.2347
	2010	0.2635	0.3322	0.2831
	2011	0.3164	0.4018	0.3413
	2012	0.3517	0.4488	0.3799
	2013	0.3832	0.4884	0.4132
	2014	0.4023	0.5143	0.4343
	2015	0.4014	0.5168	0.4351
	2016	0.3179	0.4819	0.3784
	2017	0.3348	0.4993	0.3948

Table A2. *Cont.*

The Element Layer	Year	The Entropy Weight Method	The Rough Set Method	Rough Set + Entropy Weight Method
The Response Layer	2009	0.4360	0.3566	0.3826
	2010	0.2824	0.2236	0.2643
	2011	0.2958	0.2230	0.2168
	2012	0.2298	0.1773	0.2116
	2013	0.7305	0.6101	0.5560
	2014	0.9405	0.7977	0.8637
	2015	0.5843	0.4679	0.4654
	2016	0.5711	0.4668	0.4176
	2017	0.3891	0.3223	0.3717
The Sustainable Layer	2009	0.1625	0.6836	0.4310
	2010	0.2574	0.6138	0.3955
	2011	0.1781	0.6455	0.5787
	2012	0.2459	0.6007	0.5277
	2013	0.1162	0.6836	0.6509
	2014	0.2937	0.8182	0.7775
	2015	0.1504	0.6749	0.6223
	2016	0.3844	0.4363	0.3958
	2017	0.5014	0.3675	0.4527

Table A3. Comparison of comprehensive values calculated by different methods for Liaoning.

The Element Layer	Year	The Entropy Weight Method	The Rough Set Method	Rough Set + Entropy Weight Method
The Pressure Layer	2009	0.6649	0.7194	0.5918
	2010	0.5201	0.5750	0.4549
	2011	0.6700	0.7353	0.6026
	2012	0.5629	0.6250	0.4997
	2013	0.5923	0.6518	0.5299
	2014	0.5966	0.6442	0.5176
	2015	0.6447	0.6922	0.5740
	2016	0.4519	0.4791	0.4060
	2017	0.2979	0.3183	0.2573
The State Layer	2009	0.3147	0.3227	0.2804
	2010	0.2814	0.2903	0.2397
	2011	0.4655	0.4774	0.4158
	2012	0.5076	0.5217	0.4525
	2013	0.4985	0.5137	0.4383
	2014	0.4017	0.4182	0.3386
	2015	0.5032	0.5211	0.4406
	2016	0.4505	0.4927	0.4091
	2017	0.5008	0.5421	0.4563

Table A3. *Cont.*

The Element Layer	Year	The Entropy Weight Method	The Rough Set Method	Rough Set + Entropy Weight Method
The Response Layer	2009	0.5174	0.4573	0.4277
	2010	0.2486	0.2045	0.1937
	2011	0.3711	0.3268	0.3013
	2012	0.2556	0.2181	0.2102
	2013	0.6836	0.5979	0.5163
	2014	0.8349	0.7132	0.6495
	2015	0.5190	0.4393	0.3890
	2016	0.5600	0.4905	0.4256
	2017	0.3785	0.3232	0.3046
The Sustainable Layer	2009	0.3605	0.5470	0.3951
	2010	0.4691	0.5913	0.4575
	2011	0.3333	0.4802	0.3908
	2012	0.4482	0.5543	0.4677
	2013	0.3444	0.5244	0.4417
	2014	0.5337	0.6879	0.5885
	2015	0.3522	0.5181	0.4301
	2016	0.5235	0.5173	0.4584
	2017	0.4586	0.3693	0.3572

References

1. Barbier, E.B. The concept of sustainable economic development. *Environ. Conserv.* **1987**, *14*, 101–110. [CrossRef]
2. Spangenberg, J.H. Institutional sustainability indicators: an analysis of the institutions in Agenda 21 and a draft set of indicators for monitoring their effectivity. *Sustain. Dev.* **2002**, *10*, 103–115. [CrossRef]
3. State Council of the People's Republic of China. *The 21st Century Agenda of China: White Paper on Population, Environment and Development in the 21st Century*; China Environmental Science Press: Beijing, China, 1994.
4. Liu, L. Sustainability: Living within One's Own Ecological Means. *Sustainability* **2009**, *1*, 1412–1430. [CrossRef]
5. Yang, W.; Li, L. Energy Efficiency, Ownership Structure, and Sustainable Development: Evidence from China. *Sustainability* **2017**, *9*, 912. [CrossRef]
6. Li, L.; Yang, W. Total Factor Efficiency Study on China's Industrial Coal Input and Wastewater Control with Dual Target Variables. *Sustainability* **2018**, *10*, 2121. [CrossRef]
7. Novas, N.; Gázquez, J.A.; MacLennan, J.; García, R.M.; Fernández-Ros, M.; Manzano-Agugliaro, F. A real-time underground environment monitoring system for sustainable tourism of caves. *J. Clean. Prod.* **2017**, *142*, 2707–2721. [CrossRef]
8. Opoku, A. Biodiversity and the built environment: Implications for the Sustainable Development Goals (SDGs). *Resour. Conserv. Recycl.* **2019**, *141*, 1–7. [CrossRef]
9. Yang, W.; Li, L. Efficiency evaluation of industrial waste gas control in China: A study based on data envelopment analysis (DEA) model. *J. Clean. Prod.* **2018**, *179*, 1–11. [CrossRef]
10. Santillán Soto, N.; García Cueto, O.R.; Lambert Arista, A.A.; Ojeda Benítez, S.; Cruz Sotelo, S.E. Comparative Analysis of Two Urban Microclimates: Energy Consumption and Greenhouse Gas Emissions. *Sustainability* **2019**, *11*, 2045. [CrossRef]
11. Yeh, H.F.; Hsu, H.L. Using the Markov Chain to Analyze Precipitation and Groundwater Drought Characteristics and Linkage with Atmospheric Circulation. *Sustainability* **2019**, *11*, 1817. [CrossRef]
12. Yang, W.; Yuan, G.; Han, J. Is China's air pollution control policy effective? Evidence from Yangtze River Delta cities. *J. Clean. Prod.* **2019**, *220*, 110–133. [CrossRef]

13. Rodriguez, Y.R. Great Smog of London. In *Encyclopedia of Toxicology*, 3rd ed.; Wexler, P., Ed.; Academic Press: Oxford, UK, 2014; pp. 796–797. ISBN 978-0-12-386455-0.
14. Ashraf, A.; Butt, A.; Khalid, I.; Alam, R.U.; Ahmad, S.R. Smog analysis and its effect on reported ocular surface diseases: A case study of 2016 smog event of Lahore. *Atmos. Environ.* **2019**, *198*, 257–264. [CrossRef]
15. Yuan, G.; Yang, W. Evaluating China's Air Pollution Control Policy with Extended AQI Indicator System: Example of the Beijing-Tianjin-Hebei Region. *Sustainability* **2019**, *11*, 939. [CrossRef]
16. Zhang, D.; Lei, L.; Ji, Q.; Kutan, A.M. Economic policy uncertainty in the US and China and their impact on the global markets. *Econ. Model.* **2019**, *79*, 47–56. [CrossRef]
17. Xu, X.; Li, D.D.; Zhao, M. "Made in China" matters: Integration of the global labor market and the global labor share decline. *China Econ. Rev.* **2018**, *52*, 16–29. [CrossRef]
18. Yang, W.; Li, L. Analysis of Total Factor Efficiency of Water Resource and Energy in China: A Study Based on DEA-SBM Model. *Sustainability* **2017**, *9*, 1316. [CrossRef]
19. Springer, C.; Evans, S.; Lin, J.; Roland-Holst, D. Low carbon growth in China: The role of emissions trading in a transitioning economy. *Appl. Energy* **2019**, *235*, 1118–1125. [CrossRef]
20. Wang, Y.; Wang, N. The role of the port industry in China's national economy: An input–output analysis. *Transp. Policy* **2019**, *78*, 1–7. [CrossRef]
21. Yang, W.; Li, L. Efficiency Evaluation and Policy Analysis of Industrial Wastewater Control in China. *Energies* **2017**, *10*, 1201. [CrossRef]
22. National Bureau of Statistics of China. *China Statistical Yearbook, 2017–2018*; China Statistic Press: Beijing, China, 2018.
23. Wang, J.; Zhang, X.; Yang, Q.; Zhang, K.; Zheng, Y.; Zhou, G. Pollution characteristics of atmospheric dustfall and heavy metals in a typical inland heavy industry city in China. *J. Environ. Sci.* **2018**, *71*, 283–291. [CrossRef]
24. Zhang, Q. Energy and resource conservation and air pollution abatement in China's iron and steel industry. *Resour. Conserv. Recycl.* **2019**, *147*, 67–84. [CrossRef]
25. Yang, H.; Tao, W.; Liu, Y.; Qiu, M.; Liu, J.; Jiang, K.; Yi, K.; Xiao, Y.; Tao, S. The contribution of the Beijing, Tianjin and Hebei region's iron and steel industry to local air pollution in winter. *Environ. Pollut.* **2019**, *245*, 1095–1106. [CrossRef] [PubMed]
26. Xie, L.; Flynn, A.; Tan-Mullins, M.; Cheshmehzangi, A. The making and remaking of ecological space in China: The political ecology of Chongming Eco-Island. *Political Geogr.* **2019**, *69*, 89–102. [CrossRef]
27. Dai, L.; Li, S.; Zhou, W.; Qi, L.; Zhou, L.; Wei, Y.; Li, J.; Shao, G.; Yu, D. Opportunities and challenges for the protection and ecological functions promotion of natural forests in China. *For. Ecol. Manag.* **2018**, *410*, 187–192. [CrossRef]
28. Yang, Q.; Liu, G.; Hao, Y.; Coscieme, L.; Zhang, J.; Jiang, N.; Casazza, M.; Giannetti, B.F. Quantitative analysis of the dynamic changes of ecological security in the provinces of China through emergy-ecological footprint hybrid indicators. *J. Clean. Prod.* **2018**, *184*, 678–695. [CrossRef]
29. Zhang, Q.; Liao, H.; Hao, Y. Does one path fit all? An empirical study on the relationship between energy consumption and economic development for individual Chinese provinces. *Energy* **2018**, *150*, 527–543. [CrossRef]
30. Cheng, X.; Li, N.; Mu, H.; Guo, Y.; Jiang, Y. Study on Total-Factor Energy Efficiency in Three Provinces of Northeast China Based on SBM Model. *Energy Procedia* **2018**, *152*, 131–136. [CrossRef]
31. Zhen, G.; Qieyi, L.; Xiaoxu, W. On Development Model based on Intra-county Cyclic Economy under Low-carbon Economy for Northeast China. *Energy Procedia* **2011**, *5*, 1553–1557. [CrossRef]
32. Zhang, C.; Liu, Y.; Qiao, H. An empirical study on the spatial distribution of the population, economy and water resources in Northeast China. *Phys. Chem. Earth Parts A/B/C* **2015**, *79–82*, 93–99. [CrossRef]
33. Zhang, P.; Yuan, H.; Bai, F.; Tian, X.; Shi, F. How do carbon dioxide emissions respond to industrial structural transitions? Empirical results from the northeastern provinces of China. *Struct. Chang. Econ. Dyn.* **2018**, *47*, 145–154. [CrossRef]
34. Zeng, X.-W.; Qian, Z.M.; Vaughn, M.G.; Nelson, E.J.; Dharmage, S.C.; Bowatte, G.; Perret, J.; Chen, D.-H.; Ma, H.; Lin, S.; et al. Positive association between short-term ambient air pollution exposure and children blood pressure in China–Result from the Seven Northeast Cities (SNEC) study. *Environ. Pollut.* **2017**, *224*, 698–705. [CrossRef] [PubMed]

35. Li, X.; Sun, Y.; An, Y.; Wang, R.; Lin, H.; Liu, M.; Li, S.; Ma, M.; Xiao, C. Air pollution during the winter period and respiratory tract microbial imbalance in a healthy young population in Northeastern China. *Environ. Pollut.* **2019**, *246*, 972–979. [CrossRef] [PubMed]

36. Yin, S.; Wang, X.; Zhang, X.; Zhang, Z.; Xiao, Y.; Tani, H.; Sun, Z. Exploring the effects of crop residue burning on local haze pollution in Northeast China using ground and satellite data. *Atmos. Environ.* **2019**, *199*, 189–201. [CrossRef]

37. National Bureau of Statistics of China. *China Statistical Yearbook, 2011–2017*; China Statistic Press: Beijing, China, 2018.

38. Lai, X.; Liu, J.; Georgiev, G. Low carbon technology integration innovation assessment index review based on rough set theory—An evidence from construction industry in China. *J. Clean. Prod.* **2016**, *126*, 88–96. [CrossRef]

39. Xue, Y.; Chen, Y.; Hu, Y.; Chen, H. Fuzzy Rough Set algorithm with Binary Shuffled Frog-Leaping (BSFL-FRSA): An innovative approach for identifying main drivers of carbon exchange in temperate deciduous forests. *Ecol. Indic.* **2017**, *83*, 41–52. [CrossRef]

40. Zhao, X.; Qi, Q.; Li, R. The establishment and application of fuzzy comprehensive model with weight based on entropy technology for air quality assessment. *Procedia Eng.* **2010**, *7*, 217–222. [CrossRef]

41. Chen, L.; Gao, S.; Zhang, H.; Sun, Y.; Ma, Z.; Vedal, S.; Mao, J.; Bai, Z. Spatiotemporal modeling of PM2.5 concentrations at the national scale combining land use regression and Bayesian maximum entropy in China. *Environ. Int.* **2018**, *116*, 300–307. [CrossRef] [PubMed]

42. Liu, W.; Jiao, F.; Ren, L.; Xu, X.; Wang, J.; Wang, X. Coupling coordination relationship between urbanization and atmospheric environment security in Jinan City. *J. Clean. Prod.* **2018**, *204*, 1–11. [CrossRef]

43. Ou, C.; Liu, W. Developing a sustainable indicator system based on the pressure–state–response framework for local fisheries: A case study of Gungliau, Taiwan. *Ocean Coast. Manag.* **2010**, *53*, 289–300. [CrossRef]

44. Guo, M.; Bu, Y.; Cheng, J.; Jiang, Z. Natural Gas Security in China: A Simulation of Evolutionary Trajectory and Obstacle Degree Analysis. *Sustainability* **2018**, *11*, 96. [CrossRef]

45. Zhao, Y.W.; Zhou, L.Q.; Dong, B.Q.; Dai, C. Health assessment for urban rivers based on the pressure, state and response framework—A case study of the Shiwuli River. *Ecol. Indic.* **2019**, *99*, 324–331. [CrossRef]

46. Li, S.; Li, R. Evaluating Energy Sustainability Using the Pressure-State-Response and Improved Matter-Element Extension Models: Case Study of China. *Sustainability* **2019**, *11*, 290. [CrossRef]

47. Zhang, X.C.; Ma, C.; Zhan, S.F.; Chen, W.P. Evaluation and simulation for ecological risk based on emergy analysis and Pressure-State-Response Model in a coastal city, China. *Procedia Environ. Sci.* **2012**, *13*, 221–231. [CrossRef]

48. Huang, Z.; Xu, M.; Chen, W.; Lin, X.; Cao, C.; Singh, R.P. Postseismic Restoration of the Ecological Environment in the Wenchuan Region Using Satellite Data. *Sustainability* **2018**, *10*, 3990. [CrossRef]

49. Neri, A.C.; Dupin, P.; Sánchez, L.E. A pressure–state–response approach to cumulative impact assessment. *J. Clean. Prod.* **2016**, *126*, 288–298. [CrossRef]

50. Deng, M.; Chen, J.; Huang, J.; Niu, W. Agricultural Drought Risk Evaluation Based on an Optimized Comprehensive Index System. *Sustainability* **2018**, *10*, 3465. [CrossRef]

51. Bai, X.; Tang, J. Ecological Security Assessment of Tianjin by PSR Model. *Procedia Environ. Sci.* **2010**, *2*, 881–887. [CrossRef]

52. Chen, Y.; Wang, P.; Zhong, B.; Ouyang, G.; Bai, B.; Du, J. Coarse-to-fine visual tracking with PSR and scale driven expert-switching. *Neurocomputing* **2018**, *275*, 1456–1467. [CrossRef]

53. Sun, B.; Tang, J.; Yu, D.; Song, Z.; Wang, P. Ecosystem health assessment: A PSR analysis combining AHP and FCE methods for Jiaozhou Bay, China1. *Ocean Coast. Manag.* **2019**, *168*, 41–50. [CrossRef]

54. Grossman, G.M.; Krueger, A.B. Economic Growth and the Environment. *Q. J. Econ.* **1995**, *110*, 353–377. [CrossRef]

55. Guo, Q.; Wang, J.; Yin, H.; Zhang, G. A comprehensive evaluation model of regional atmospheric environment carrying capacity: Model development and a case study in China. *Ecol. Indic.* **2018**, *91*, 259–267. [CrossRef]

56. Wang, Q.; Li, S.; Li, R. Evaluating water resource sustainability in Beijing, China: Combining PSR model and matter-element extension method. *J. Clean. Prod.* **2019**, *206*, 171–179. [CrossRef]

57. National Bureau of Statistics of China. *China Statistical Yearbook, 2009–2017*; China Statistic Press: Beijing, China, 2018.

58. Liu, F.; Zhao, S.; Weng, M.; Liu, Y. Fire risk assessment for large-scale commercial buildings based on structure entropy weight method. *Saf. Sci.* **2017**, *94*, 26–40. [CrossRef]
59. Xu, H.; Ma, C.; Lian, J.; Xu, K.; Chaima, E. Urban flooding risk assessment based on an integrated k-means cluster algorithm and improved entropy weight method in the region of Haikou, China. *J. Hydrol.* **2018**, *563*, 975–986. [CrossRef]
60. Järvinen, J.; Kovács, L.; Radeleczki, S. Defining rough sets using tolerances compatible with an equivalence. *Inf. Sci.* **2019**. [CrossRef]
61. Han, S.E. Roughness measures of locally finite covering rough sets. *Int. J. Approx. Reason.* **2019**, *105*, 368–385. [CrossRef]
62. Fang, Y.; Min, F. Cost-sensitive approximate attribute reduction with three-way decisions. *Int. J. Approx. Reason.* **2019**, *104*, 148–165. [CrossRef]
63. Kunz, M.; Puchta, A.; Groll, S.; Fuchs, L.; Pernul, G. Attribute quality management for dynamic identity and access management. *J. Inf. Secur. Appl.* **2019**, *44*, 64–79. [CrossRef]
64. Wang, Q.; Qian, Y.; Liang, X.; Guo, Q.; Liang, J. Local neighborhood rough set. *Knowl.-Based Syst.* **2018**, *153*, 53–64. [CrossRef]
65. Han, S.E. Covering rough set structures for a locally finite covering approximation space. *Inf. Sci.* **2019**, *480*, 420–437. [CrossRef]
66. Zhang, L.; Jin, Z.; Zheng, Y.; Jiang, R. Mechanical Product Ecological Design Knowledge Reduction Based on Rough Set. *Procedia CIRP* **2019**, *80*, 33–38. [CrossRef]
67. Wang, C.; Huang, Y.; Shao, M.; Fan, X. Fuzzy rough set-based attribute reduction using distance measures. *Knowl.-Based Syst.* **2019**, *164*, 205–212. [CrossRef]
68. Zhang, S.; Xu, S.; Zhang, W.; Yu, D.; Chen, K. A hybrid approach combining an extended BBO algorithm with an intuitionistic fuzzy entropy weight method for QoS-aware manufacturing service supply chain optimization. *Neurocomputing* **2018**, *272*, 439–452. [CrossRef]
69. Zhou, M.; Liu, X.; Yang, J.; Chen, Y.; Wu, J. Evidential reasoning approach with multiple kinds of attributes and entropy-based weight assignment. *Knowl.-Based Syst.* **2019**, *163*, 358–375. [CrossRef]
70. Zhou, J.; Chen, L.; Chen, C.L.P.; Zhang, Y.; Li, H.-X. Fuzzy clustering with the entropy of attribute weights. *Neurocomputing* **2016**, *198*, 125–134. [CrossRef]
71. Wang, K.; Wang, Y. Entropy numbers of functions on [−1,1] with Jacobi weights. *J. Math. Anal. Appl.* **2017**, *445*, 985–997. [CrossRef]
72. Jing, Y.; Li, T.; Fujita, H.; Wang, B.; Cheng, N. An incremental attribute reduction method for dynamic data mining. *Inf. Sci.* **2018**, *465*, 202–218. [CrossRef]
73. Konecny, J.; Krajča, P. On attribute reduction in concept lattices: The polynomial time discernibility matrix-based method becomes the CR-method. *Inf. Sci.* **2019**, *491*, 48–62. [CrossRef]
74. Chen, J.; Mi, J.; Xie, B.; Lin, Y. A fast attribute reduction method for large formal decision contexts. *Int. J. Approx. Reason.* **2019**, *106*, 1–17. [CrossRef]
75. Qiu, Y. *Management Decision Entropy and Its Application*; China Electric Power Press: Beijing, China, 2011.
76. Virto, L.R. A preliminary assessment of the indicators for Sustainable Development Goal (SDG) 14 "Conserve and sustainably use the oceans, seas and marine resources for sustainable development". *Mar. Policy* **2018**, *98*, 47–57. [CrossRef]
77. Kılkış, Ş.; Krajačić, G.; Duić, N.; Rosen, M.A.; Al-Nimr, M.A. Advancements in sustainable development of energy, water and environment systems. *Energy Convers. Manag.* **2018**, *176*, 164–183. [CrossRef]
78. Ye, J. Multicriteria fuzzy decision-making method using entropy weights-based correlation coefficients of interval-valued intuitionistic fuzzy sets. *Appl. Math. Model.* **2010**, *34*, 3864–3870. [CrossRef]
79. Ji, Y.; Huang, G.H.; Sun, W. Risk assessment of hydropower stations through an integrated fuzzy entropy-weight multiple criteria decision making method: A case study of the Xiangxi River. *Expert Syst. Appl.* **2015**, *42*, 5380–5389. [CrossRef]
80. Cui, Y.; Feng, P.; Jin, J.; Liu, L. Water Resources Carrying Capacity Evaluation and Diagnosis Based on Set Pair Analysis and Improved the Entropy Weight Method. *Entropy* **2018**, *20*, 359. [CrossRef]
81. Heilongjiang Provincial Government. *Implementation Rules of Air Pollution Prevention Action Plan in Heilongjiang*; Heilongjiang Provincial Government: Harbin, Heilongjiang, China, 2014.
82. Heilongjiang Provincial Government. *Special Action Plan for Air Pollution Prevention and Control in Heilongjiang Province, 2016–2018*; Heilongjiang Provincial Government: Harbin, Heilongjiang, China, 2016.

Sustainability **2019**, *11*, 3793

83. The 12th People's Congress of Heilongjiang Province. *Air Pollution Prevention and Control Regulations of Heilongjiang Province*; People's Congress of Heilongjiang Province: Harbin, Heilongjiang, China, 2017.

84. Xinhua Net Ministry of Ecology and Environment: 65 People in 4 Cities of Heilongjiang Province Are Reproaches for Their Ineffectively Control Haze. Available online: http://www.xinhuanet.com/2018-05/29/c_129882781.htm (accessed on 1 July 2019).

85. Yang, Y.; Yang, W. Does Whistleblowing Work for Air Pollution Control in China? A Study Based on Three-party Evolutionary Game Model under Incomplete Information. *Sustainability* **2019**, *11*, 324. [CrossRef]

sustainability

MDPI

Article

A Novel Linear Time-Varying GM(1,N) Model for Forecasting Haze: A Case Study of Beijing, China

Pingping Xiong [1,*], Jia Shi [1], Lingling Pei [2] and Song Ding [3]

1   College of Mathematics and Statistics, Nanjing University of Information Science and Technology, Nanjing 210044, China
2   School of Business Administration, Zhejiang University of Finance and Economics, Hangzhou 310018, China
3   School of Economics, Zhejiang University of Finance and Economics, Hangzhou 310018, China
*   Correspondence: xiongpingping@nuist.edu.cn

Received: 29 May 2019; Accepted: 10 July 2019; Published: 13 July 2019

Abstract: Haze is the greatest challenge facing China's sustainable development, and it seriously affects China's economy, society, ecology and human health. Based on the uncertainty and suddenness of haze, this paper proposes a novel linear time-varying grey model (GM)(1,N) based on interval grey number sequences. Because the original GM(1,N) model based on interval grey number sequences has constant parameters, it neglects the dynamic change characteristics of parameters over time. Therefore, this novel linear time-varying GM(1,N) model, based on interval grey number sequences, is established on the basis of the original GM(1,N) model by introducing a linear time polynomial. To verify the validity and practicability of this model, this paper selects the data of PM_{10}, SO_2 and NO_2 concentrations in Beijing, China, from 2008 to 2018, to establish a linear time-varying GM(1,3) model based on interval grey number sequences, and the prediction results are compared with the original GM(1,3) model. The result indicates that the prediction effect of the novel model is better than that of the original model. Finally, this model is applied to forecast PM_{10} concentration for 2019 to 2021 in Beijing, and the forecast is made to provide a reference for the government to carry out haze control.

Keywords: haze; linear time-varying GM(1,N) model; interval grey number; Beijing; forecasting

1. Introduction

In recent years, the process of urbanization and industrialization has been accelerating in China, but air pollution is seriously increasingly. At present, haze is one of the greatest atmospheric, environmental pollution problems in China. Haze not only has adverse effects on the ecological environment [1–3] but also poses a major threat to human health [4,5]. As one of the largest developing countries in the world, China is facing the biggest challenge of sustainable development. In 2018, China's government officially issued the 'Three-Year Plan of Action to Win the Blue Sky Defense War'. Therefore, an accurate study of haze is of great importance.

Beijing is one of key cities for haze control in China. 16 haze events occurred in Beijing from November, 2012 to January, 2013, and the minimum visibility was even 667 m during this period [6]. Besides, the haze event in January, 2013 may have been the cause of 690 deaths in Beijing, which could lead to 253.8 million dollars losses [7]. To reduce the haze event, China's government implemented the 'Air Pollution Prevention and Control Action Plan' in September, 2013. Since then, an increasing number of scholars started to research haze from different aspects. Current haze studies mainly focused on health impact [3,4,7], economic loss [7], chemical composition [8], statistical characteristics [9], trend prediction [10], formation mechanism [11], and the rest. Above researches mainly based on hourly or daily data, but the goal of Blue Sky Defense War is to reduce the average annual concentration of pollutants. Therefore, using annual data to forecast haze is necessary. Due to the uncertainty and suddenness of haze, the data of its related indices are uncertain. In the system research, grey number refers to the uncertain number in an interval or a general number set [12]. In other words, due to the constraints of data acquisition tools, acquisition conditions and errors by the acquisition personnel, the value of haze may contain inevitable measurement errors in a certain range, thus, haze has obvious grey number features. Therefore, Wu et al. established a grey prediction model to forecast annual pollutant concentrations effecting haze by limited data for the first time [13]. Grey prediction model is an important part of grey system. As an emerging uncertain system theory, grey system theory is characterized by small data modeling to obtain accurate results, and widely applied in all professions and trades. The grey system theory was established by Professor Deng, and assists with solving an uncertain system of small data and poor information, mainly through the deep mining of the inherent law in the existing information research system. Besides, grey number is the basic unit of grey system [12]. To summarize, grey prediction model is an effective tool for studying haze.

At present, the original grey model (GM)(1,N) based on interval grey number sequences is used to estimate structural parameters by the least squares method, and the result of its structural parameters are fixed values that are not related to time. Although this method is simple, it ignores the dynamic change characteristics of the parameters over time, which may lead the model to have a high precision fit, but not an ideal predictive effect. Therefore, this paper selects the data of PM_{10} concentration (PM_{10} refers to particles with aerodynamic equivalent diameter less than or equal to 10 microns in ambient air, called inhalable particulate matter), SO_2 concentration and NO_2 concentration in Beijing, China, from 2008 to 2018, and establishes a novel linear time-varying GM(1,3) model based on interval grey number sequences to perform simulations and make predictions. Meanwhile, we establish an original GM(1,3) model based on interval grey number sequences as the contrast model, and then we compare the results of these two models.

This paper is organized as follows: the literature reviews of studying on haze and grey prediction model are presented in Section 2; the modeling algorithms and model testing methods of the linear time-varying GM(1,N) model based on interval grey number sequences are illustrated in Section 3; the linear time-varying GM(1,N) model based on interval grey number sequences is established to forecast PM_{10} concentration of Beijing in Section 4; the main conclusions of this paper are summarized in Section 5.

2. Literature Reviews

2.1. Study of Haze

To control haze more scientifically and effectively, many scholars at home and abroad have carried out a series of extensive studies. $PM_{2.5}$ ($PM_{2.5}$ refers to particles with an aerodynamic equivalent diameter less than or equal to 2.5 microns in ambient air, called fine particulate matter) and PM_{10} are main factors causing haze, which not only affect the quality of atmospheric environment and visibility, but also endanger human health. Voukantsis et al. used the neural network method of multilayer perceptive structure to predict the daily change of PM_{10} concentration in Thessaloniki and Helsinki [14]. Kumar et al. used a principal component regression method to predict short-term air quality index days in Delhi [15]. Lang et al. generalized the season autoregressive integrated moving average (SARIMA) model to the small-scale time sequences and used this model to make a short-term prediction of $PM_{2.5}$ concentration at 10 stations in Hangzhou [16]. Mishra et al. studied the meteorological factors and atmospheric pollutants affecting $PM_{2.5}$ in Delingha, and predicted the $PM_{2.5}$ concentration under haze conditions in Delingha using a combination of a neural network and fuzzy logic [17]. Konovalov et al. predicted future PM_{10} values in Europe by combining the deterministic prediction method and the linear regression method [18]. Based on the data of $PM_{2.5}$ concentration in 285 cities in China from 2001 to 2012, Cheng et al. used a dynamic space panel model to analyze the main driving factors affecting air pollution [19]. Wu et al. established an input-oriented zero sum gains-data envelopment analysis (ZSG-DEA) model, and studied the allocation efficiency of $PM_{2.5}$ emissions in China's provinces under fixed target amount conditions [20]. The novel air quality early warning system was proposed by Xu et al., which included an evaluation module, a prediction module and a feature estimation module, and this literature used a novel dynamic fuzzy comprehensive evaluation method to determine the air quality level and major pollutants in the study area [21]. Han et al. predicted the spatial distribution of lung cancer in males induced by $PM_{2.5}$ in China from 2010 to 2015, and the spatial autocorrelation method was used to evaluate the spatial relationship between the incidence of lung cancer and the atmospheric level of satellite-derived $PM_{2.5}$ from 2006 to 2009 [22].

Nowadays, there are some scholars studying haze with grey system theory. Xiong et al. established the multivariate grey model (MGM) (1,M) based on interval grey number sequences to predict the visibility and relative humidity during a haze period in Nanjing [23]. Gong et al. combined the GM(1,1) model with both the Markoff chain model and the residual error correction model to establish a modified grey Markoff chain model, and used this model to predict the $PM_{2.5}$ concentration in Shanghai [24]. Wang studied the distribution characteristics of $PM_{2.5}$ concentration in Huaian with the non-parametric hypothesis test, and predicted the $PM_{2.5}$ concentration of Huaian in the next five years by using GM(1,1) model [25]. Chen et al. predicted the hourly $PM_{2.5}$ and PM_{10} in Taichung's Dali area by using several GM(1,1) models and back-propagation artificial neural network, and compared their predictive performances [26]. Wang et al. established a novel grey correlation degree model, and used this model to dynamically analyze the influencing factors of haze in southern China [27].

2.2. Study of Grey Prediction Model

The grey prediction model has been widely applied in many fields, and it occupies an important position in the grey system theory. The GM(1,1) model that has only one variable is the most widely used prediction model in the grey system, but it is most suitable for the time sequences of monotonic increasing or monotonic decreasing. Hence, many scholars have worked to improve it. Wu et al. proposed a new GM(1,1) model with the fractional order accumulation, and this model had a better predicted performance than the traditional model [28]. Focusing on why the discrete grey model simulated the constant value growth rate, Zhang et al. established a linear time-varying discrete grey model by introducing a linear time polynomial [29]. Wang introduced the time polynomial function into the GM(1,1) power model and optimized the power exponent of the model [30]. Zeng et al. built a self-adapting intelligent grey prediction model to predict the natural gas demand of China [31].

However, the GM(1,1) model ignores the effect of related factors on the system behavior data, and then, Professor Deng proposed the GM(1,N) model, which has one system behavior sequence and N-1 related factors sequences [12]. The GM(1,N) model, as an extension of the GM(1,1) model, fully considers the impact of related factors on the system behavior data. Therefore, the GM(1,N) model and its optimization models are increasing in researches and applications. Ding et al. established a novel GM(1,N) model combined with the changing trend of the driving term, and this model was applied to predict CO_2 emissions from fuel combustion in China [32]. Zeng et al. proposed the optimal background-value GM(1,N) model through optimizing the background-value coefficient with the particle swarm optimization algorithm [33]. Wang et al. constructed a nonlinear GM(1,N) model by introducing the power exponent of the related factors, and this model was used to predict the carbon emissions of fossil energy consumption in China [34]. The above studies on grey prediction models were mainly based on real number sequence. Recently, increasingly more scholars have begun to explore the modeling problem of grey prediction models based on interval grey number sequences. Ye et al. fully explored and expanded the axiom of generalized non-decreasing grey degrees and established a prediction model for interval grey number sequences [35]. Luo et al. established a discrete GM(1,1) model for kernel and measure sequences, and then restored the predicted value of the interval grey number [36]. Yang et al. established a prediction model for the normal distribution based on interval grey number sequences in the context of the normal distribution of uncertain information [37]. A new nonlinear GM(1,N) model was established based on interval grey number sequences by Xiong et al., which was applied to the prediction of air quality index in haze period [38].

3. Methodology

3.1. The Introduction of Interval Grey Number

In this section, we will mainly introduce the basic concepts of interval grey number, including the definitions of interval grey number, kernel and grey radius.

Definition 1. [12] grey number that has both lower bound a_k and upper bound b_k is called interval grey number, denoted as $\otimes_k \in [a_k, b_k]$.

Definition 2. [12] suppose that interval grey number $\otimes_k$ is a continuous function, $\widetilde{\otimes} = (a_k + b_k)/2$ is called a kernel of interval grey number $\otimes_k$.

Definition 3. [12] when $\otimes_k$ is a continuous interval grey number, $r(k) = (b_k - a_k)/2$ is called a grey radius of interval grey number $\otimes_k$.

3.2. Linear Time-Varying GM(1,N) Model Based on Interval Grey Number Sequences

In this section, the modeling mechanism of the linear time-varying GM(1,N) model based on interval grey number sequences will be introduced. Besides, this model will be constructed from the kernel and grey radius sequences, respectively. Finally, the kernel and grey radius sequences will be restored to the interval grey number sequences. To illustrate the model more clearly, we will show the modeling steps of the linear time-varying GM(1,N) model based on interval grey number sequences in Figure 1.

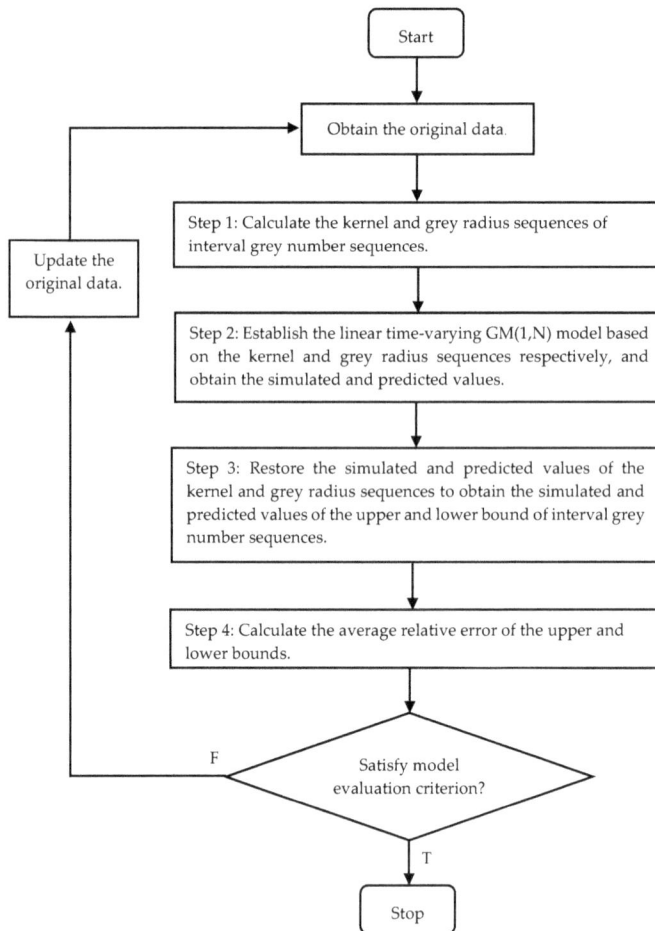

Figure 1. Modeling flow chart of the linear time-varying GM(1,N) model based on interval grey number sequences.

Then, this section will mainly introduce the linear time-varying GM(1,N) model based on kernel sequences as follows:

Definition 4. Suppose that the system behavior sequence is $\widetilde{\otimes}_1^{(0)} = \left(\widetilde{\otimes}_1^{(0)}(1), \widetilde{\otimes}_1^{(0)}(2), \cdots, \widetilde{\otimes}_1^{(0)}(n) \right)$, and the related factor sequences is $\widetilde{\otimes}_i^{(0)} = \left(\widetilde{\otimes}_i^{(0)}(1), \widetilde{\otimes}_i^{(0)}(2), \cdots, \widetilde{\otimes}_i^{(0)}(n) \right), i = 2, 3, \cdots, N$, where $\widetilde{\otimes}_i^{(1)}$ is the first order accumulating generation sequence of $\widetilde{\otimes}_i^{(0)}$, $i = 1, 2, \cdots, N; z_1^{(1)}(k) = 0.5 \left(\widetilde{\otimes}_1^{(1)}(k) + \widetilde{\otimes}_1^{(1)}(k-1) \right)$ is the mean sequence generated by consecutive neighbors of $\widetilde{\otimes}_1^{(1)}$; thus, the linear time-varying GM(1,N) model based on kernel sequences is shown as follows:

$$\widetilde{\otimes}_1^{(0)}(k) + a z_1^{(1)}(k) = \sum_{i=2}^{N} (b_{i1} + b_{i2}k) \widetilde{\otimes}_i^{(1)}(k). \tag{1}$$

Also, the whitening equation of the linear time-varying model based on kernel sequences is shown as follows:

$$\frac{d\widetilde{\otimes}_1^{(1)}}{dt} + a\widetilde{\otimes}_1^{(1)} = \sum_{i=2}^{N}(b_{i1} + b_{i2}t)\widetilde{\otimes}_i^{(1)}. \tag{2}$$

Theorem 1. The undetermined coefficient vector of the linear time-varying GM(1,N) model based on kernel sequences is $\hat{a} = [a, b_{21}, \cdots, b_{N1}, b_{22}, \cdots, b_{N2}]^T$, which can be estimated by utilizing the least squares method as follows:

(1) When $n - 1 = 2N - 1$, that is $n = 2N$, and $|B| \neq 0$, $\hat{a} = B^{-1}Y$;

(2) When $n - 1 > 2N - 1$, that is $n > 2N$, and $|B^T B| \neq 0$, $\hat{a} = (B^T B)^{-1} B^T Y$;

(3) When $n - 1 < 2N - 1$, that is $n < 2N$, and $|B^T B| \neq 0$, $\hat{a} = B^T (BB^T)^{-1} Y$;

where $B = \begin{bmatrix} -z_1^{(1)}(2) & \widetilde{\otimes}_2^{(1)}(2) & \cdots & \widetilde{\otimes}_N^{(1)}(2) & 2\widetilde{\otimes}_2^{(1)}(2) & \cdots & 2\widetilde{\otimes}_N^{(1)}(2) \\ -z_1^{(1)}(3) & \widetilde{\otimes}_2^{(1)}(3) & \cdots & \widetilde{\otimes}_N^{(1)}(3) & 3\widetilde{\otimes}_2^{(1)}(3) & \cdots & 3\widetilde{\otimes}_N^{(1)}(3) \\ \vdots & \vdots & & \vdots & \vdots & & \vdots \\ -z_1^{(1)}(n) & \widetilde{\otimes}_2^{(1)}(n) & \cdots & \widetilde{\otimes}_N^{(1)}(n) & n\widetilde{\otimes}_2^{(1)}(n) & \cdots & n\widetilde{\otimes}_N^{(1)}(n) \end{bmatrix}$, $Y = \begin{bmatrix} \widetilde{\otimes}_1^{(0)}(2) \\ \widetilde{\otimes}_1^{(0)}(3) \\ \vdots \\ \widetilde{\otimes}_1^{(0)}(n) \end{bmatrix}$

The proof is similar to literature [39].

Theorem 2. After calculating the coefficient vector $\hat{a}$, the solution of the linear time-varying GM(1,N) model based on kernel sequences is shown as follows:

$$\hat{\otimes}_1^{(1)}(k+1) = \frac{1}{a}\sum_{i=2}^{N}(b_{i1} + b_{i2}k)\widetilde{\otimes}_i^{(1)}(k+1) + e^{-ak}\left[\widetilde{\otimes}_1^{(1)}(0) - \frac{1}{a}\sum_{i=2}^{N}(b_{i1} + b_{i2}k)\widetilde{\otimes}_i^{(1)}(k+1)\right], \tag{3}$$

where the solution can be obtained using the initial condition $\widetilde{\otimes}_1^{(1)}(0) = \widetilde{\otimes}_1^{(0)}(1)$. Also, the inverse accumulating reduction equation is shown as follows:

$$\hat{\otimes}_1^{(0)}(k+1) = \hat{\otimes}_1^{(1)}(k+1) - \hat{\otimes}_1^{(1)}(k). \tag{4}$$

The modeling mechanism of the linear time-varying GM(1,N) model based on grey radius sequences are the same as the linear time-varying GM(1,N) model based on kernel sequences, thus we will not present the linear time-varying GM(1,N) model based on grey radius sequences repeatedly.

After obtaining the values of kernel and grey radius respectively, we will calculate the upper and lower bounds of the interval grey number sequences as follows [40]:

$$\begin{cases} \hat{a}_{k+1} = \hat{\otimes}_1^{(0)}(k+1) - \hat{r}_1^{(0)}(k+1) \\ \hat{b}_{k+1} = \hat{\otimes}_1^{(0)}(k+1) + \hat{r}_1^{(0)}(k+1) \end{cases}. \tag{5}$$

3.3. Model Evaluation Criterion

To analyze the reliability and credibility of the prediction model, we will show the model evaluation criterion for testing the model accuracy in this section. By comparing the relative error and average relative error of the upper and lower bounds of the interval grey number sequences to test the prediction model, the testing equations are shown as follows:

The relative error of the upper and lower bounds of the interval grey number sequences are shown as follows:

$$\begin{cases} \Delta_k = \frac{|\hat{a}_k - a_k|}{a_k} \times 100\% \\ \overline{\Delta}_k = \frac{|\hat{b}_k - b_k|}{b_k} \times 100\% \end{cases}, k = 1, 2, \cdots, n. \tag{6}$$

The average relative error of the upper and lower bounds of the interval grey number sequences are shown as follows:

$$\begin{cases} \overline{\hat{a}_k} = \frac{1}{n} \sum_{k=1}^{n} \Delta_k \\ \overline{\hat{b}_k} = \frac{1}{n} \sum_{k=1}^{n} \overline{\Delta_k} \end{cases}, k = 1, 2, \cdots, n. \qquad (7)$$

Prediction accuracy is an important criterion for measuring the reliability of the prediction model. Therefore, this paper provides the prediction accuracy corresponding to the average relative error in Table 1.

Table 1. The average relative error criterion for testing model [41].

Average Relative Error	Prediction Accuracy
<10%	High prediction
10–20%	Good prediction
20–50%	Reasonable prediction
>50%	Weak prediction

4. Empirical Results and Discussion

In this section, the linear time-varying GM(1,N) model and the original GM (1,N) model, based on interval grey number sequences, will be established to simulate the development trend of haze in Beijing, and the model with high prediction accuracy will be selected to forecast the haze situation in Beijing from 2019 to 2021.

4.1. Data Selection and Processing

PM_{10} can be formed by the interaction of sulfur oxides, nitrogen oxides and other compounds in the ambient air. Additionally, PM_{10} is highly correlated with SO_2 and NO_2 in 31 cities of China [42]. Therefore, PM_{10} concentration in Beijing from 2008 to 2018 is selected as the system behavior sequence, and SO_2 concentration and NO_2 concentration are selected as related factor sequences. In addition, the data are from the annual report of air quality in Beijing. In data processing, the maximum and minimum observed values over the previous three values are the upper and lower bounds of the third interval grey number. Moreover, the interval grey number sequences for 2008 to 2014 are used as the modeling data, and the interval grey number sequences for 2015 to 2018 are used as the prediction data. Besides, we denote the interval grey number sequences of PM_{10} concentration, SO_2 concentration and NO_2 concentration as $X_1(\otimes)$, $X_2(\otimes)$, $X_3(\otimes)$ respectively. The original data is shown in Table 2.

Table 2. The interval grey number sequences of PM_{10} concentration, SO_2 concentration and NO_2 concentration

k	Year	$X_1(\otimes)(\mu g/m^3)$	$X_2(\otimes)(\mu g/m^3)$	$X_3(\otimes)(\mu g/m^3)$
1	2008	[122,161]	[36,53]	[49,66]
2	2009	[121,148]	[34,47]	[49,66]
3	2010	[114,122]	[32,36]	[49,55]
4	2011	[114,121]	[28,34]	[53,55]
5	2012	[109,121]	[28,32]	[52,55]
6	2013	[108,114]	[27,28]	[52,56]
7	2014	[108,116]	[22,28]	[52,59]
8	2015	[102,116]	[14,27]	[50,59]
9	2016	[100,116]	[10,22]	[48,59]
10	2017	[95,108]	[8,14]	[46,50]
11	2018	[91,100]	[6,10]	[42,48]

4.2. Establishment and Comparison of Model

Step 1: on the basis of the data from 2008 to 2014 in Table 2, the kernel and grey radius sequences of PM_{10} concentration, SO_2 concentration and NO_2 concentration are calculated according to Definition 2 and 3, and the results are shown in Table 3.

Table 3. The kernel and grey radius sequences of PM_{10} concentration, SO_2 concentration and NO_2 concentration

k	$\tilde{\otimes}_1(k)$	$\tilde{\otimes}_2(k)$	$\tilde{\otimes}_3(k)$	$r_1(k)$	$r_2(k)$	$r_3(k)$
1	141.5	44.5	57.5	19.5	8.5	8.5
2	134.5	40.5	57.5	13.5	6.5	8.5
3	118	34	52	4	2	3
4	117.5	31	54	3.5	3	1
5	115	30	53.5	6	2	1.5
6	111	27.5	54	3	0.5	2
7	112	25	55.5	4	3	3.5

Step 2: after calculating model parameters by least squares method, a linear time-varying GM(1,3) model for the PM_{10} concentration kernel sequence is established as follows:

$$\tilde{\otimes}_1^{(0)}(k) + 1.9972z_1^{(1)}(k) = (1.5419 - 0.0591k)\tilde{\otimes}_2^{(1)}(k) + (3.7449 - 0.0014k)\tilde{\otimes}_3^{(1)}(k) \qquad (8)$$

Similarly, a linear time-varying GM(1,3) model for the PM_{10} concentration grey radius sequence is established as follows:

$$r_1^{(0)}(k) + 1.6900z_1^{(1)}(k) = (-6.9833 + 2.2742k)r_2^{(1)}(k) + (9.5542 - 2.0577k)r_3^{(1)}(k). \qquad (9)$$

According to the linear time-varying GM(1,3) model of PM_{10} concentration kernel and grey radius sequences, the simulated values of PM_{10} concentration kernel and grey radius sequences can be respectively obtained. The SO_2 concentration and NO_2 concentration sequences that are predicted by the GM(1,1) model are used as the related factor sequences for 2015 to 2018. Then, a linear time-varying GM(1,3) model is used to obtain the predicted values of PM_{10} concentration kernel and grey radius sequences.

Step 3: according to Equation (5), the simulated and predicted values of the upper and lower bounds over the PM_{10} concentration interval grey number sequence are obtained by restoring the simulated and predicted values of kernel and grey radius sequences. The results are shown in Table 4.

Step 4: the relative error and average relative error of the PM_{10} concentration upper and lower bounds are calculated according to Equation (6) and Equation (7). The results are shown in Table 4.

At the same time, the original GM(1,N) model based on interval grey number sequences is selected as the contrast model. Establishing the GM(1,3) model based on interval grey number sequences to simulate and forecast PM_{10} concentration in Beijing from 2008 to 2018, and the relative error and average relative error of upper and lower bounds are calculated. The results are shown in Table 4.

Table 4. The simulated and predicted values and the average relative error of the PM$_{10}$ concentration upper and lower bounds.

		Linear Time-Varying GM(1,3) Model			GM(1,3) Model		
Year	Actual Value (µg/m³)	Simulated Value (µg/m³)	Lower Bound Relative Error (%)	Upper Bound Relative Error (%)	Simulated Value (µg/m³)	Lower Bound Relative Error (%)	Upper Bound Relative Error (%)
2008	[122,161]	[122.00,161.00]	0.00	0.00	[122.00,161.00]	0.00	0.00
2009	[121,148]	[107.05,130.05]	11.54	12.13	[105.05,126.89]	13.18	14.26
2010	[114,122]	[124.80,141.14]	9.47	15.69	[124.73,138.16]	9.41	13.25
2011	[114,121]	[121.66,123.27]	6.72	1.88	[115.38,125.57]	1.21	3.78
2012	[109,121]	[110.83,121.07]	1.68	0.06	[111.76,119.48]	2.53	1.26
2013	[108,114]	[110.18,115.51]	2.02	1.32	[110.10,114.59]	1.95	0.52
2014	[108,116]	[105.85,118.84]	1.99	2.45	[104.30,117.56]	3.43	1.35
Average simulated relative error (%)			4.78	4.79		4.53	4.92
Year	Actual Value (µg/m³)	Predicted Value (µg/m³)	Lower Bound Relative Error (%)	Upper Bound Relative Error (%)	Predicted Value (µg/m³)	Lower Bound Relative Error (%)	Upper Bound Relative Error (%)
2015	[102,116]	[103.77,109.13]	1.74	5.92	[103.88,106.75]	1.85	7.98
2016	[100,116]	[100.99,106.51]	0.99	8.18	[101.46,103.48]	1.46	10.80
2017	[95,108]	[98.52,104.11]	3.70	3.60	[99.14,100.56]	4.36	6.89
2018	[91,100]	[96.31,101.91]	5.84	1.91	[96.95,97.95]	6.53	2.05
Average predicted relative error (%)			3.07	4.90		3.55	6.93

This paper will compare the simulated and predicted values, relative errors of upper and lower bounds of the PM$_{10}$ concentration interval grey number sequence. According to Table 4, the average simulation relative error of upper and lower bounds of linear time-varying GM(1,3) model are 4.79% and 4.78%, respectively, and the average simulation relative error of upper and lower bounds of GM(1,3) model are 4.92% and 4.53%, respectively. The average simulation relative errors of those two models are very close, and both belong to high simulation accuracy. However, the average predicted relative error of upper and lower bounds of linear time-varying GM(1,3) model are 4.90% and 3.07% respectively, which are less than those of GM(1,3) model.

For ease of comparison, we make comparison maps as shown in Figures 2–5. For simplicity, we briefly note the linear time-varying GM(1,N) model based on interval grey number sequences as LTVGM(1,N), and briefly note the GM(1,N) model based on interval grey number sequences as GM(1,N). From Figure 2 to Figure 5, the novel linear time-varying GM(1,3) model based on interval grey number sequences can more accurately describe the upper and lower bounds of PM$_{10}$ concentration in Beijing, which especially has the better prediction effect.

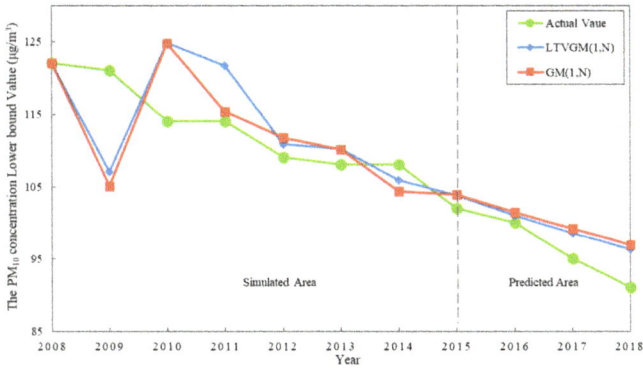

Figure 2. Comparison map of the PM_{10} concentration lower bound values for LTVGM(1,N) model and GM(1,N) model.

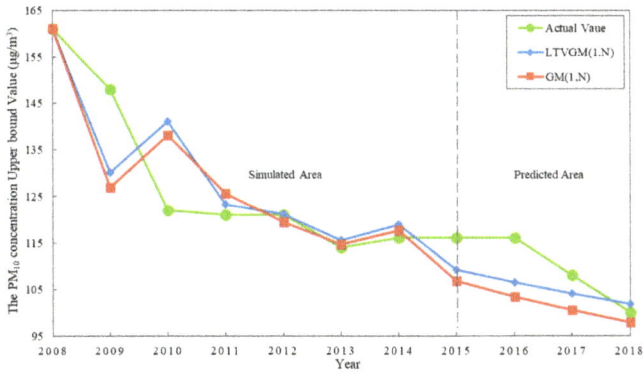

Figure 3. Comparison map of the PM_{10} concentration upper bound values for LTVGM(1,N) model and GM(1,N) model.

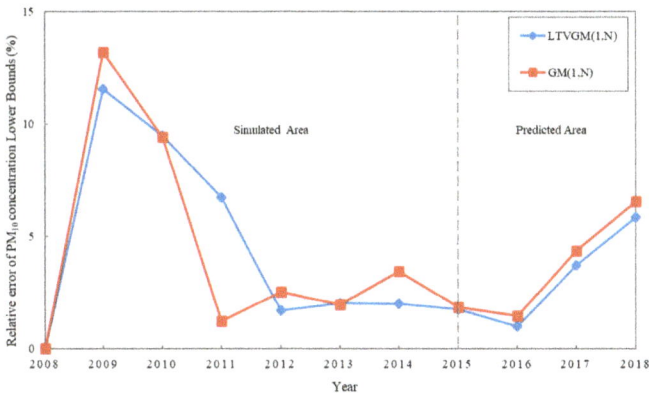

Figure 4. Comparison map of the relative errors of PM_{10} concentration lower bound for LTVGM(1,N) model and GM(1,N) model.

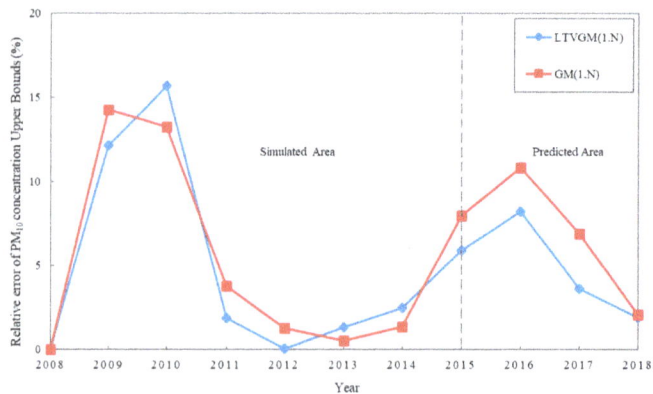

Figure 5. Comparison map of the relative errors of PM_{10} concentration upper bound for LTVGM(1,N) model and GM(1,N) model.

4.3. Forecast Results and Discussion

To understand the trend of haze in Beijing in the future, the linear time-varying GM(1,3) model proposed in this paper is used to forecast PM_{10} concentration in Beijing for 2019 to 2021, the forecast results are shown in Table 5. According to Table 5, PM_{10} concentration in Beijing will decrease slowly from 2019 to 2021, but still exceed the China's environmental air quality standard [13].

Table 5. The predicted value of PM_{10} concentration in Beijing for 2019 to 2021.

Year	2019	2020	2021
PM_{10} concentration ($\mu g/m^3$)	[94.36,99.88]	[92.62,98.07]	[91.06,96.29]

According to all the results considered, we will perform a discussion as follows: the average predicted relative errors of the linear time-varying GM(1,3) model are less than those of the original GM(1,3) model, which is attributed to the improved adaptability of the model to the dynamic change characteristics data. The model proposed in this paper is simple in the modeling method and convenient for calculation and applications. In addition, this novel model expands the range of predicted values from the real number to interval grey number, which can broaden the range of applications. To a certain extent, it can make up for the errors caused by data acquisition tools, acquisition conditions and acquisition personnel. However, this novel model still has certain limitations. It only considers that the related factor sequences are air pollutants, and it neglects the influence of meteorological factors on haze. Therefore, the future research will select both air pollutants and meteorological factors as the related factor sequences to study haze.

5. Conclusions

Aiming at forecasting haze of Beijing more accurately, this paper introduced a linear time polynomial into the GM(1,N) model based on interval grey number sequences, and established a novel linear time-varying GM(1,N) model based on interval grey number sequences. The data of PM_{10} concentration, SO_2 concentration and NO_2 concentration in Beijing, China, were selected as modeling data to establish a linear time-varying GM(1,3) model from 2008 to 2018. The results indicate that the prediction model proposed in this paper has a higher prediction accuracy than the original model, and both the corresponding prediction errors are less than 5%, which proves the validity and practicability of this model. When using this novel model to make a forecast of PM_{10} concentration in Beijing, the forecast shows that PM_{10} concentration will be a downward trend for 2019 to 2021 in Beijing. This is because PM_{10} concentration is not only determined by air pollutants, but also related to meteorological

Sustainability **2019**, *11*, 3832

factors, such as wind speed, precipitation, and so forth. Specifically, when wind speed or precipitation increases, haze will decrease in Beijing [43]. Meteorological factors vary from year to year, and the extreme weather event such as El Niño cannot be accurately forecasted a year in advance. When the extreme weather event occurs, the actual PM_{10} concentration in Beijing may be different with our forecast for 2019 to 2021. If there is no extreme weather event in that year, the haze predicted value of this paper will be accurate. Therefore, this model can provide decision support for the government when working toward greater haze control.

The model proposed in this paper had a high accuracy of prediction, but there is much room for improvement. This paper only selected two pollutants as related factors, which were SO_2 and NO_2, in order to forecast PM_{10} concentration in Beijing. In future research, we will consider a few meteorological factors as related factors to construct model, such as wind speed, relative humidity, air pressure, temperature, precipitation, etc. Besides, we will apply this novel model to several key cities in the Beijing-Tianjin-Hebei region of China. On this basis, we will use data mining technology to compare and analyze the haze of different cities in the Beijing-Tianjin-Hebei region.

Author Contributions: Conceptualization, J.S.; methodology, P.X.; software, J.S.; validation, J.S. and S.D.; formal analysis, P.X.; investigation, L.P.; resources, S.D.; data curation, L.P.; writing—original draft preparation, P.X. and J.S.; writing—review and editing, P.X. and J.S.; visualization, L.P. and S.D.; supervision, P.X.; project administration, P.X.; funding acquisition, P.X.

Funding: National Natural Science Foundation of China (71701105, 41505118); Major Program of the National Social Science Fund of China (17ZDA092); Ministry of Education Humanities and Social Sciences Research Youth Subsidy Project in China (17YJC630182, 17YJC630123); Key Research Project of Philosophy and Social Sciences in Universities of Jiangsu Province (2018SJZDI111).

Conflicts of Interest: The authors declare no conflict of interest.

References

1. Tao, M.; Chen, L.F.; Xiong, X.Z.; Zhang, M.G.; Ma, P.F.; Tao, J.H.; Wang, Z.F. Formation process of the widespread extreme haze pollution over northern China in January 2013: Implications for regional air quality and climate. *Atmos. Environ.* **2014**, *98*, 417–425. [CrossRef]
2. Lu, Y.L.; Wang, Y.; Zuo, J.; Jiang, H.Q.; Huang, D.C.; Rameezdeen, R. Characteristics of public concern on haze in China and its relationship with air quality in urban areas. *Sci. Total Environ.* **2018**, *637*, 1597–1606. [CrossRef] [PubMed]
3. Behera, S.N.; Cheng, J.P.; Huang, X.; Zhu, Q.Y.; Liu, P.; Balasubramanian, R. Chemical composition and acidity of size-fractionated inorganic aerosols of 2013-14 winter haze in Shanghai and associated health risk of toxic elements. *Atmos. Environ.* **2015**, *122*, 259–271. [CrossRef]
4. Yin, H.; Pizzol, M.; Jacobsen, J.B.; Xu, L.Y. Contingent valuation of health and mood impacts of $PM_{2.5}$ in Beijing, China. *Sci. Total Environ.* **2018**, *630*, 1269–1282. [CrossRef] [PubMed]
5. Faridi, S.; Shamsipour, M.; Krzyzanowski, M.; Künzli, N.; Amini, H.; Azimi, F.; Malkawi, M.; Momeniha, F.; Gholampour, A.; Hassanvand, M.S.; et al. Long-term trends and health impact of $PM_{2.5}$ and O_3 in Tehran, Iran, 2006–2015. *Environ. Int.* **2018**, *114*, 37–49. [CrossRef] [PubMed]
6. Zhang, Q.; Quan, J.N.; Tie, X.X.; Li, X.; Liu, Q.; Gao, Y.; Zhao, D.L. Effects of meteorology and secondary particle formation on visibility during heavy haze events in Beijing, China. *Sci. Total Environ.* **2015**, *502*, 578–584. [CrossRef]
7. Gao, M.; Guttikunda, S.K.; Carmichael, G.R.; Wang, Y.S.; Liu, Z.R.; Stanier, C.O.; Saide, P.E.; Yu, M. Health impacts and economic losses assessment of the 2013 severe haze event in Beijing area. *Sci. Total Environ.* **2015**, *511*, 553–561. [CrossRef]
8. Tian, S.L.; Pan, Y.P.; Liu, Z.R.; Wen, T.X.; Wang, Y.S. Size-resolved aerosol chemical analysis of extreme haze pollution events during early 2013 in urban Beijing, China. *J. Hazard. Mater.* **2014**, *279*, 452–460. [CrossRef]
9. Zheng, Z.F.; Xu, G.R.; Yang, Y.J.; Wang, Y.T.; Li, Q.C. Statistical characteristics and the urban spillover effect of haze pollution in the circum-Beijing region. *Atmos. Pollut. Res.* **2018**, *9*, 1062–1071. [CrossRef]
10. Zhu, S.L.; Lian, X.Y.; Liu, H.X.; Hu, J.M.; Wang, Y.Y.; Che, J.X. Daily air quality index forecasting with hybrid models: A case in China. *Environ. Pollut.* **2017**, *231*, 1232–1244. [CrossRef]

11. Liu, X.G.; Li, J.; Qu, Y.; Han, T.; Hou, L.; Gu, J.; Chen, C.; Yang, Y.; Liu, X.; Yang, T.; et al. Formation and evolution mechanism of regional haze: A case study in the megacity Beijing, China. *Atmos. Chem. Phys.* **2013**, *13*, 4501–4514. [CrossRef]

12. Liu, S.F.; Dang, Y.G.; Fang, Z.G.; Xie, N.M. *Grey Systems Theory and Its Applications*, 5th ed.; Science Press: Beijing, China, 2010.

13. Wu, L.F.; Nu, L.; Yang, Y.J. Prediction of air quality indicators for the Beijing-Tianjin-Hebei region. *J. Clean. Prod.* **2018**, *196*, 682–687. [CrossRef]

14. Voukantsis, D.; Karatzas, K.; Kukkonen, J.; Räsänen, T.; Karppinen, A.; Kolehmainen, M. Intercomparison of air quality data using principal component analysis, and forecasting of PM_{10} and $PM_{2.5}$ concentrations using artificial neural networks, in Thessaloniki and Helsinki. *Sci. Total Environ.* **2011**, *409*, 1266–1276. [CrossRef] [PubMed]

15. Kumar, A.; Goyal, P. Forecasting of air quality in Delhi using principal component regression technique. *Atmos. Pollut. Res.* **2011**, *2*, 436–444. [CrossRef]

16. Lang, Y.C.; Xiao, L.; George, C. Forecasting concentrations of $PM_{2.5}$ in main urban area of Hangzhou and mapping using SARIMA model and ordinary Kringing method. *Acta Sci. Circum.* **2018**, *1*, 62–70.

17. Mishra, D.; Goyal, P.; Upadhyay, A. Artificial intelligence based approach to forecast $PM_{2.5}$, during haze episodes: A case study of Delhi, India. *Atmos. Environ.* **2015**, *102*, 239–248. [CrossRef]

18. Konovalov, I.B.; Beekmann, M.; Meleux, F.; Dutot, A.; Foret, G. Combining deterministic and statistical approaches for PM_{10}, forecasting in Europe. *Atmos. Environ.* **2009**, *43*, 6425–6434. [CrossRef]

19. Cheng, Z.H.; Li, L.S.; Liu, J. Identifying the spatial effects and driving factors of urban $PM_{2.5}$, pollution in China. *Ecol. Indic.* **2017**, *82*, 61–75. [CrossRef]

20. Wu, X.H.; Tan, L.; Guo, J.; Wang, Y.Y.; Liu, H.; Zhu, W.W. A study of allocative efficiency of $PM_{2.5}$, emission rights based on a zero sum gains data envelopment model. *J. Clean. Prod.* **2016**, *113*, 1024–1031. [CrossRef]

21. Xu, Y.Z.; Du, P.; Wang, J.Z. Research and application of a hybrid model based on dynamic fuzzy synthetic evaluation for establishing air quality forecasting and early warning system: A case study in China. *Environ. Pollut.* **2017**, *223*, 435–448. [CrossRef]

22. Han, X.; Liu, Y.Q.; Gao, H.; Ma, J.M.; Mao, X.X.; Wang, Y.T.; Ma, X.D. Forecasting $PM_{2.5}$, induced male lung cancer morbidity in China using satellite retrieved $PM_{2.5}$, and spatial analysis. *Sci. Total Environ.* **2017**, *607*, 1009–1017. [CrossRef] [PubMed]

23. Xiong, P.P.; Zhang, Y.; Yao, T.X.; Zeng, B. Multivariate grey prediction model based on interval grey number sequences. *Math. Pract. Theor.* **2017**, *7*, 310–319.

24. Gong, M.; Ye, C.M. Prediction of the $PM_{2.5}$ concentration in Shanghai Municipality based on Modified Grey Markov chain model. *J. Nat. Disasters* **2016**, *5*, 97–104.

25. Wang, Z.X. Distribution characteristics and grey prediction model of $PM_{2.5}$ in Huai'an. *Math. Pract. Theor.* **2018**, *48*, 286–291.

26. Chen, L.; Pai, T.Y. Comparisons of GM(1,1) and BPNN for predicting hourly particulate matter in Dali area of Taichung City, Taiwan. *Atmos. Pollut. Res.* **2015**, *6*, 572–580. [CrossRef]

27. Wang, J.J.; Hipel, K.W.; Dang, Y.G. An improved grey dynamic trend incidence model with application to factors causing smog weather. *Expert Syst. Appl.* **2017**, *87*, 240–251. [CrossRef]

28. Wu, L.F.; Liu, S.F.; Yao, L.G.; Yan, S.L.; Liu, D.L. Grey system model with the fractional order accumulation. *Commun. Nonlinear Sci.* **2013**, *18*, 1775–1785. [CrossRef]

29. Zhang, K.; Liu, S.F. Linear time-varying parameters discrete grey forecasting model. *Syst. Eng. Theor. Pract.* **2010**, *30*, 1650–1657.

30. Wang, Z.X. GM(1,1) power model with time-varying parameters and its application. *Control Decis.* **2014**, *10*, 1828–1832.

31. Zeng, B.; Li, C. Forecasting the natural gas demand in China using a self-adapting intelligent grey model. *Energy* **2016**, *112*, 810–825. [CrossRef]

32. Ding, S.; Dang, Y.G.; Li, X.M.; Wang, J.J.; Zhao, K. Forecasting Chinese CO_2 emissions from fuel combustion using a novel grey multivariable model. *J. Clean. Prod.* **2017**, *162*, 1527–1538. [CrossRef]

33. Zeng, B.; Li, C. Improved multi-variable grey forecasting model with a dynamic background-value coefficient and its application. *Comput. Ind. Eng.* **2018**, *118*, 278–290. [CrossRef]

34. Wang, Z.X.; Ye, D.J. Forecasting Chinese carbon emissions from fossil energy consumption using non-linear grey multivariable models. *J. Clean. Prod.* **2017**, *142*, 600–612. [CrossRef]

35. Ye, J.; Dang, Y.G.; Ding, S. Grey prediction model of interval grey numbers based on axiom of generalized non-decrease grey degree. *Control Decis.* **2016**, *31*, 1831–1836.

36. Luo, D.; Li, L. Interval grey number prediction model based on kernel and measure. *Math. Pract. Theor.* **2014**, *44*, 96–100.

37. Yang, J.W.; Xiao, X.P.; Guo, J.H. Grey prediction model for interval grey number of normal distribution. *Control Decis.* **2015**, *9*, 1711–1716.

38. Xiong, P.P.; Yin, Y.; Shi, J.; Gao, H. Nonlinear multivariable GM(1,N) model based on interval grey number sequence. *J. Grey Syst.* **2018**, *30*, 33–47.

39. Zeng, B.; Duan, H.M.; Zhou, Y.F. A new multivariable grey prediction model with structure compatibility. *Appl. Math. Model.* **2019**, *75*, 385–397. [CrossRef]

40. Liu, J.F.; Liu, S.F.; Fang, Z.G. Continuous interval grey number prediction model based on kernel and grey radius. *Syst. Eng.* **2013**, *2*, 61–64.

41. Lewis, C. *Industrial and Business Forecasting Methods: A Practical Guide to Exponential Smoothing and Curve Fitting*; Butterworth Scientific: Oxford, UK, 1982.

42. Xie, Y.Y.; Zhao, B.; Zhang, L.; Luo, R. Spatiotemporal variations of $PM_{2.5}$ and PM_{10} concentrations between 31 Chinese cities and their relationships with SO_2, NO_2, CO and O_3. *Particuology* **2015**, *20*, 141–149. [CrossRef]

43. Chang, L.Y.; Xu, J.M.; Tie, X.X.; Wu, J.B. Impact of the 2015 El Nino event on winter air quality in China. *Sci. Rep.* **2016**, *6*, 34275. [CrossRef] [PubMed]

![sustainability logo] *sustainability*

MDPI

Article

Long-Term Cointegration Relationship between China's Wind Power Development and Carbon Emissions

Wenhui Zhao [1], Ruican Zou [1], Guanghui Yuan [2,3,*], Hui Wang [1] and Zhongfu Tan [4]

[1] College of Economics and Management, Shanghai University of Electric Power, Shanghai 200090, China
[2] Fintech Research Institute, Shanghai University of Finance and Economics, Shanghai 200433, China
[3] School of Information Management and Engineering, Shanghai University of Finance and Economics, Shanghai 200433, China
[4] North China Electric Power University, Beijing 102206, China
*   Correspondence: guanghuiyuan@outlook.com; Tel.: +86-21-6590-4028

Received: 2 August 2019; Accepted: 21 August 2019; Published: 26 August 2019

Abstract: Faced with the deterioration of the environment and resource shortages, countries have turned their attention to renewable energy and have actively researched and applied renewable energy. At present, a large number of studies have shown that renewable energy can effectively improve the environment and control the reduction of resources. However, there are few studies on how renewable energy improves the environment through its influencing factors. Therefore, this paper mainly analyses the relationship between wind energy and carbon emissions in renewable energy and uses Chinese data as an example for the case analysis. Based on the model and test methods, this paper uses the 1990–2018 data from the China Energy Statistical Yearbook to study and analyse the correlation between wind energy and carbon emissions and finally gives suggestions for wind energy development based on environmental improvements.

Keywords: sustainable development; wind power development; carbon emissions

1. Introduction

With the rapid development of the economy and rapid growth of the population, human behaviour has an increasing impact on the global environment, such as resource shortages and environmental pollution [1–4]. Therefore, sustainable development has become the focus of all countries. In recent years, global fog and haze have become severe, the environment has drastically deteriorated, and carbon emissions have remained high [5–7]. These issues have become the main targets of the environmental governance of all the countries in the world. Governing the environment and alleviating the energy crisis have become the top priorities. The concept of sustainable development was proposed by the World Commission for Environment and Development, which was chaired by Mrs. Brundtland in 1987, and was adopted at the 1992 United Nations Conference on Environment and Development. The latter marks the entry of countries from all over the world into an environmentally friendly and sustainable development stage. Sustainable development is the core content of the scientific concept of development. Sustainable development refers to development that meets the needs of the present without compromising the ability of future generations to meet their needs. The theoretical core of the concept of sustainable development mainly includes two aspects: One is the harmonious coexistence between man and nature, and the other is the coordination of the relationships between people. Due to serious environmental pollution and poor environmental quality around the world, carbon emissions control is an important issue for sustainable development. Countries around the world have made relevant agreements to ensure that carbon emissions can be effectively controlled, such as the Kyoto Protocol and the Paris Agreement.

It is necessary to vigorously develop clean energy, since the total amount of carbon emissions remains high. It is also important to alleviate the continued environmental deterioration to improve its future. Among them, wind power generation is one of the most mature, large-scale developmental goals and commercialization prospects of clean energy.

After reviewing a large amount of literature, we found that the previous research mainly focused on carbon emissions, economic growth and energy consumption [8–10]. There are few scholars who have studied the relationship between carbon emissions and renewable energy. Since China has completed wind power generation data, this paper takes China as an example to analyse the relationship between wind energy and carbon emissions. The article concludes with recommendations for sustainable development in countries around the world. The research methods in this paper are as follows. (1) We establish an autoregressive distributed lag model, which is usually defined as an ARDL model to test whether there is a long-term co-integration relationship between wind power development and total carbon emissions. (2) We use the Granger causality test to verify the causal relationship between wind power development and total carbon emissions.

This research has far-reaching significance. It can provide a theoretical reference for environmental sustainability analysis, and provide a scientific basis for sustainable development strategies, which is conducive to future wind power construction planning. The structure of the remainder of this paper is as follows. First, it introduces the relevant research background and a large amount of literature that is related to the disciplines we studied. Second, the literature review describes the development of wind power, the spatial and temporal distributions of carbon emissions, and the application of the ARDL model and the Granger causality test. Third, the research areas and data are introduced. Fourth, we take China as an example and use the ARDL model and the Granger causality test method to analyse the case. Fifth, the results are obtained, and relevant recommendations and prospects for future research are given.

2. Literature Review

2.1. Development and Application of Wind Energy

Many countries are turning their attention to renewable energy to alleviate the status quo, since the significant global warming greenhouse effect; excessive energy shortages and the inability of traditional energy sources such as oil, coal and natural gas cannot meet the growing global energy demand. Furthermore, the relatively low power density of renewable energy, the returns on energy investment, and the trends towards energy conservation and emissions reductions will all contribute to the widespread use of renewable energy. Wind energy has wide application prospects due to its small impact on the local environment and easy management [11–13].

Wind energy resources are abundant, inexhaustible and sustainable. The total amount of wind energy in the world is approximately 130 million megawatts, of which 20 million megawatts of wind energy is available. This is 10 times more than the total amount of hydroelectric power that can be developed and utilized on the earth, which is up to 53 billion MWh per year. Wind power is currently the most mature renewable energy. Therefore, in order to achieve sustainable social and economic development, wind power development is a key goal. Countries have knowledge gaps in their existing wind energy abilities. To understand the distribution of wind energy resources, scholars have combined the advantages of large wind speed observation networks and grid-based advantages to geographically analyse the distribution and intensity of terrestrial wind resources on the global, continental and national scales. Better use of wind energy resources can achieve economic growth, energy conservation, emission reductions, and sustainable development [14–17].

Electricity is closely related to people's lives. However, the traditional power industry is dominated by thermal power. The environmental pollution is very serious, causing a global greenhouse effect, which is not conducive to sustainable development. Therefore, energy conservation and emission reduction in the power industry is imperative. Due to environmental impacts and fluctuations in fossil fuel prices, the application of renewable energy in the power industry has received great attention. Wind power significantly reduces the power system life cycle costs, energy costs, greenhouse gas emissions costs and annual load loss costs, effectively alleviating global environmental degradation [18,19].

Offshore wind energy resources are the most abundant wind energy resources in all regions. Therefore, the most aspect of the technology is to understand the characteristics of the best locations for installing offshore wind power plants. To achieve this goal, it is necessary to fully understand wind resources, marine life and environmental protection, and the related conflict management [20].

2.2. Time and Spatial Distributions of Carbon Emissions and the Reduction of Renewable Energy

The carbon emissions of fast-growing cities around the world are becoming increasingly severe. To alleviate the greenhouse gas pollution, scholars usually focus on the changes in the amounts of carbon emissions. For the spatial and temporal characteristics of carbon emissions, meteorological models have difficulties simulating their transportation and diffusion, partly due to the increased surface heterogeneity and fine spatial and temporal scales. The Eulerian model has been used to observe the air or to generate a high-precision time-space distribution map using a comprehensive coefficient model of the carbon density factor and the corresponding land cover type. It is necessary for the government to understand the spatial and temporal distributions of carbon emissions and take corresponding measures to reduce carbon emissions to make the energy economy sustainable [21–24].

The decarbonization of the power industry is critical to achieve the goal of reducing greenhouse gas emissions in countries around the world and addressing fossil fuel shortages. Among them, intermittent clean energy such as wind energy and photovoltaic solar power are major components of low-carbon power systems. Scholars have studied whether the use of clean energy to generate electricity is reducing carbon emissions and if its productivity is increasing. Studies have shown that clean energy generation can not only reduce the marginal costs of power generation but also reduce carbon dioxide emissions, preventing the greenhouse effect that is caused by carbon emissions from becoming more severe [25–27].

Wind power plays a vital role in mitigating climate change and responding to environmental pollution. However, the difficulty and instability of wind farm scheduling hinder the development of wind power. To meet the ever-increasing energy demand and energy-savings emissions reduction targets, scholars have conducted relevant research on these problems. This research includes combining wind turbines with compressed air energy storage; combining water energy and wind energy, and using deterministic dynamic programming in long-term power generation. The extended model was optimized and the hourly power system simulation tool was used to evaluate the five additional supplements of the lowest cost integration batch to maintain system stability [28–31].

2.3. Related Research on the ARDL Model

The ARDL co-integration test method was developed by Charemza and Deadman. It was proposed by Pesaran and the ARDL co-integration test is a model for checking whether there are stable relationships between variables [32,33].

The ARDL model has the following points: It only requires that the time sequence monotony does not exceed one without pre-checking whether the time series has first-order monotonicity; the ARDL boundary test is also robust enough in small sample cases; and when the explanatory variable is an endogenous variable, the ARDL model can also be derived from unbiased and effective estimates. Therefore, it has attracted the attention of many scholars [34–36].

Many experts have studied the application of the ARDL model in the correlation analysis. (1) Some studied the relationship between Ghana's carbon dioxide emissions, GDP, energy use and population growth, and added renewable energy technologies to Ghana's energy structure, which is conducive to solving Ghana's climate change and environmental problems. There was evidence of bidirectional causality running from energy use to GDP and a unidirectional causality running from carbon dioxide emissions to energy use, carbon dioxide emissions to GDP, carbon dioxide emissions to population, and population to energy use. As a policy implication, the addition of renewable energy and clean energy technologies into Ghana's energy mix can help mitigate climate change and its impact in the future. [37]. (2) The causal relationship between transportation consumption, fuel price, added value of the transportation sector and carbon emissions was studied. The results show that the increase in transportation consumption technology, the increase in fuel prices, and the increase in the value added of the transportation sector are all conducive to the reduction of carbon emissions. Providing new ideas to policy makers to improve the transportation environment and other issues [38]. (3) Related researchers have studied the relationship between carbon emissions and economic growth, and given corresponding policy recommendations [39–42].

2.4. Granger Causality Test and Its Application

The Granger causality test is a statistical hypothesis testing method that examines whether a set of time series *x* is related to another set of time series *y*. It is based on an autoregressive model from the regression analysis. On the basis of this, scholars have conducted in-depth research, have made great progress in the research on the nonlinearity and robustness of causality tests, and have tested extensive applications of the causal tests in economics and finance [43,44].

There have been several applications of Granger causality tests in the research and analysis of correlation. (1) Some researchers have investigated the relationship between carbon dioxide emissions, energy consumption, actual production, actual production squared, trade openness, urbanization and financial development. In the long run, energy consumption and urbanization will exacerbate environmental degradation, while financial development has no impact on it, and trade will bring about environmental improvements. Therefore, effective energy policy recommendations are given in the literature to help reduce carbon dioxide emissions without affecting the actual output [45]. (2) Related field experts have studied the relationships between carbon dioxide, energy consumption, foreign direct investment and economic growth, and concluded that the use of clean technology for foreign investment is essential to reducing carbon dioxide emissions while maintaining economic development [46]. (3) Scholars have studied the relationship between carbon emissions, energy consumption, and economic growth in various countries, and provided useful advice to relevant policy makers to ensure the reduction of carbon emissions while promoting sustainable economic development [47–51].

3. Materials and Methods

3.1. Data Processing

This paper uses the annual data from 1990–2018 from the Energy Statistics Yearbook for the empirical analysis. We estimate China's carbon emissions using the following formula:

$$Y = \sum_i q_i \times \beta_i \tag{1}$$

Here, *Y* is the total annual carbon emissions, q_i is the annual consumption of China's primary energy, and β_i is the carbon emissions coefficient. By consulting the relevant data, the carbon emissions coefficient of energy consumption is found, and the average values is used to calculate the carbon emissions coefficient of each type of energy (see Table 1). We input the data into Equation (1) to obtain China's annual carbon emissions.

Table 1. Carbon emissions coefficients of different types of energy.

Data Sources	Coal Consumption Carbon Emissions Coefficient $t(c)/t$	Petroleum Consumption Carbon Emissions Coefficient $t(c)/t$	Natural Gas Consumption Carbon Emissions Coefficient $t(c)/t$
DOE/EIA	0.702	0.478	0.389
Japan Institute of Energy Economics	0.756	0.586	0.449
National Science and Technology Commission Climate Change Project	0.726	0.583	0.409
Xu Guoquan	0.7476	0.5825	0.4435
average value	0.7329	05574	0.4226

The total carbon emissions in China from 1990 to 2018 are shown in Table 2.

Table 2. The total carbon emissions in China from 1990 to 2018.

Years	Total Amount	Years	Total Amount
1990	65,131.425	2005	151,096.135
1991	68,652.9008	2006	165,792.045
1992	72,093.6518	2007	179,477.125
1993	76,201.8913	2008	184,448.433
1994	80,354.932	2009	193,867.63
1995	85,513.0048	2010	202,395.535
1996	91,068.6173	2011	242,899.904
1997	89,110.1581	2012	248,151.036
1998	87,583.4832	2013	255,020.596
1999	90,768.2709	2014	256,274.833
2000	93,169.9358	2015	255,275.396
2001	95,090.6289	2016	254,395.482
2002	100,890.264	2017	259,093.115
2003	117,681.65	2018	263,041.807
2004	136,325.105		

Wind power development is measured using the total annual installed capacity of wind power every year and is recorded as X. To eliminate the possible heteroscedasticity in the original data and not change the co-integration relationships between them, this paper takes the logarithms of the variables and records them separately as $\ln X$ and $\ln Y$. The newly installed capacity and total carbon emissions of China's wind power in 1990–2018 are shown in Figure 1.

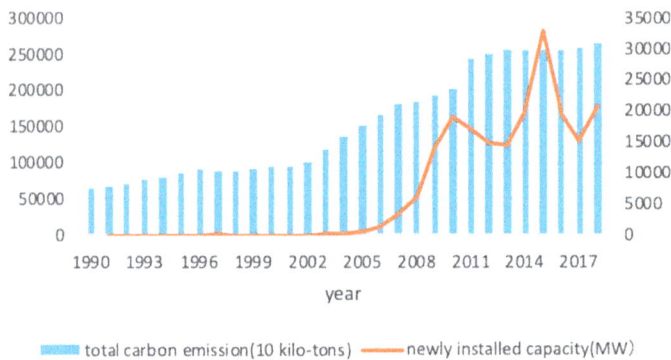

Figure 1. Annual carbon emissions and newly installed wind power capacity.

3.2. ARDL Cointegration Test

The ARDL co-integration test is a method that was proposed by Charemza and Deadman and also Pesaran [52,53]. The main idea is to determine whether there is a long-term stable relationship between variables using the boundary test method, and then estimate the correlation coefficient between the variables under the premise of the existence of a co-integration relationship. Combined with the research content of this paper, the main calculation steps of the ARDL co-integration test are briefly introduced.

First, we construct an unconstrained error correction model (UECM).

$$\Delta \ln Y_t = a_{10} + \sum_{i=1}^{n} a_{11i}\Delta \ln Y_{t-i} + \sum_{i=1}^{n} a_{12i}\Delta \ln X_{t-i} + a_{13}\ln Y_{t-1} + a_{14} \ln X_{t-1} + \varepsilon_{1t} \qquad (2)$$

$$\Delta \ln X_t = a_{20} + \sum_{i=1}^{n} a_{21i}\Delta \ln X_{t-i} + \sum_{i=1}^{n} a_{12i}\Delta \ln Y_{t-i} + a_{23}\ln X_{t-1} + a_{24}\ln Y_{t-1} + \varepsilon_{2t} \qquad (3)$$

Here, $a_{i0}(i = 1,2)$ is a constant, Δ is the first-order difference of the variable, $\varepsilon_{it}(i = 1,2)$ is the stable white noise sequence, n is the lag order, and t is the year. The optimal lag order of each difference term in Equations (1) and (2) is determined using a priori information or related information criteria (AIC, SBC or another). Then, the F statistic is used to test the following joint hypothesis.

$$\begin{cases} \text{Null hypothesis} H_0 : a_{i3} = a_{i4} = 0, i = 1,2 \\ \text{Hypothesis} H_1 : a_{i3} \neq or \ a_{i4} \neq 0, i = 1,2 \end{cases} \qquad (4)$$

If the null hypothesis is accepted, it means that there is no co-integration relationship between the variables; otherwise, it means that there is a co-integration relationship between the variables. The relevant literature gives the upper and lower thresholds of the F statistic. If the calculated F value is greater than the upper critical value, the original hypothesis is rejected. If the calculated F value is less than the lower critical value, the original hypothesis is accepted. If the calculated F value falls between the upper and lower thresholds, the unit root test is needed. If the test results show that all variables are first-order and single-order and the sequence is an $I(1)$ process, the conclusion is based on the upper bound. If the test result indicates that all variables are 0-order and the sequence is an $I(0)$ process, the conclusion is drawn according to the lower bound.

Under the premise that there is a co-integration relationship between variables, the correlation coefficient between variables is estimated. The parameters of the long-term dynamic equation are estimated under the condition that the lag order is determined.

Compared with the traditional co-integration test method, the biggest advantage of the ARDL test is that regardless of whether the time series in the model is an $I(0)$ or $I(1)$ process, the ARDL results are relatively robust, especially for small sample estimations. However, if the variable's single order is more than one, the F statistic regarding whether there is a co-integration relationship between the test variables will be invalid [54]. Considering the characteristics of the ARDL and the number of samples in this study, the method will be used to test the co-integration relationship between variables.

3.3. Granger Causality Test

The Granger causality test is the most common method of determining whether a change in one variable is the cause of another variable's change. The main calculation steps are briefly given with respect to the research content of this paper.

(1) Construct a lagged term regression equation containing explanatory variables and interpreted variables. If there is a cointegration relationship between the variables, the corresponding ECM model can be constructed, and, based on this, the Granger causality test is used to analyse the long-term causal dynamic relationship between the variables. Assuming that there is a cointegration relationship

between wind power development and carbon emissions, the following formal ECM models can be used.

$$\Delta \ln Y_t = b_{10} + \sum_{i=1}^{m} b_{11i} \Delta \ln Y_{t-i} + \sum_{i=1}^{m} b_{12i} \Delta \ln X_{t-i} + \lambda_1 ECM_{t-1} + v_{1t} \tag{5}$$

$$\Delta \ln X_t = b_{20} + \sum_{i=1}^{m} b_{21i} \Delta \ln X_{t-i} + \sum_{i=1}^{m} b_{22i} \Delta \ln Y_{t-i} + \lambda_2 ECM_{t-1} + v_{2t} \tag{6}$$

Here, $b_{i0}(i = 1, 2)$ is a constant, $v_{it}(i = 1, 2)$ is a stable white noise sequence, and ECM_t is an error correction term, which is the residual of the cointegration equation. The following steps only use equation (3) as an example.

(2) Regress the current $\Delta \ln Y_t$ for all the lagged terms $\Delta \ln Y_{t-i}(i = 1, 2, \cdots, m)$ and other variables (if any), but this regression does not include the lagged term of $\Delta \ln X_t$. It returns the sum of the squared residuals.

(3) Add the lagged term of $\Delta \ln X_t$ to the regression equation of step (2) and then perform the regression to obtain the sum of the squared residuals.

(4) Test the null hypothesis using the calculated squared sum of the F values of the two residuals in steps (2) and (3).

$$H_0 : b_{12i} = 0, (i = 1, 2, \cdots, m) \tag{7}$$

If the F value that is calculated at the set significance level exceeds the critical value, the null hypothesis H_0 is rejected. This means that $\Delta \ln X_t$ belongs to this regression, indicating that $\Delta \ln X_t$ is the *Granger* cause of $\Delta \ln Y_t$, and the magnitude of the causal relationship can be the pair of F values.

It should be pointed out that the *Granger* causality test requires that the time series be covariance stable; otherwise, the pseudo-regression problem may occur. Therefore, the unit root test is needed to assess the covariance stationarity of each time series. If there is no unit root in the test, then the time series non-covariance is stable, and the first-order difference test must be conducted before the *Granger* causality test. In addition, the *Granger* causality test is sometimes sensitive to the choice of the lagged order of the regression model and needs to be screened.

4. Model Results and Analysis

4.1. Unit Root Test

As mentioned above, if the variable order is more than one, the F statistic that uses the *ARDL* model to check for the existence of a co-integration relationship between variables will be invalid. Therefore, we first perform a unit root test on the variables. Since the sample size that is selected in this paper is small, the $DF - GLS$ test method is proposed. The selection of the lagged order is based on the AIC criterion, and the maximum lagged order is eight. The results are shown in Table 3.

Table 3. Unit root test result.

Null Hypothesis: D(LNY) Has A Unit Root				
Exogenous: Constant				
Lag Length: 0 (Automatic—Based on SIC, Maxlag = 8)				
			t-Statistic	Prob. *
Augmented Dickey-Fuller	test statistic		−3.020011	0.0456
Test critical values:	1% level		−3.699871	
	5% level		−2.976263	
	10% level		−2.627420	
Variable	Coefficient	Std. Error	t-Statistic	Prob.
D(LNY(−1))	−0.543901	0.180099	−3.020011	0.0058
C	0.026425	0.012796	2.065096	0.0494

It can be seen from the above table that the intercept of ln X and lnY plus the delay term, the intercept only, and the P value without the intercept and the delay term are all greater than 0.05; therefore, it is an unstable sequence, and Δ ln X and ΔlnY do not contain the intercept. When the delay term is P, the value is less than 0.05, which is a stationary sequence. The single integer order of the two variables is more than the first order, and so the ARDL model can be used.

4.2. Co-Integration Test Based on the ARDL Model

This section may be divided by subheadings. It will provide a concise and precise description of the experimental results and their interpretations and the experimental conclusions that can be drawn. We use the Eviews software to run the F tests, and the results are shown in Table 4.

Table 4. F test results.

EC = LNX − (6.1556 *LNY − 68.9407)				
F-Bounds Test		Null Hypothesis: No level relationship		
Test Statistic	Value	Signif.	I (0)	I (1)
		Asymptotic: n = 1000		
F-statistic	5.501287	10%	3.02	3.51
K	1	5%	3.62	4.16
		2.5%	4.18	4.79
		1%	4.94	5.58

If the F statistic is greater than the upper critical value, the null hypothesis is rejected, indicating that there is a co-integration relationship between the variables. If the F statistic is less than the lower critical value, then there is no co-integration relationship between the variables. It can be seen in Table 4 that at the 5% significance level, the F statistic is greater than the upper critical value. Therefore, when wind power development is the dependent variable, the variable has a long-term co-integration relationship.

On the basis of this model, the stability of the selected model is tested by using the recursive residual accumulation sum test (CUSUM) and the recursive residual cumulative square sum test (CUSUM of squares). The results are shown in Figures 2 and 3.

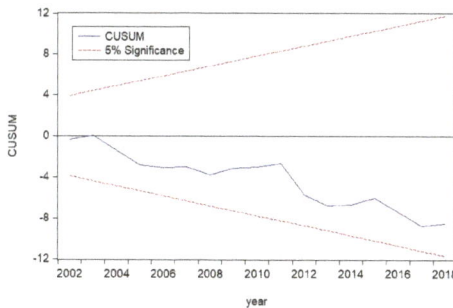

Figure 2. CUSUM TEST.

It can be seen from Figures 2 and 3 that both statistics are within the 5% significance lines, indicating that the parameters in the ARDL model are stable. From Table 4, the long-term co-integration equations for wind power development and carbon emissions can be obtained.

$$\ln X = 6.1556 \times \ln Y + 68.9407 \tag{8}$$

From the above formula, it can be seen that for every 1% increase in carbon emissions, 1% of the growth rate of the newly installed of wind power capacity will increase by 6.1556%. It can be seen that the increase in carbon emissions will accelerate the development of wind power.

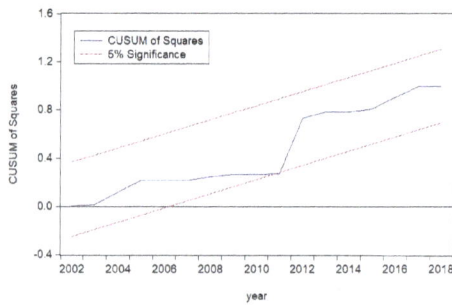

Figure 3. CUSUM of squares.

4.3. Granger Causality Analysis

Since there is a long-term co-integration relationship between China's wind power development and carbon emissions, an error correction model can be used to analyse the Granger causal relationship between the two. We run the analysis using the Eviews software, and the results are shown in Table 5.

Table 5. ECM-based Granger causality test results.

Pairwise Granger Causality Tests			
Sample: 1990 2018 Lags: 4			
Null Hypothesis	Obs	F-Statistic	Prob.
LNY does not Granger Cause LNX	24	3.42253	0.0354
LNX does not Granger Cause LNY		0.38472	0.8162

It can be seen from Table 5 that carbon emissions are the cause of wind power development, and wind power development will not increase carbon emissions, but it will inhibit carbon emissions and enable the environment to be effectively improved. Therefore, the growth of carbon emissions will drive the development of wind power, but the development of wind power will also inhibit the growth of carbon emissions, thereby reducing the speed of wind power development. In the long run, the final installed capacity of wind power will reach a certain value, and the amount of carbon emissions will be reduced to a certain value, achieving stable development.

5. Conclusions and Outlook

There have been a few scholars that have studied the relationship between a certain energy's development and carbon emissions. Almost no research has been conducted on the relationship between wind power development and carbon emissions. Here, the relationship between wind power development and carbon emissions are effectively assessed using models and causality tests. The data we used are open to the public and published by the government, and the data source is authoritative and reliable.

Results are as follows.

1. The ARDL model verifies the long-term stable relationship between wind power development and carbon emissions reductions. The two interact with each other, and so the development of wind power will impact carbon emissions.

2. The Granger causality test shows that an increase in carbon emissions will contribute to the annual increase in the newly installed wind power capacity. The reductions in carbon emissions will also inhibit the development of wind power. In the long run, the two will eventually stabilize in a certain region and steadily develop. However, at present, global carbon emissions are always increasing, which will certainly drive the development of wind power. Therefore, countries around

the world should increase investments related to wind power, promote wind power development, effectively curb the increase in carbon emissions, protect the environment, and conserve resources.

3. The excessive use of fossil energy makes greenhouse gas emissions harmful to the environment. The rapid growth of carbon emissions is one of the driving forces for our development of clean energy. To promote the sustainable development of China's energy and alleviate the energy consumption crisis, China should promote the development of wind power and promote the utilization of renewable energy.

First, in order to promote sustainable energy development, alleviate environmental pollution and reduce carbon emissions, the Chinese government should vigorously develop wind power, increase investments in wind power development, and formulate relevant subsidy policies to promote the development of renewable energy. Second, the Chinese government is also facing the problem of wind power consumption, which was solved by the government through administrative means. Currently, China is in the stage of the power market reform, and the problem of wind power consumption should be solved through market means. This will be the focus of our future research.

Author Contributions: W.Z., R.Z., G.Y., and H.W. devised the research. G.Y. and W.Z. calculated the results. Z.T. was responsible for the overall editing of the paper. G.Y. was responsible for analysing the results. Z.R. provided the numerical results and gave valuable information for calculating carbon emissions.

Funding: This work was supported by The Ministry of Education of Humanities and Social Science project and The National Natural Science Foundation of China, grant numbers 18YJAZH138 and 71403163.

Acknowledgments: The authors thank the National Energy Administration and the National Bureau of Statistics for their data support (http://www.stats.gov.cn/). We are very grateful for the anonymous reviewers' insightful comments that significantly increased the clarity of this work.

Conflicts of Interest: The authors declare no conflict of interest.

References

1. Gao, H.; Yang, W.; Yang, Y.; Yuan, G. Analysis of the Air Quality and the Effect of Governance Policies in China's Pearl River Delta, 2015–2018. *Atmosphere* **2019**, *10*, 412. [CrossRef]
2. Yang, W.X.; Li, L.G. Energy Efficiency, Ownership Structure, and Sustainable Development: Evidence from China. *Sustainability* **2017**, *9*, 912. [CrossRef]
3. Yang, W.; Li, L. Analysis of Total Factor Efficiency of Water Resource and Energy in China: A Study Based on DEA-SBM Model. *Sustainability* **2017**, *9*, 1316. [CrossRef]
4. Yuan, G.; Yang, W. Evaluating China's Air Pollution Control Policy with Extended AQI Indicator System: Example of the Beijing-Tianjin-Hebei Region. *Sustainability* **2019**, *11*, 939. [CrossRef]
5. Li, L.; Yang, W. Total Factor Efficiency Study on China's Industrial Coal Input and Wastewater Control with Dual Target Variables. *Sustainability* **2018**, *10*, 2121. [CrossRef]
6. Yang, W.; Li, L. Efficiency Evaluation and Policy Analysis of Industrial Wastewater Control in China. *Energies* **2017**, *10*, 1201. [CrossRef]
7. Yang, W.; Li, L. Efficiency evaluation of industrial waste gas control in China: A study based on data envelopment analysis (DEA) model. *J. Clean. Prod.* **2018**, *179*, 1–11. [CrossRef]
8. Yang, W.; Yuan, G.; Han, J. Is China's air pollution control policy effective? Evidence from Yangtze River Delta cities. *J. Clean. Prod.* **2019**, *220*, 110–133. [CrossRef]
9. Yuan, G.; Yang, W. Study on optimization of economic dispatching of electric power system based on Hybrid Intelligent Algorithms (PSO and AFSA). *Energy* **2019**, *183*, 926–935. [CrossRef]
10. Yang, Y.; Yang, W. Does Whistleblowing Work for Air Pollution Control in China? A Study Based on Three-party Evolutionary Game Model under Incomplete Information. *Sustainability* **2019**, *11*, 324. [CrossRef]
11. Gnatowska, R.; Moryń-Kucharczyk, E. Current status of wind energy policy in Poland. *Renew. Energy* **2019**, *135*, 232–237. [CrossRef]
12. Jefferson, M. Renewable and low carbon technologies policy. *Energy Policy* **2018**, *123*, 367–372. [CrossRef]
13. Wei, X.; Duan, Y.; Liu, Y.; Jin, S.; Sun, C. Onshore-offshore wind energy resource evaluation based on synergetic use of multiple satellite data and meteorological stations in Jiangsu Province, China. *Front. Earth Sci.* **2019**, *13*, 132–150. [CrossRef]

14. Adefarati, T.; Bansal, R.C. Reliability, economic and environmental analysis of a microgrid system in the presence of renewable energy resources. *Appl. Energy* **2019**, *236*, 1089–1114. [CrossRef]

15. Dalla Longa, F.; Strikkers, T.; Kober, T.; van der Zwaan, B. Advancing Energy Access Modelling with Geographic Information System Data. *Environ. Model. Assess.* **2018**, *23*, 627–637. [CrossRef]

16. Ebrahimi, M.; Rahmani, D. A five-dimensional approach to sustainability for prioritizing energy production systems using a revised GRA method: A case study. *Renew. Energy* **2019**, *135*, 345–354. [CrossRef]

17. Liu, Z.; Liu, Y.; He, B.-J.; Xu, W.; Jin, G.; Zhang, X. Application and suitability analysis of the key technologies in nearly zero energy buildings in China. *Renew. Sustain. Energy Rev.* **2019**, *101*, 329–345. [CrossRef]

18. Bandoc, G.; Pravalie, R.; Patriche, C.; Degeratu, M. Spatial assessment of wind power potential at global scale. A geographical approach. *J. Clean. Prod.* **2018**, *200*, 1065–1086. [CrossRef]

19. Liu, F.; Sun, F.; Liu, W.; Wang, T.; Wang, H.; Wang, X.; Lim, W.H. On wind speed pattern and energy potential in China. *Appl. Energy* **2019**, *236*, 867–876. [CrossRef]

20. deCastro, M.; Costoya, X.; Salvador, S.; Carvalho, D.; Gomez-Gesteira, M.; Javier Sanz-Larruga, F.; Gimeno, L. An overview of offshore wind energy resources in Europe under present and future climate. *Ann. N. Y. Acad. Sci.* **2019**, *1436*, 70–97. [CrossRef]

21. Li, X.; Feng, D.; Li, J.; Zhang, Z. Research on the Spatial Network Characteristics and Synergetic Abatement Effect of the Carbon Emissions in Beijing–Tianjin–Hebei Urban Agglomeration. *Sustainability* **2019**, *11*, 1444. [CrossRef]

22. Liu, Y.; Hu, X.; Wu, H.; Zhang, A.; Feng, J.; Gong, J. Spatiotemporal Analysis of Carbon Emissions and Carbon Storage Using National Geography Census Data in Wuhan, China. *ISPRS Int. J. Geo-Inf.* **2019**, *8*, 7. [CrossRef]

23. Martin, C.R.; Zeng, N.; Karion, A.; Mueller, K.; Ghosh, S.; Lopez-Coto, I.; Gurney, K.R.; Oda, T.; Prasad, K.; Liu, Y.; et al. Investigating sources of variability and error in simulations of carbon dioxide in an urban region. *Atmos. Environ.* **2019**, *199*, 55–69. [CrossRef]

24. Xu, Q.; Dong, Y.; Yang, R.; Zhang, H.; Wang, C.; Du, Z. Temporal and spatial differences in carbon emissions in the Pearl River Delta based on multi-resolution emission inventory modeling. *J. Clean. Prod.* **2019**, *214*, 615–622. [CrossRef]

25. Brouwer, A.S.; van den Broek, M.; Zappa, W.; Turkenburg, W.C.; Faaij, A. Least-cost options for integrating intermittent renewables in low-carbon power systems. *Appl. Energy* **2016**, *161*, 48–74. [CrossRef]

26. Hanak, D.P.; Anthony, E.J.; Manovic, V. A review of developments in pilot-plant testing and modelling of calcium looping process for CO_2 capture from power generation systems. *Energy Environ. Sci.* **2015**, *8*, 2199–2249. [CrossRef]

27. Hertwich, E.G.; Gibon, T.; Bouman, E.A.; Arvesen, A.; Suh, S.; Heath, G.A.; Bergesen, J.D.; Ramirez, A.; Vega, M.I.; Shi, L. Integrated life-cycle assessment of electricity-supply scenarios confirms global environmental benefit of low-carbon technologies. *Proc. Natl. Acad. Sci. USA* **2015**, *112*, 6277–6282. [CrossRef] [PubMed]

28. Foley, A.M.; Leahy, P.G.; Li, K.; McKeogh, E.J.; Morrison, A.P. A long-term analysis of pumped hydro storage to firm wind power. *Appl. Energy* **2015**, *137*, 638–648. [CrossRef]

29. Guandalini, G.; Campanari, S.; Romano, M.C. Power-to-gas plants and gas turbines for improved wind energy dispatchability: Energy and economic assessment. *Appl. Energy* **2015**, *147*, 117–130. [CrossRef]

30. Lund, P.D.; Mikkola, J.; Ypya, J. Smart energy system design for large clean power schemes in urban areas. *J. Clean. Prod.* **2015**, *103*, 437–445. [CrossRef]

31. Sun, H.; Luo, X.; Wang, J. Feasibility study of a hybrid wind turbine system—Integration with compressed air energy storage. *Appl. Energy* **2015**, *137*, 617–628. [CrossRef]

32. Boluk, G.; Mert, M. The renewable energy, growth and environmental Kuznets curve in Turkey: An ARDL approach. *Renew. Sustain. Energy Rev.* **2015**, *52*, 587–595. [CrossRef]

33. Shahbaz, M.; Loganathan, N.; Muzaffar, A.T.; Ahmed, K.; Jabran, M.A. How urbanization affects CO_2 emissions of STIRPAT model in Malaysia? The application. *Renew. Sustain. Energy Rev.* **2016**, *57*, 83–93. [CrossRef]

34. Asumadu-Sarkodie, S.; Owusu, P.A. The relationship between carbon dioxide and agriculture in Ghana: A comparison of VECM and ARDL model. *Environ. Sci. Pollut. Res.* **2016**, *23*, 10968–10982. [CrossRef]

35. Cerdeira Sento, J.P.; Moutinho, V. CO_2 emissions, non-renewable and renewable electricity production, economic growth, and international trade in Italy. *Renew. Sustain. Energy Rev.* **2016**, *55*, 142–155. [CrossRef]

36. Shahbaz, M.; Loganathan, N.; Zeshan, M.; Zaman, K. Does renewable energy consumption add in economic growth? An application of auto-regressive distributed lag model in Pakistan. *Renew. Sustain. Energy Rev.* **2015**, *44*, 576–585. [CrossRef]

37. Asumadu-Sarkodie, S.; Owusu, P.A. Carbon dioxide emissions, GDP, energy use, and population growth: A multivariate and causality analysis for Ghana, 1971-2013. *Environ. Sci. Pollut. Res.* **2016**, *23*, 13508–13520. [CrossRef]

38. Shahbaz, M.; Khraief, N.; Ben Jemaa, M.M. On the causal nexus of road transport CO_2 emissions and macroeconomic variables in Tunisia: Evidence from combined cointegration tests. *Renew. Sustain. Energy Rev.* **2015**, *51*, 89–100. [CrossRef]

39. Abbasi, F.; Riaz, I. CO_2 emissions and financial development in an emerging economy: An augmented VAR approach. *Energy Policy* **2016**, *90*, 102–114. [CrossRef]

40. Ahmad, N.; Du, L. Effects of energy production and CO_2 emissions on economic growth in Iran: ARDL approach. *Energy* **2017**, *123*, 521–537. [CrossRef]

41. Liu, X.; Bae, J. Urbanization and industrialization impact of CO_2 emissions in China. *J. Clean. Prod.* **2018**, *172*, 178–186. [CrossRef]

42. Sun, C.; Zhang, F.; Xu, M. Investigation of pollution haven hypothesis for China: An ARDL approach with breakpoint unit root tests. *J. Clean. Prod.* **2017**, *161*, 153–164. [CrossRef]

43. Li, N.; Chen, M.Z.; Chen, M. Nonlinear Progress and Application Research of Granger Causality Test. *J. Appl. Stat. Manag.* **2017**, *36*, 891–905.

44. Liu, H.; Wang, Y.L.; Chen, D.K. Analysis of the Efficacy and Robustness of Mixed Granger Causality Test. *J. Quant. Tech. Econ.* **2017**, *34*, 144–161.

45. Dogan, E.; Turkekul, B. CO_2 emissions, real output, energy consumption, trade, urbanization and financial development: Testing the EKC hypothesis for the USA. *Environ. Sci. Pollut. Res.* **2016**, *23*, 1203–1213. [CrossRef]

46. Tang, C.F.; Tan, B.W. The impact of energy consumption, income and foreign direct investment on carbon dioxide emissions in Vietnam. *Energy* **2015**, *79*, 447–454. [CrossRef]

47. Ben Jebli, M.; Ben Youssef, S.; Ozturk, I. Testing environmental Kuznets curve hypothesis: The role of renewable and non-renewable energy consumption and trade in OECD countries. *Ecol. Indic.* **2016**, *60*, 824–831. [CrossRef]

48. Dogan, E. The relationship between economic growth and electricity consumption from renewable and non-renewable sources: A study of Turkey. *Renew. Sustain. Energy Rev.* **2015**, *52*, 534–546. [CrossRef]

49. Seker, F.; Ertugrul, H.M.; Cetin, M. The impact of foreign direct investment on environmental quality: A bounds testing and causality analysis for Turkey. *Renew. Sustain. Energy Rev.* **2015**, *52*, 347–356. [CrossRef]

50. Wang, Y.; Chen, L.; Kubota, J. The relationship between urbanization, energy use and carbon emissions: Evidence from a panel of Association of Southeast Asian Nations (ASEAN) countries. *J. Clean. Prod.* **2016**, *112*, 1368–1374. [CrossRef]

51. Yousefi-Sahzabi, A.; Sasaki, K.; Yousefi, H.; Pirasteh, S.; Sugai, Y. GIS aided prediction of CO_2 emission dispersion from geothermal electricity production. *J. Clean. Prod.* **2011**, *19*, 1982–1993. [CrossRef]

52. Charenza, W.W.; Deadman, D.F. *New Directions in Econometric Analysis*; Oxford University Press: Oxford, UK, 1997.

53. Pesaran, M.H.; Shin, Y.; Smith, R.J. Bounds testing approaches to the analysis of level relationships. *J. Appl. Econom.* **2001**, *16*, 289–326. [CrossRef]

54. Ouattara, B. *Foreign Aid and Fiscal Policy in Senegal*; Mimeo University of Manchester: Manchester, UK, 2004.

MDPI

St. Alban-Anlage 66

4052 Basel

Switzerland

Tel. +41 61 683 77 34

Fax +41 61 302 89 18

www.mdpi.com

Sustainability Editorial Office

E-mail: sustainability@mdpi.com

www.mdpi.com/journal/sustainability